K. Omasa, I. Nouchi, L.J. De Kok (Eds.)
**Plant Responses to Air Pollution and Global Change**

K. Omasa, I. Nouchi,
L.J. De Kok (Eds.)

# Plant Responses to Air Pollution and Global Change

With 100 Figures, Including 2 in Color

Springer

Kenji Omasa
Professor, Graduate School of Agricultural and Life Sciences
The University of Tokyo
1-1-1 Yayoi, Bunkyo-ku, Tokyo 113-8657, Japan

Isamu Nouchi
Head, Agro-Meteorology Group
National Institute for Agro-Environmental Sciences
3-1-3 Kannondai, Tsukuba, Ibaraki 305-8604, Japan

Luit J. De Kok
Professor, Laboratory of Plant Physiology
University of Groningen
P.O. Box 14, 9750 AA Haren, The Netherlands

Library of Congress Control Number: 2006921340

ISBN-10  4-431-31013-4 Springer Tokyo Berlin Heidelberg New York
ISBN-13  978-4-431-31013-6 Springer Tokyo Berlin Heidelberg New York

This work is subject to copyright. All rights are reserved, whether the whole or part of the material is concerned, specifically the rights of translation, reprinting, reuse of illustrations, recitation, broadcasting, reproduction on microfilms or in other ways, and storage in data banks.
The use of registered names, trademarks, etc. in this publication does not imply, even in the absence of a specific statement, that such names are exempt from the relevant protective laws and regulations and therefore free for general use.

Springer is a part of Springer Science+Business Media
springer.com
© Springer 2005
Second printing 2007
Printed in Japan

Typesetting: Camera-ready by the editors
Printing and binding: Nikkei, Japan

Printed on acid-free paper

# Preface

The main force behind climate change is the elevated concentration of $CO_2$ in the atmosphere. Carbon dioxide and air pollutants come mostly from the same industrial sources and diffuse globally, so that air pollution is also part of global change in the present era. The impacts on plants and plant ecosystems have complex interrelationships and lead to global change in a circular manner as changes in land cover and atmospheric and soil environments. Plant metabolism of $CO_2$ and air pollutants and their gas fluxes in plant ecosystems influence the global gaseous cycles as well as the impacts on plants.

The 6th International Symposium on Plant Responses to Air Pollution and Global Changes was held at the Tsukuba Center for Institutes and Epochal Tsukuba, in Tsukuba, Japan, October 19–22, 2004. The aim of the symposium series is to bring together scientists of various disciplines who are actively involved in research on responses of plant metabolism to air pollution and global change. The previous symposia were held in Oxford, UK, 1982 (1st), in Munich, Germany, 1987 (2nd), in Blacksburg, USA, 1992 (3rd), in Egmond aan Zee, The Netherlands, 1997 (4th), and in Pulawy, Poland, 2001 (5th).

This book is one of three publications (this volume and special issues of *Phyton* and the *Journal of Agricultural Meteorology*) coming out of the symposium and contains a selection of invited papers. It also includes current topics on plant metabolism of air pollutants and elevated $CO_2$, responses of whole plants and plant ecosystems, genetics and molecular biology for functioning improvement, experimental ecosystems and climate change research, global carbon-cycle monitoring in plant ecosystems, and remote sensing and modeling of climate change impacts, with additional topics in risk assessment and protection against air pollution and global change in East Asia. Because the authors are researchers from 18 countries, coming from Europe, the United States, Australia, and East Asia, readers can obtain information on current research in those regions as well as finding a source of expert knowledge about the topics that are included.

The publication of this volume has been made possible by a grant from the Commemorative Organization for the Japan World Exposition ('70).

Kenji Omasa
Isamu Nouchi
Luit J.De Kok

# Contents

Preface ··················································································· V

Contributors ············································································· X I

## I. Plant Responses to Air Pollution

**Metabolism of atmospheric sulfur gases in onion** ···························· 3
Mark Durenkamp, Freek S. Posthumus, C. Elisabeth E. Stuiver,
and Luit J. De Kok

**Impact of atmospheric $NH_3$ deposition on plant growth and functioning
– a case study with *Brassica oleracea* L.** ···································· 13
Ana Castro, Ineke Stulen, and Luit J. De Kok

**How sensitive are forest trees to ozone? - New research on an old issue** ········ 21
Rainer Matyssek, Gerhard Wieser, Angela J. Nunn, Markus Löw,
Christiane Then, Karin Herbinger, Manuela Blumenröther, Sascha Jehnes,
Ilja M. Reiter, Christian Heerdt, Nina Koch, Karl-Heinz Häberle,
Kris Haberer, Herbert Werner, Michael Tausz, Peter Fabian,
Heinz Rennenberg, Dieter Grill and Wolfgang Oßwald

**Northern conditions enhance the susceptibility of birch (*Betula pendula* Roth)
to oxidative stress caused by ozone** ············································ 29
Elina Oksanen

**Physiological responses of trees to air pollutants at high elevation sites** ········ 37
Dieter Grill, Hardy Pfanz, Bohumir Lomsky, Andrzej Bytnerowicz,
Nancy E. Grulke, and Michael Tausz

**Complex assessment of forest condition under air pollution impacts** ··········· 45
Tatiana A. Mikhailova, Nadezhda S. Berezhnaya, Olga V. Ignatieva,
and Larisa V. Afanasieva

**Evaluation of the ozone-related risk for Austrian forests** ···················· 53
Friedl Herman, Stefan Smidt, Wolfgang Loibl,
and Harald R. Bolhar-Nordenkampf

Causes of differences in response of plant species to nitrogen supply
and the ecological consequences · · · · · · · · · · · · · · · · · · · · · · · · · · · · · · · · · · · · · 63
David W. Lawlor

## II. Plant Responses to Climate Change

Long-term effects of elevated $CO_2$ on sour orange trees · · · · · · · · · · · · · · · · · · · · · · 73
Bruce A. Kimball, and Sherwood B. Idso

Plant responses to climate change: impacts and adaptation · · · · · · · · · · · · · · · · · · 81
David W Lawlor

Effects of elevated carbon dioxide concentration on wood structure
and formation in trees · · · · · · · · · · · · · · · · · · · · · · · · · · · · · · · · · · · · · · · · · · · · · · · · 89
Ken'ichi Yazaki, Yutaka Maruyama, Shigeta Mori, Takayoshi Koike,
and Ryo Funada

## III. Plant Responses to Combination of Air Pollution and Climate Change

Carbon dioxide and ozone affect needle nitrogen and abscission
in *Pinus ponderosa* · · · · · · · · · · · · · · · · · · · · · · · · · · · · · · · · · · · · · · · · · · · · · · · · · · · 101
David M. Olszyk, David T. Tingey, William E. Hogsett, and E. Henry Lee

Effects of air pollution and climate change on forests of the Tatra Mountains,
Central Europe · · · · · · · · · · · · · · · · · · · · · · · · · · · · · · · · · · · · · · · · · · · · · · · · · · · · · · · 111
Peter Fleischer, Barbara Godzik, Svetlana Bicarova,
and Andrzej Bytnerowicz

## IV. Genetics and Molecular Biology for Functioning Improvement

MAPK signalling and plant cell survival in response
to oxidative environmental stress · · · · · · · · · · · · · · · · · · · · · · · · · · · · · · · · · · · · · · · · 125
Marcus A. Samuel, Godfrey P. Miles, and Brian E. Ellis

Expression of cyanobacterial *ictB* in higher plants enhanced photosynthesis
and growth · · · · · · · · · · · · · · · · · · · · · · · · · · · · · · · · · · · · · · · · · · · · · · · · · · · · · · · · · · · 133
Judy Lieman-Hurwitz, Leonid Asipov, Shimon Rachmilevitch,
Yehouda Marcus, and Aaron Kaplan

Improvement of photosynthesis in higher plants · · · · · · · · · · · · · · · · · · · · · · · · · · · · 141
Masahiro Tamoi and Shigeru Shigeoka

**Modification of $CO_2$ fixation of photosynthetic prokaryote** ·················149
  Akira Wadano, Manabu Tsukamoto, Yoshihisa Nakano,
  and Toshio Iwaki

**Specificity of diatom Rubisco** ········································157
  Richard P. Haslam, Alfred J. Keys, P John Andralojc,
  Pippa J. Madgwick, Inger Andersson, Anette Grimsrud,
  Hans C. Eilertsen, and Martin A.J. Parry

**Regulation of $CO_2$ fixation in non-sulfur purple photosynthetic bacteria** ······165
  Simona Romagnoli and F. Robert Tabita

## V. Experimental Ecosystem and Climate Change Research

**Experimental ecosystem and climate change research in controlled environments : lessons from the Biosphere 2 Laboratory 1996-2003** ····················173
  Barry Osmond

**Importance of air movement for promoting gas and heat exchanges between plants and atmosphere under controlled environments** ··············185
  Yoshiaki Kitaya

**Pros and cons of $CO_2$ springs as experimental sites** ·······················195
  Elena Paoletti, Hardy Pfanz, and Antonio Raschi

## VI. Global Carbon Cycles in Ecosystem and Assessment of Climate Change Impacts

**Carbon dynamics in response to climate and disturbance: Recent progress from multi-scale measurements and modeling in AmeriFlux** ················205
  Beverly Law

**Synthetic analysis of the $CO_2$ fluxes at various forests in East Asia** ···········215
  Susumu Yamamoto, Nobuko Saigusa, Shohei Murayama, Minoru Gamo,
  Yoshikazu Ohtani, Yoshiko Kosugi, and Makoto Tani

**3-D remote sensing of woody canopy height and carbon stocks by helicopter-borne scanning lidar** ·································227
  Kenji Omasa and Fumiki Hosoi

**Assessments of climate change impacts on the terrestrial ecosystem in Japan using the Bio-Geographical and GeoChemical (BGGC) Model** ···············235
  Yo Shimizu, Tomohiro Hajima, and Kenji Omasa

## VII. Air Pollution and Global Change in Asia

**Establishing critical levels of air pollutants for protecting East Asian vegetation – A challenge** ··············································································243
Yoshihisa Kohno, Hideyuki Matsumura, Takashi Ishii, and Takeshi Izuta

**Major activities of acid deposition monitoring network in East Asia (EANET) and related studies** ································································251
Tsumugu Totsuka, Hiroyuki Sase and Hideyuki Shimizu

**Land degradation and blown-sand disaster in China** ···························261
Pei-Jun Shi, Hideyuki Shimizu, Jing-Ai Wang, Lian-You Liu, Xiao-Yan Li, Yi-Da Fan, Yun-Jiang Yu, Hai-Kun Jia, Yanzhi Zhao, Lei Wang, and Yang Song

**Impact of meteorological fields and surface conditions on Asian dust** ········271
Seiji Sugata, Masataka Nishikawa, Nobuo Sugimoto, Ikuko Mori, and Atsushi Shimizu

**A case study on combating desertification at a small watershed in the hills-gully area of loess plateau, China** ·······································277
Junliang Tian, Puling Liu, Hideyuki Shimizu, and Shinobu Inanaga

**A recipe for sustainable agriculture in drylands** ······························285
Shinobu Inanaga, A. Egrinya Eneji, Ping An, and Hideyuki Shimizu

Index ·················································································295

# Contributors

**Afanasieva, Larisa V.,** Institute of General and Experimental Biology Siberian Branch of the Russian Academy of Sciences, Sahjanova, 6, Ulan-Ude, 670042, Russia

**An, Ping,** Arid Land Research Center, Tottori University, Hamasaka 1390, Tottori 681-0001, Japan

**Andersson, Inger,** Department of Molecular Biology, Swedish University of Agricultural Sciences, Uppsala Biomedical Centre, Box 590, S-751 24 Uppsala, Sweden

**Andralojc, P. John,** Crop Performance and Improvement Division, Rothamsted Research, Harpenden, Hertfordshire AL5 2JQ, UK

**Asipov, Leonid,** Department of Plant and Environmental Sciences, The Hebrew University of Jerusalem, 91904 Jerusalem, Israel

**Berezhnaya, Nadezhda S.,** Siberian Institute of Plant Physiology and Biochemistry, Siberian Branch of the Russian Academy of Sciences, Lermontova, 132, Irkutsk, 664033, Russia

**Bicarova, Svetlana,** Institute of Geophysics, Slovak Academy of Sciences, Meteorological Observatory, Stara Lesna, 059 60 Tatranska Lomnica, Slovakia

**Blumenröther, Manuela,** Pathology of Woody Plants, Technische Universität München, Am Hochanger 13, 85354 Freising, Germany

**Bolhar-Nordenkampf, Harald R.,** Institute of Ecology and Conservation Biology, University of Vienna, Althanstraße 14, A-1090 Vienna, Austria

**Bytnerowicz, Andrzej,** USDA Forest Service, Pacific Southwest Research Station, 4955 Canyon Crest Drive, Riverside, CA 92507-6090, USA

**Castro, Ana,** Laboratory of Plant Physiology, University of Groningen, P.O. Box 14, 9750 AA Haren, The Netherlands

**De Kok, Luit J.,** Laboratory of Plant Physiology, University of Groningen, P.O. Box 14, 9750 AA Haren, The Netherlands

**Durenkamp, Mark,** Laboratory of Plant Physiology, University of Groningen, P.O. Box 14, 9750 AA Haren, The Netherlands

**Eilertsen, Hans C.,** Norwegian College of Fisheries Science (NFH), University of Tromsø, N-9037 Tromsø, Norway

**Ellis, Brian E.,** Michael Smith Laboratories, University of British Columbia, Vancouver BC V6T 1Z4 Canada

**Eneji, A. Egrinya,** Arid Land Research Center, Tottori University, Hamasaka 1390, Tottori 681-0001, Japan

**Fabian, Peter,** Ecoclimatology, Technische Universität München, Am Hochanger 13, 85354 Freising, Germany

**Fan, Yi-Da,** National Disaster Reduction Center of China, Ministry of Civil Affairs, Bai Guang Lu No.7, Beijing 100053,China

**Fleischer, Peter,** Research Station of Tatra National Park, State Forest of TANAP, 059 60 Tatranska Lomnica, Slovakia

**Funada, Ryo,** Faculty of Agriculture, Tokyo University of Agriculture and Technology, Saiwai-cho 3-5-8, Fuchu, Tokyo 183-8509, Japan

**Gamo, Minoru,** Institute for Environmental Management Technology, National Institute of Advanced Industrial Sciences and Technology, Onogawa, 16-1, Tsukuba, Ibaraki 305-8506, Japan

**Godzik, Barbara,** Institute of Botany, Polish Academy of Sciences, Lubicz 46, 31 512 Krakow, Poland

**Grill, Dieter,** Institut für Pflanzenwissenschaften, Karl-Franzens-Universität Graz, Schubertstraße 51, A-8010 Graz, Austria

**Grimsrud, Anette,** Norwegian College of Fisheries Science (NFH), University of Tromsø, N-9037 Tromsø, Norway

**Grulke, Nancy E.,** USDA Forest Service, Pacific Southwest Research Station, 4955 Canyon Crest Drive, Riverside, CA 92507-6090, USA

**Haberer, Kris,** Forest Botany and Tree Physiology, Albert-Ludwigs Universität Freiburg i. Br., Georges-Köhler-Allee 53/54, 79110 Freiburg, Germany

**Häberle, Karl-Heinz,** Ecophysiology of Plants, Technische Universität München, Am Hochanger 13, 85354 Freising, Germany

**Hajima, Tomohiro,** Graduate School of Agricultural and Life Sciences, The University of Tokyo, Yayoi 1-1-1, Bunkyo-ku, Tokyo 113-8657, Japan

**Haslam, Richard P.,** Crop Performance and Improvement Division, Rothamsted Research, Harpenden, Hertfordshire AL5 2JQ, UK

**Heerdt, Christian,** Ecoclimatology, Technische Universität München, Am Hochanger 13, 85354 Freising, Germany

**Herbinger, Karin,** Institut für Pflanzenwissenschaften, Karl-Franzens Universität Graz, Schubertstraße 51, 8010 Graz, Austria

**Herman, Friedl,** Federal Office and Research Centre for Forests, Seckendorff-Gudent Weg 8, A-1130 Vienna, Austria

**Hogsett, William E.,** US Environmental Protection Agency, National Health and Environmental Effects Laboratory, Western Ecology Division, 200 SW 35th Street, Corvallis, OR 97333, USA

**Hosoi, Fumiki,** Graduate School of Agricultural and Life Sciences, The University of Tokyo, Yayoi 1-1-1, Bunkyo-ku, Tokyo 113-8657, Japan

**Idso, Sherwood B.,** Center for the Study of Carbon Dioxide and Global Change, Tempe, AZ 85285, USA

**Ignatieva, Olga V.,** Siberian Institute of Plant Physiology and Biochemistry, Siberian Branch of the Russian Academy of Sciences, Lermontova, 132, Irkutsk, 664033, Russia

**Inanaga, Shinobu,** Arid Land Research Center, Tottori University, Hamasaka 1390, Tottori 681-0001, Japan

**Ishii, Takashi,** Environmental Science Research Laboratory, Central Research Institute of Electric Power Industry, Abiko 1646, Abiko City, Chiba 270-1194, Japan

**Iwaki, Toshio,** Department of Applied Biochemistry, Osaka Prefecture University, Gakuencho 1-1, Sakai, Osaka, 599-8531, Japan

**Izuta, Takeshi,** Institute of Symbiotic Science and Technology, Tokyo University of Agriculture and Technology, Saiwai-cho 3-5-8, Fuchu, Tokyo 183-8509, Japan

**Jehnes, Sascha,** Forest Botany and Tree Physiology, Albert-Ludwigs Universität Freiburg i. Br., Georges-Köhler-Allee 53/54, 79110 Freiburg, Germany

**Jia, Hai-Kun,** College of Resources Science and Technology, Beijing Normal University, No.19 Xinjiekouwai Street, Beijing 100875, China

**Kaplan, Aaron,** Department of Plant and Environmental Sciences, The Hebrew University of Jerusalem, 91904 Jerusalem, Israel

**Keys, Alfred J.,** Crop Performance and Improvement Division, Rothamsted Research, Harpenden, Hertfordshire AL5 2JQ, UK

**Kimball, Bruce A.,** U.S. Water Conservation Laboratory, USDA, Agricultural Research Service 4331 East Broadway Road, Phoenix, AZ 85040, USA

**Kitaya, Yoshiaki,** Graduate School of Life and Environmental Sciences, Osaka Prefecture University, Gakuen-cho 1-1, Sakai, Osaka 599-8531, Japan

**Koch, Nina,** Ecophysiology of Plants, Technische Universität München, Am Hochanger 13, 85354 Freising, Germany

**Kohno, Yoshihisa,** Environmental Science Research Laboratory, Central Research Institute of Electric Power Industry, Abiko 1646, Abiko City, Chiba 270-1194, Japan

**Koike, Takayoshi,** Field Science Center for Northern Biosphere, Hokkaido University, Kita-9, Nishi-9, Kita-ku, Sapporo, Hokkaido 060-0809, Japan

**Kosugi, Yoshiko,** Graduate School of Agriculture, Kyoto University, Kitashirakawa, Sakyo-ku, Kyoto 606-8502, Japan

**Law, Beverly,** College of Forestry, Oregon State University, Corvallis, OR 97331-5752, USA

**Lawlor, David W.,** Crop Performance and Improvement Division, Rothamsted Research, Harpenden, Hertfordshire, AL5 2JQ, UK

**Lee, E. Henry,** US Environmental Protection Agency, National Health and Environmental Effects Laboratory, Western Ecology Division, 200 SW 35th Street, Corvallis, OR 97333, USA

**Li, Xiao-Yan,** College of Resources Science and Technology, Beijing Normal University, No.19 Xinjiekouwai Street, Beijing 100875, China

**Lieman-Hurwitz, Judy,** Department of Plant and Environmental Sciences, The Hebrew University of Jerusalem, 91904 Jerusalem, Israel

**Liu, Lian-You,** College of Resources Science and Technology, Beijing Normal University, No.19 Xinjiekouwai Street, Beijing 100875, China

**Liu, Puling,** Institute of Soil and Water Conservation, Chinese Academy of Sciences and Ministry of Water Resources, 26# Xinong Rd. Yangling, Shaanxi 712100, China

**Loibl, Wolfgang,** ARC Systems Research, Austrian Research Centers, A-2444 Seibersdorf, Austria

**Lomsky, Bohumir,** VULHM, Jiloviste, Strnady, 15604 Zbraslav-Praha, Czechia

**Löw, Markus,** Ecophysiology of Plants, Technische Universität München, Am Hochanger 13, 85354 Freising, Germany

**Madgwick, Pippa J.**, Crop Performance and Improvement Division, Rothamsted Research, Harpenden, Hertfordshire AL5 2JQ, UK

**Marcus, Yehouda**, Department of Plant Sciences, Tel Aviv University, 69978 Tel Aviv, Israel

**Maruyama, Yutaka**, Forestry and Forest Products Research Institute, Matsunosato 1, Tsukuba, Ibaraki 305-8687, Japan

**Matsumura, Hideyuki**, Environmental Science Research Laboratory, Central Research Institute of Electric Power Industry, Abiko 1646, Abiko City, Chiba 270-1194, Japan

**Matyssek, Rainer**, Ecophysiology of Plants, Technische Universität München, Am Hochanger 13, 85354 Freising, Germany

**Mikhailova, Tatiana A.**, Siberian Institute of Plant Physiology and Biochemistry, Siberian Branch of the Russian Academy of Sciences, Lermontova, 132, Irkutsk, 664033, Russia

**Miles, Godfrey P.**, Michael Smith Laboratories, University of British Columbia, Vancouver BC V6T 1Z4 Canada

**Mori, Ikuko,** National Institute for Environmental Studies, Onogawa 16-2, Tsukuba, Ibaraki, 305-8506, Japan

**Mori, Shigeta,** Tohoku Research Center, Forestry and Forest Products Research Institute Nabeyashiki 92-25, Shimo-Kuriyagawa, Morioka, Iwate 020-0123, Japan

**Murayama, Shohei,** Institute for Environmental Management Technology, National Institute of Advanced Industrial Sciences and Technology, Onogawa 16-1, Tsukuba, Ibaraki 305-8569, Japan

**Nakano, Yoshihisa,** Department of Applied Biochemistry, Osaka Prefecture University, Gakuencho 1-1, Sakai, Osaka, 599-853, Japan

**Nishikawa, Masataka,** National Institute for Environmental Studies, Onogawa 16-2, Tsukuba, Ibaraki, 305-8506, Japan

**Nunn, Angela J.**, Ecophysiology of Plants, Technische Universität München, Am Hochanger 13, 85354 Freising, Germany

**Ohtani, Yoshikazu,** Forestry and Forest Products Research Institute, Matsunosato 1, Tsukuba, Ibaraki 305-8687, Japan

**Oksanen, Elina,** Department of Biology, University of Joensuu, POB 111, 80101 Joensuu, Finland

**Olszyk, David M.,** US Environmental Protection Agency, National Health and Environmental Effects Laboratory, Western Ecology Division, 200 SW 35th Street, Corvallis, OR 97333, USA

**Omasa, Kenji,** Graduate School of Agricultural and Life Sciences, The University of Tokyo, Yayoi 1-1-1, Bunkyo-ku, Tokyo 113-8657, Japan

**Osmond, Barry,** School of Biochemistry and Molecular Biology, Australian National University, P.O. Box 3252 Weston Creek ACT 2611, Australia

**Oßwald, Wolfgang,** Pathology of Woody Plants, Technische Universität München, Am Hochanger 13, 85354 Freising, Germany

**Paoletti, Elena,** IPP-CNR, Via Madonna del Piano, 50019 Sesto Fiorentino, Italy

**Parry, Martin A. J.,** Crop Performance and Improvement Division, Rothamsted Research, Harpenden, Hertfordshire AL5 2JQ, UK

**Pfanz, Hardy,** Institut für Angewandte Botanik, Universität Duisburg-Essen, Campus Essen, Universitätsstraße 5, 45117 Essen, Germany

**Posthumus, Freek S.,** Laboratory of Plant Physiology, University of Groningen, P.O. Box 14, 9750 AA Haren, The Netherlands

**Rachmilevitch, Shimon,** Department of Plant and Environmental Sciences, The Hebrew University of Jerusalem, 91904 Jerusalem, Israel

**Raschi, Antonio,** IBIMET-CNR, Via Madonna del Piano, 50019 Sesto Fiorentino, Italy

**Reiter, Ilja M.,** Ecophysiology of Plants, Technische Universität München, Am Hochanger 13, 85354 Freising, Germany

**Rennenberg, Heinz,** Forest Botany and Tree Physiology, Albert-Ludwigs Universität Freiburg i. Br., Georges-Köhler-Allee 53/54, 79110 Freiburg, Germany

**Romagnoli, Simona,** Department of Microbiology, The Ohio State University, 484 West 12th Avenue, Columbus, OH 43210 USA

**Saigusa, Nobuko,** Institute for Environmental Management Technology, National Institute of Advanced Industrial Sciences and Technology, Onogawa 16-1, Tsukuba, Ibaraki 305-8506, Japan

**Samuel, Marcus A.,** Michael Smith Laboratories, University of British Columbia, Vancouver BC V6T 1Z4 Canada

**Sase, Hiroyuki,** Acid Deposition and Oxidant Research Center, Sowa 1182, Niigata 950-2144, Japan

**Shi, Pei-Jun,** College of Resources Science and Technology, Beijing Normal University, No.19 Xinjiekouwai Street, Beijing 100875, China

**Shigeoka, Shigeru,** Faculty of Agriculture, Kinki University, Nakamachi 3327-204, Nara 631-8505, Japan

**Shimizu, Atsushi,** National Institute for Environmental Studies, Onogawa 16-2, Tsukuba, Ibaraki, 305-8506, Japan

**Shimizu, Hideyuki,** National Institute for Environmental Studies, Onogawa 16-2, Tsukuba, Ibaraki 305-8506 Japan

**Shimizu, Yo,** Graduate School of Agricultural and Life Sciences, The University of Tokyo, Yayoi 1-1-1, Bunkyo-ku, Tokyo 113-8657, Japan

**Smidt, Stefan,** Federal Office and Research Centre for Forests, Seckendorff-Gudent Weg 8, A-1130 Vienna, Austria

**Song, Yang,** College of Resources Science and Technology, Beijing Normal University, No.19 Xinjiekouwai Street, Beijing 100875, China

**Stuiver, C. Elisabeth E.,** Laboratory of Plant Physiology, University of Groningen, P.O. Box 14, 9750 AA Haren, The Netherlands

**Stulen, Ineke,** Laboratory of Plant Physiology, University of Groningen, P.O. Box 14, 9750 AA Haren, The Netherlands

**Sugata, Seiji,** National Institute for Environmental Studies, Onogawa 16-2, Tsukuba, Ibaraki, 305-8506, Japan

**Sugimoto, Nobuo,** National Institute for Environmental Studies, Onogawa 16-2, Tsukuba, Ibaraki, 305-8506, Japan

**Tabita, Robert,** Department of Microbiology, The Ohio State University, 484 West 12th Avenue, Columbus, OH 43210 USA

**Tamoi, Masahiro,** Faculty of Agriculture, Kinki University, Nakamachi 3327-204, Nara 631-8505, Japan

**Tani, Makoto,** Graduate School of Agriculture, Kyoto University, Kitashirakawa, Sakyo-ku, Kyoto 606-8502, Japan

**Tausz, Michael,** Institut für Pflanzenwissenschaften, Karl-Franzens Universität Graz, Schubertstraße 51, A-8010 Graz, Austria

**Then, Christiane,** Ecophysiology of Plants, Technische Universität München, Am Hochanger 13, 85354 Freising, Germany

**Tian, Junliang,** Institute of Soil and Water Conservation, Chinese Academy of Sciences and Ministry of Water Resources, 26# Xinong Rd. Yangling, Shaanxi 712100, China

**Tingey, David T.,** US Environmental Protection Agency, National Health and Environmental Effects Laboratory, Western Ecology Division, 200 SW 35th Street, Corvallis, OR 97333, USA

**Totsuka, Tsumugu,** Acid Deposition and Oxidant Research Center, Sowa 1182, Niigata 950-2144, Japan

**Tsukamoto, Manabu,** Department of Applied Biochemistry, Osaka Prefecture University, Gakuencho 1-1, Sakai, Osaka, 599-8531, Japan

**Wadano, Akira,** Department of Applied Biochemistry, Osaka Prefecture University, Gakuencho 1-1, Sakai, Osaka, 599-8531, Japan

**Wang, Jing-Ai,** College of Geography and Remote Sensing Science, Beijing Normal University, No.19 Xinjiekouwai Street, Beijing 100875, China

**Wang, Lei,** College of Resources Science and Technology, Beijing Normal University, No.19 Xinjiekouwai Street, Beijing 100875, China

**Werner, Herbert,** Ecoclimatology, Technische Universität München, Am Hochanger 13, 85354 Freising, Germany

**Wieser, Gerhard,** Federal Office and Research Centre for Forests, Alpine Timberline Ecophysiology, Rennweg 1, 6020 Innsbruck, Austria

**Yamamoto, Susumu,** Institute for Environmental Management Technology, National Institute of Advanced Industrial Sciences and Technology, Onogawa 16-1, Tsukuba, Ibaraki 305-8569, Japan

**Yazaki, Ken'ichi,** Forestry and Forest Products Research Institute, Matsunosato 1, Tsukuba, Ibaraki 305-8687, Japan

**Yu, Yun-Jiang,** College of Resources Science and Technology, Beijing Normal University, No.19 Xinjiekouwai Street, Beijing 100875, China

**Zhao, Yanzhi,** College of Resources Science and Technology, Beijing Normal University, No.19 Xinjiekouwai Street, Beijing 100875, China

# I. Plant Responses to Air Pollution

# Metabolism of atmospheric sulfur gases in onion

Mark Durenkamp, Freek S. Posthumus, C. Elisabeth E. Stuiver, and Luit J. De Kok

Laboratory of Plant Physiology, University of Groningen, P.O. Box 14, 9750 AA Haren, The Netherlands

**Summary.** The impact of atmospheric sulfur gases was studied in onion (*Allium cepa* L.). The occurrence of toxic effects of $H_2S$ in onion depended not only on the atmospheric $H_2S$ level but also on the duration of the exposure. Prolonged exposure of onion to $\geq 0.3$ µl l$^{-1}$ $H_2S$ resulted in a strong reduction in shoot biomass production. $H_2S$ exposure resulted in a decrease in the organic N/S ratio at all levels (0.15 to 0.6 µl l$^{-1}$), which could be attributed to an increase in the pool of secondary sulfur compounds and not to changes in the sulfolipid content. The latter even decreased upon $H_2S$ exposure when expressed on a lipid basis. $SO_2$ exposure resulted in an enhanced content of sulfate and total sulfur in the shoot, whereas roots were not affected. In contrast to exposure to $H_2S$, $SO_2$ exposure did not result in an increase in non-protein organic (secondary) sulfur compounds, which showed that these compounds only were a sink pool for reduced atmospheric sulfur, when both the uptake of sulfate by the roots and its reduction in the shoot were bypassed.

**Key words.** *Allium cepa*, $H_2S$, $SO_2$, Sulfolipids, Sulfur metabolism

## 1. Introduction

Generally, sulfate taken up by the roots is used as the main source of sulfur for plants and the uptake, transport and subcellular distribution of sulfate are mediated by specific sulfate transporter proteins (Hawkesford 2003; Hawkesford et al. 2003; Buchner et al. 2004). The uptake of sulfate by the roots and its transport to other plant parts are highly regulated and the affinity of the sulfate transporters towards sulfate is high; a maximum uptake and transport rate is generally already reached at $\leq 0.1$ mM sulfate (Hawkesford and Wray 2000; Durenkamp and De Kok 2004; Buchner et al. 2004). The expression and activity of the sulfate transporter proteins, as well as the activity of the enzymes of the sulfate reduction pathway, strongly depend on the sulfur nutritional status of the plant (Buchner et al. 2004). Prior to its incorporation into organic compounds, sulfate needs to be reduced to sulfide, a process that primarily takes place in the chloroplasts. Subsequently, sulfide is incorporated into cysteine, the precursor for most other organic sulfur compounds (Fig. 1). In most plants the predominant proportion of the organic sulfur is present in the protein fraction as cysteine and methionine residues (up to 70 % of total S), however, species like onion also may contain high amounts of secondary sulfur compounds. Part of the organic sulfur is present in the lipid fraction; in general sulfoquinovosyldiacylglycerol (SQDG) appears to be the predominant plant sulfolipid and it accounts for 1 to 6 % of total S (Heinz 1993; De Kok et al. 1997; Benning 1998; Harwood and

---

*Plant Responses to Air Pollution and Global Change*
Edited by K. Omasa, I. Nouchi, and L. J. De Kok ( Springer-Verlag Tokyo 2005 )

Okanenko 2003).

In spite of their potential phytotoxic effects, foliarly deposited atmospheric sulfur gases as $H_2S$ and $SO_2$ can also be used as sulfur source for growth, and they even may be beneficial if the sulfate supply to the roots is limited (De Kok et al. 2000, 2002a,b; Durenkamp and De Kok 2004). Due to the impermeability of the cuticle, $H_2S$ and $SO_2$ are taken up via the stomates and their uptake is both dependent on the stomatal conductance

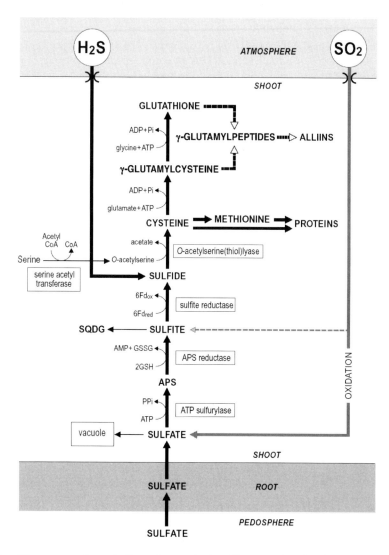

**Fig. 1.** Possible patterns of metabolism of atmospheric sulfur gases in onion (adapted from De Kok et al. 2002a). APS, adenosine 5'-phosphosulfate; Fd, ferredoxin; GSH, glutathione; SQDG, sulfoquinovosyldiacylglycerol.

and the internal (mesophyll) resistance towards these gases (De Kok et al. 1998, 2002a,b). The uptake of $H_2S$ is largely determined by the internal resistance, *viz.* the rate of metabolism of the absorbed sulfide into cysteine (Fig. 1). The rate of uptake depends on the activity of *O*-acetylserine(thiol)lyase and the availability of its substrate *O*-acetylserine (Stuiver and De Kok 2001) and it shows saturation kinetics with the atmospheric $H_2S$ level, which can be described by Michaelis-Menten kinetics (De Kok et al. 1998; Stuiver and De Kok 2001; Durenkamp and De Kok 2002). In contrast to $H_2S$, the uptake of $SO_2$ is largely determined by the stomatal conductance, since the internal resistance to $SO_2$ is low due to its high solubility and hydration in the cell sap. In general, there is a linear relation between the uptake of $SO_2$ and the level in the atmosphere (De Kok and Tausz 2001). Although $SO_2$, via sulfite, can directly be used in the sulfate reduction pathway, the greater part is oxidized to sulfate and transferred into the vacuole, especially at levels exceeding the sulfur requirement for growth (Fig. 1). Atmospheric sulfur gases have shown to be a useful tool to study sulfate uptake and sulfur assimilation by providing an extra source of sulfur taken up by the shoot, beyond the existing controls of sulfate uptake by the roots.

*Allium cepa* (onion) is one of the most important horticultural crops in the world. Secondary sulfur compounds (γ-glutamyl peptides and alliins) and their degradation products are responsible for the important role of *Allium* species in the food and phytopharmaceutical industry. The γ-glutamyl peptides are thought to act as precursors for the synthesis of alliins and they might have a function in the storage of sulfur and nitrogen (Randle and Lancaster 2002; Jones et al. 2004). The likely precursors for the synthesis of γ-glutamyl peptides and alliins are the thiol compounds γ-glutamyl cysteine and glutathione, which are products of the sulfur assimilation pathway (Fig. 1). In onion $H_2S$ exposure resulted in an increase in sulfate, thiols and other organic sulfur compounds in the shoot. The estimated N/S ratio of the latter compounds appeared to be 2 or less (Durenkamp and De Kok 2002, 2003, 2004), indicating that the increase could not be explained by an increase in the protein fraction (N/S ratio of proteins is generally around 40). It needs to be evaluated whether the increase in organic sulfur compounds upon $H_2S$ exposure was due to an accumulation of secondary sulfur compounds (γ-glutamyl peptides and alliins) and/or sulfolipids (Durenkamp and De Kok 2002, 2003, 2004). In addition, it needs to be assessed to what extent the observed accumulation of sulfur compounds is specific for $H_2S$ or the consequence of by-passing the regulatory control of the uptake of sulfate by the roots. In the present paper the impact of $H_2S$ and $SO_2$ on growth and sulfur metabolism has been compared. The significance of sulfolipids and secondary sulfur compounds as possible pool for excessive deposited atmospheric sulfur and the possible down-regulation of the sulfate reduction pathway upon $H_2S$ exposure are discussed.

## 2. Atmospheric $H_2S$: toxin vs. nutrient

Atmospheric sulfur gases are potentially phytotoxic, however, there is a large variation between species in the susceptibility towards these gases and the mechanisms of toxicity are still not completely understood. Like cyanide, sulfide complexes with high affinity to metallo groups in proteins (for instance heme-containing NADH oxidizing enzymes) and this reaction is probably the primary biochemical basis for the phytotoxicity of $H_2S$ (Maas and De Kok 1988; De Kok et al. 1998, 2002b). Mutagenic effects of accumulated thiol compounds (Glatt et al. 1983) or sulfide itself might also be a cause for the phytotoxicity of $H_2S$, since exposure to $H_2S$ resulted in an increase in chromosomal aberrations in apical meristems and root tips (Wonisch et al. 1999a,b; Stulen et al. 2000). In general, dicotyledons are more susceptible to $H_2S$ than monocotyledons, since in the latter $H_2S$ hardly has direct access to the vegetation point (Stulen et al. 2000).

Onion and related *Allium* species, as monocotyledons, were not very susceptible to the toxic effects of $H_2S$ (Durenkamp and De Kok 2002, 2003, 2004). A one-week exposure up to 0.6 µl $l^{-1}$ $H_2S$, a level which by far exceeds the sulfur requirement for growth, did

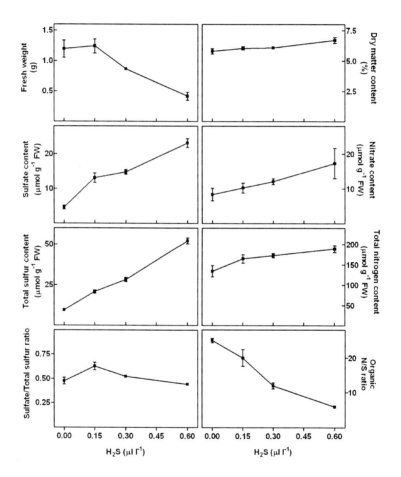

**Fig. 2.** Impact of prolonged $H_2S$ exposure on growth and sulfur and nitrogen metabolism of onion shoots (*Allium cepa* L.). Seedlings were grown in vermiculite for two weeks and subsequently transferred to a regular potting soil and exposed to 0, 0.15, 0.3 and 0.6 µl $l^{-1}$ $H_2S$ for 38 days. Fresh weight (g), dry matter content (%), metabolite contents (µmol $g^{-1}$ FW), sulfate/total sulfur ratio and organic N/S ratio of the shoot were determined as described in Durenkamp and De Kok (2002, 2004). Data represent the mean of three measurements with five plants in each (± SD).

not result in a reduction of growth in onion (Durenkamp and De Kok 2004). However, prolonged exposure to the same range of $H_2S$ levels for 38 days resulted in a substantial decrease in biomass production and a slight increase in dry matter content in onion shoots at levels $\geq 0.3$ µl l$^{-1}$ $H_2S$ (Fig. 2). Apparently, the occurrence of toxic effects of $H_2S$ in onion depended not only on the atmospheric $H_2S$ level but also on the duration of the exposure. The latter might be due to a cumulative effect of sulfide or produced toxic metabolites for instance in meristematic tissue. Prolonged exposure to $H_2S$ resulted in an increased content of sulfate and other sulfur-containing compounds, as illustrated by a maximal five-fold increase in the total sulfur content of the shoot upon exposure up to 0.6 µl l$^{-1}$ $H_2S$ (Fig. 2). The organic N/S ratio was decreased at all levels of $H_2S$ exposure, independent of the effects of $H_2S$ phytotoxicity (Fig. 2). The decrease in the organic N/S ratio could be attributed to an increase in non-protein organic (secondary) sulfur compounds, which pool might be a sink for reduced sulfur (Durenkamp and De Kok 2002, 2003, 2004). Prolonged $H_2S$ exposure also resulted in an enhancement of nitrogen-containing compounds in the shoot, which possibly was the consequence of a disturbed metabolism and/or an alteration in tissue and shoot development.

Atmospheric $H_2S$ could be used as a sulfur source for growth in onion, especially when the sulfate supply to the roots was deprived (Durenkamp and De Kok 2004). However, upon prolonged exposure $H_2S$ appeared to be phytotoxic and it reduced biomass production.

## 3. Impact of H₂S exposure on sulfolipids

The main plant sulfolipid sulfoquinovosyldiacylglycerol (SQDG) is synthesized from UDP-sulfoquinovose and diacylglycerol with sulfite as the likely sulfur precursor (Sanda et al. 2001; Harwood and Okanenko 2003). Sulfite is synthesized from APS by APS reductase and this enzyme is the predominant site of regulatory control of the sulfate reduction pathway (De Kok et al. 2002a; Vauclare et al. 2002). The sulfolipid content of the shoot (expressed on a lipid basis) decreased upon exposure to $H_2S$ (Table 1), which could be caused by a down-regulation of the sulfate reduction pathway and by a subsequent decrease in sulfite production, the sulfur precursor of SQDG (Sanda et al. 2001). This suggestion is supported by observations in *Brassica oleracea*, where a similar decrease in sulfolipid content (expressed on a lipid basis) was observed upon $H_2S$ exposure (De Kok et al. 1997). The sulfate reduction pathway is known to be down-regulated via APS reductase upon $H_2S$ exposure in *B. oleracea* (Westerman et al. 2001b). Since the sulfolipid content was not increased upon exposure to $H_2S$, sulfolipids did not act as a sink pool for atmospheric reduced sulfur.

The total lipid content of the shoot was increased upon exposure to $H_2S$, which could not be explained by an increase in either sulfolipid or pigment content (Table 1). It needs to be evaluated to what extent this increase in lipid content upon $H_2S$ exposure can be attributed to changes in the overall structure and/or composition of membranes. Another option for the increase in total lipid content could be the formation of vesicles containing secondary sulfur compounds (as suggested by Turnbull et al. 1981). The possible enhancement of secondary sulfur compounds content in the shoot might be accompanied with a subsequent increase in vesicle formation resulting in an increase in the total lipid content. The latter was not observed in *Brassica oleracea* (De Kok et al. 1997), since in

**Table 1.** Impact of short-term $H_2S$ exposure on pigment content in shoot and lipid content in shoot and roots of onion (*Allium cepa* L.). Seedlings were grown in vermiculite for two weeks and subsequently transferred to a 25% Hoagland nutrient solution. Four-week-old seedlings were transferred to a fresh nutrient solution and exposed to 0.3 µl l$^{-1}$ $H_2S$ for one week. Total lipid content and sulfolipid content in shoot and roots were determined as described by De Kok et al. (1997) and the content of chlorophylls and carotenoids in the shoot was measured as described by Lichtenthaler (1987). Data represent the mean of three measurements with 12 plants in each (± SD).

| | 0 µl l$^{-1}$ $H_2S$ | 0.3 µl l$^{-1}$ $H_2S$ |
|---|---|---|
| **Shoot** | | |
| Total lipid content (mg g$^{-1}$ FW) | 3.60 ± 0.09 | 4.26 ± 0.17** |
| Sulfolipid content (nmol g$^{-1}$ FW) | 89.0 ± 6.1 | 86.6 ± 2.6 |
| Sulfolipid content (nmol mg$^{-1}$ total lipids) | 24.7 ± 2.1 | 20.7 ± 0.2* |
| Sulfolipid content (nmol mg$^{-1}$ chlorophyll) | 189 ± 5 | 187 ± 10 |
| Total chlorophyll content (mg g$^{-1}$ FW) | 0.47 ± 0.03 | 0.46 ± 0.01 |
| Total carotenoid content (mg g$^{-1}$ FW) | 0.11 ± 0.00 | 0.11 ± 0.00 |
| **Root** | | |
| Total lipid content (mg g$^{-1}$ FW) | 1.44 ± 0.12 | 1.52 ± 0.15 |
| Sulfolipid content (nmol g$^{-1}$ FW) | 36.5 ± 4.2 | 37.3 ± 4.2 |
| Sulfolipid content (nmol mg$^{-1}$ total lipids) | 25.3 ± 1.0 | 24.5 ± 0.3 |

*$P<0.05$; **$P<0.01$ vs 0 µl l$^{-1}$ $H_2S$; Student's *t*-test.

this species an accumulation of secondary sulfur compounds was absent upon $H_2S$ exposure (Westerman et al. 2001a).

The observed increase in the non-protein organic sulfur content upon $H_2S$ exposure (Durenkamp and De Kok 2002, 2003, 2004) could not be attributed to changes in the content of sulfolipids. Therefore, secondary sulfur compounds appeared to be the most likely pool for excessive deposited atmospheric sulfur in onion.

## 4. Impact of atmospheric $SO_2$ on sulfur metabolism: a comparison with $H_2S$

In general, plant exposure to $SO_2$ results in an increase in the sulfate content and a slight increase in the thiol content (mainly glutathione) of the shoot since part of the $SO_2$ can be assimilated into organic sulfur compounds via sulfite (De Kok and Tausz 2001; Tausz et al. 2003; Yang et al. 2003).

Growth of onion was not affected upon exposure to 0.3 µl l$^{-1}$ $SO_2$ (Table 2). An increase in the sulfate and total sulfur content of the shoot was observed upon exposure to $SO_2$ in both sulfate-sufficient and sulfate-deprived plants, whereas the content in the roots was not affected (Table 2). The increase in the total sulfur content of the shoot in sulfate-sufficient plants could solely be explained by an increase in the sulfate content (Table 2). Apparently, $SO_2$ was for the greater part oxidized to sulfate and transferred into the vacuole (Fig. 1). In contrast to exposure to $H_2S$, $SO_2$ exposure did not result in a significant

**Table 2.** Impact of sulfate nutrition and short-term SO$_2$ exposure on growth and sulfur metabolism in shoot and roots of onion (*Allium cepa* L.). Seedlings were grown in vermiculite for two weeks and transferred to a 25% Hoagland nutrient solution. Four-week-old seedlings were transferred to a fresh nutrient solution with 0 (-S) or 0.5 (+S) mM sulfate and exposed to 0 (-SO$_2$) or 0.3 (+SO$_2$) µl l$^{-1}$ SO$_2$ for one week. Fresh weight (g), sulfate and total sulfur content (µmol g$^{-1}$ FW) and sulfate/total sulfur ratio in shoot and roots were determined as described in Durenkamp and De Kok (2002, 2004). Data represent the mean of four measurements with 12 or 24 (initial) plants in each (± SD). Different letters indicate significant differences between treatments (P<0.05, Student's *t*-test).

|  | Initial | -S | -S +SO$_2$ | +S | +S +SO$_2$ |
|---|---|---|---|---|---|
| **Shoot** | | | | | |
| Fresh weight | 0.48 ± 0.05 | 1.10 ± 0.04$^a$ | 1.12 ± 0.06$^{ab}$ | 0.98 ± 0.23$^{ab}$ | 1.27 ± 0.13$^b$ |
| Total sulfur content | 9.0 ± 0.3 | 4.0 ± 0.3$^a$ | 9.3 ± 0.3$^b$ | 8.5 ± 1.2$^b$ | 14.8 ± 1.2$^c$ |
| Sulfate content | 2.6 ± 0.2 | 0.6 ± 0.0$^a$ | 4.7 ± 0.2$^c$ | 3.6 ± 0.5$^b$ | 9.0 ± 0.5$^d$ |
| Sulfate/total sulfur | 0.29 ± 0.03 | 0.14 ± 0.03$^a$ | 0.50 ± 0.02$^c$ | 0.43 ± 0.02$^b$ | 0.61 ± 0.03$^d$ |
| **Root** | | | | | |
| Fresh weight | 0.23 ± 0.02 | 0.43 ± 0.03$^a$ | 0.42 ± 0.06$^a$ | 0.40 ± 0.08$^a$ | 0.46 ± 0.03$^a$ |
| Total sulfur content | 9.2 ± 0.7 | 4.1 ± 0.2$^a$ | 4.3 ± 0.6$^a$ | 8.9 ± 0.3$^b$ | 9.5 ± 0.4$^b$ |
| Sulfate content | 5.6 ± 0.5 | 0.9 ± 0.3$^a$ | 0.8 ± 0.3$^a$ | 5.1 ± 0.2$^b$ | 5.5 ± 0.2$^c$ |
| Sulfate/total sulfur | 0.61 ± 0.05 | 0.21 ± 0.08$^a$ | 0.18 ± 0.06$^a$ | 0.58 ± 0.02$^b$ | 0.58 ± 0.01$^b$ |

decrease in the organic N/S ratio of the shoot of sulfate-sufficient plants (27.7 ± 1.8 and 23.9 ± 3.5 at 0 and 0.3 µl l$^{-1}$ SO$_2$, respectively). As has been indicated above, a decrease in the organic N/S ratio upon H$_2$S exposure could likely be attributed to an increase in secondary sulfur compounds (Durenkamp and De Kok 2002, 2003, 2004). These compounds only seemed to be a sink for reduced atmospheric sulfur like H$_2$S, via by-passing of the sulfate uptake in the roots and its reduction in the shoot, and not for oxidized (atmospheric) sulfur like SO$_2$. The reduction of sulfate is known to be highly regulated (De Kok et al. 2002a; Vauclare et al. 2002), in contrast to the uptake of SO$_2$, which resulted in an accumulation of sulfate upon SO$_2$ exposure. Sulfate accumulation was not observed when onion was subjected to increasing levels of pedospheric sulfate, since uptake of sulfate by the roots was strictly regulated (Hawkesford and Wray 2000; Durenkamp and De Kok, 2004; Buchner et al. 2004). A combination of H$_2$S exposure and different levels of pedospheric sulfate nutrition will be used to further investigate the regulation of sulfate uptake, transport, subcellular distribution and reduction through APS reductase, since these processes predominantly control the assimilation of sulfate in plants.

# References

Benning C (1998) Biosynthesis and function of the sulfolipid sulfoquinovosyl diacylglycerol. Ann Rev Plant Physiol Plant Mol Biol 49:53-75

Buchner P, Stuiver CEE, Westerman S, Wirtz M, Hell R, Hawkesford MJ, De Kok LJ (2004) Regulation of sulfate uptake and expression of sulfate transporter genes in *Brassica oleracea* as affected by atmospheric H$_2$S and pedospheric sulfate nutrition. Plant Physiol 136:3396-3408

De Kok LJ, Tausz M (2001) The role of glutathione in plant reaction and adaptation to air pollutants. In: Grill D, Tausz M, De Kok LJ (Eds) Significance of glutathione in plant adaptation to the environment. Kluwer Academic Publishers, Dordrecht, pp 185-206

De Kok LJ, Stuiver CEE, Rubinigg M, Westerman S, Grill D (1997) Impact of atmospheric sulfur deposition on sulfur metabolism in plants: $H_2S$ as sulfur source for sulfur deprived *Brassica oleracea* L. Bot Acta 110:411-419

De Kok LJ, Stuiver CEE, Stulen I (1998) Impact of atmospheric $H_2S$ on plants. In: De Kok LJ, Stulen I (Eds) Responses of plant metabolism to air pollution and global change. Backhuys Publishers, Leiden, pp 51-63

De Kok LJ, Westerman S, Stuiver CEE, Stulen I (2000) Atmospheric $H_2S$ as plant sulfur source: interaction with pedospheric sulfur nutrition – a case study with *Brassica oleracea* L. In: Brunold C, Rennenberg H, De Kok LJ, Stulen I, Davidian J-C (Eds) Sulfur nutrition and sulfur assimilation in higher plants: molecular, biochemical and physiological aspects. Paul Haupt, Bern, pp 41-55

De Kok LJ, Castro A, Durenkamp M, Stuiver CEE, Westerman S, Yang L, Stulen I (2002a) Sulphur in plant physiology. Proc No 500, International Fertiliser Society, York, pp 1-26

De Kok LJ, Stuiver CEE, Westerman S, Stulen I (2002b) Elevated levels of hydrogen sulfide in the plant environment: nutrient or toxin. In: Omasa K, Saji H, Youssefian S, Kondo N (Eds) Air pollution and plant biotechnology – prospects for phytomonitoring and phytoremediation. Springer, Tokyo, pp 201-219

Durenkamp M, De Kok LJ (2002) The impact of atmospheric $H_2S$ on growth and sulfur metabolism of *Allium cepa* L. Phyton 42(3):55-63

Durenkamp M, De Kok LJ (2003) Impact of atmospheric $H_2S$ on sulfur and nitrogen metabolism in *Allium* species and cultivars. In: Davidian J-C, Grill D, De Kok LJ, Stulen I, Hawkesford MJ, Schnug E, Rennenberg H (Eds) Sulfur transport and assimilation in plants: regulation, interaction, signaling. Backhuys Publishers, Leiden, pp 197-199

Durenkamp M, De Kok LJ (2004) Impact of pedospheric and atmospheric sulphur nutrition on sulphur metabolism of *Allium cepa* L., a species with a potential sink capacity for secondary sulphur compounds. J Exp Bot 55:1821-1830

Glatt H, Protić-Sabljić M, Oesch F (1983) Mutagenicity of glutathione and cysteine in the Ames test. Science 220:961-963

Harwood JL, Okanenko AA (2003) Sulphoquinovosyl diacylglycerol (SQDG) – the sulpholipid of higher plants. In: Abrol YP, Ahmad A (Eds) Sulphur in plants. Kluwer Academic Publishers, Dordrecht, pp 189-219

Hawkesford MJ (2003) Transporter gene families in plants: the sulphate transporter gene family – redundancy or specialization? Physiol Plant 117:155-163

Hawkesford MJ, Wray JL (2000) Molecular genetics of sulphate assimilation. Adv Bot Res 33:159-223

Hawkesford MJ, Buchner P, Hopkins L, Howarth JR (2003) Sulphate uptake and transport. In: Abrol YP, Ahmad A (Eds) Sulphur in plants. Kluwer Academic Publishers, Dordrecht, pp 71-86

Heinz E (1993) Recent investigations on the biosynthesis of the plant sulfolipid. In: De Kok LJ, Stulen I, Rennenberg H, Brunold C, Rauser WE (Eds) Sulfur nutrition and assimilation in higher plants: regulatory, agricultural and environmental aspects. SPB Academic Publishing, The Hague, pp 163-178

Jones MG, Hughes J, Tregova A, Milne J, Tomsett AB, Collin HA (2004) Biosynthesis of the flavour precursors of onion and garlic. J Exp Bot 55:1903-1918

Lichtenthaler HK (1987) Chlorophylls and carotenoids: pigments of photosynthetic biomembranes. Methods in Enzymology 148:350-382

Maas FM, De Kok LJ (1988) In vitro NADH oxidation as an early indicator for growth reduction in spinach exposed to $H_2S$ in the ambient air. Plant Cell Physiol 29:523-526

Randle WM, Lancaster JE (2002) Sulphur compounds in *Alliums* in relation to flavour quality. In: Rabinowitch HD, Currah L (Eds) Allium crop science: recent advances. CAB International, Wallingford, pp 329-356

Sanda S, Leustek T, Theisen MJ, Garavito RM, Benning C (2001) Recombinant *Arabidopsis* SQD1 converts UDP-glucose and sulfite to the sulfolipid head group precursor UDP-sulfoquinovose in vitro. J Biol Chem 276:3941-3946

Stuiver CEE, De Kok LJ (2001) Atmospheric $H_2S$ as sulfur source for *Brassica oleracea*: kinetics of $H_2S$ uptake and activity of $O$-acetylserine (thiol)lyase as affected by sulfur nutrition. Environ Exp Bot 46:29-36

Stulen I, Posthumus F, Amâncio S, Masselink-Beltman I, Müller M, De Kok LJ (2000) Mechanisms of $H_2S$ phytotoxicity. In: Brunold C, Rennenberg H, De Kok LJ, Stulen I, Davidian J-C (Eds) Sulfur nutrition and sulfur assimilation in higher plants: molecular, biochemical and physiological aspects. Paul Haupt, Bern, pp 381-383

Tausz M, Weidner W, Wonisch A, De Kok LJ, Grill D (2003) Uptake and distribution of $^{35}$S-sulfate in needles and roots of spruce seedlings as affected by exposure to $SO_2$ and $H_2S$. Environ Exp Bot 50:211-220

Turnbull A, Galpin IJ, Smith JL, Collin HA (1981) Comparison of the onion plant (*Allium cepa*) and onion tissue culture. IV. Effect of shoot and root morphogenesis on flavour precursor synthesis in onion tissue culture. New Phytol 87:257-268

Vauclare P, Kopriva S, Fell D, Suter M, Sticher L, von Ballmoos P, Krähenbühl U, Op den Camp R, Brunold C (2002) Flux control of sulphate assimilation in *Arabidopsis thaliana*: adenosine 5'-phosphosulphate reductase is more susceptible than ATP sulphurylase to negative control by thiols. Plant J 31:729-740

Westerman S, Blake-Kalff MMA, De Kok LJ, Stulen I (2001a) Sulfate uptake and utilization by two different varieties of *Brassica oleracea* with different sulfur need as affected by atmospheric $H_2S$. Phyton 41:49-62

Westerman S, Stulen I, Suter M, Brunold C, De Kok LJ (2001b) Atmospheric $H_2S$ as sulphur source for *Brassica oleracea*: consequences for the activity of the enzymes of the assimilatory sulphate reduction pathway. Plant Physiol Biochem 39:425-432

Wonisch A, Tausz M, Müller M, Weidner W, De Kok LJ, Grill D (1999a) Low molecular weight thiols and chromosomal aberrations in *Picea omorika* upon exposure to two concentrations of $H_2S$. Phyton 39(3):167-170

Wonisch A, Tausz M, Müller M, Weidner W, De Kok LJ, Grill D (1999b) Treatment of young spruce shoots with $SO_2$ and $H_2S$: effects on fine root chromosomes in relation to changes in the thiol content and redox state. Water Air Soil Pollut 116:423-428

Yang L, Stulen I, De Kok LJ (2003) Interaction between atmospheric sulfur dioxide deposition and pedospheric sulfate nutrition in Chinese cabbage. In: Davidian J-C, Grill D, De Kok LJ, Stulen I, Hawkesford MJ, Schnug E, Rennenberg H (Eds) Sulfur transport and assimilation in plants: regulation, interaction, signaling. Backhuys Publishers, Leiden, pp 181-183

# Impact of atmospheric $NH_3$ deposition on plant growth and functioning – a case study with *Brassica oleracea* L.

Ana Castro, Ineke Stulen, and Luit J. De Kok

Laboratory of Plant Physiology, University of Groningen, P.O. Box 14, 9750 AA Haren, The Netherlands

**Summary.** *Brassica oleracea* L. (curly kale) was exposed to 0, 2, 4, 6 and 8 $\mu l\ l^{-1}$ $NH_3$ during one week and the impact on growth and N compounds was determined. Exposure to $NH_3$ increased shoot biomass production at 2 and 4 $\mu l\ l^{-1}$, but resulted in an inhibition of shoot and root growth at 6 and 8 $\mu l\ l^{-1}$. Shoot to root ratio was not affected up to 4 $\mu l\ l^{-1}$, but decreased at higher levels. Shoot total N content was increased at all levels, mainly due to the increase in free amino acids. Even at atmospheric $NH_3$ levels, at which the foliarly absorbed $NH_3$ would cover a limited proportion of N requirement there was already an enhancement of the nitrogen content of the shoots and roots. Apparently there was no direct regulatory control of and/or interaction between atmospheric and pedospheric nitrogen utilization in *B. oleracea*. It needs to be evaluated to what extent foliarly absorbed $NH_3$ is used as nitrogen source for growth.

**Key words** Ammonia, *Brassica oleracea*, Nitrogen pollutants, Nutrient, Toxin

## 1. Atmospheric N deposition in Europe

$NH_3$ is a major air pollutant, which accounts for up to 80% of the total N deposition in central Europe (Fangmeier et al. 1994; Gessler and Rennenberg 1998; Krupa 2003). Atmospheric $NH_3$ pollution is the consequence of intensive farming activities (animal manure and fertilizer use), and to a lesser extent to anthropogenic sources and natural background emissions (Leith et al. 2002; Krupa 2003; Pitcairn et al. 2003; Erisman and Schaap 2004). High $NH_3$ emissions and consequently, excessive N deposition will lead to direct phytotoxic effects, eutrophication and acidification (Stulen et al. 1998; Rennenberg and Gessler 1999; Krupa 2003). The toxic effect of $NH_3$ has often been ascribed to nutrient imbalances due to cation release (Wollenweber and Raven 1993).

While the impact of atmospheric N deposition on ecosystems such as heathlands (Van der Eerden et al. 1991; Leith et al. 2002; Sheppard and Leith 2002) and forests (Högberg et al. 1998; Rennenberg and Gessler 1999; Bassirirad 2000) has been studied in detail, fewer studies have dealt with its impact on crop plants (Van der Eerden 1982; Clement et al. 1997). In addition, there are hardly any data available on the contribution of foliar uptake of atmospheric $NH_3$ to the plant's N requirement for growth (Pérez-Soba and Van der Eerden 1993; Stulen et al. 1998).

---

*Plant Responses to Air Pollution and Global Change*
Edited by K. Omasa, I. Nouchi, and L. J. De Kok ( Springer-Verlag Tokyo 2005 )

## 2. Foliar uptake and metabolism of $NH_3$

The uptake of $NH_3$ shows a diurnal variation and is dependent on the water status of the plant, temperature, light intensity, internal $CO_2$ level and nutrient availability (Hutchinson et al. 1972; Rogers and Aneja 1980; Van Hove et al. 1987; Husted and Schjoerring 1996; Schjoerring et al. 1998). The foliar uptake of $NH_3$ is determined by the stomatal conductance and the internal (mesophyll) resistance to the gas and its uptake via the cuticle surface can be neglected (Krupa 2003). The internal resistance of the mesophyll cells appears to be the limiting factor for foliar uptake of $NH_3$ (Hutchinson et al. 1972). The internal resistance to $NH_3$ is low, since this gas is highly water-soluble and in addition it is rapidly converted into $NH_4^+$ in the aqueous phase of the mesophyll cells (Fangmeier et al. 1994). $NH_3$ uptake takes place as long as the atmospheric level exceeds the internal $NH_4^+$ level (Husted and Schjoerring 1996).

The $NH_4^+$ formed in the mesophyll cells may be assimilated by the glutamine synthetase/glutamate synthase cycle (Lea and Mifflin 1974; Pérez-Soba et al. 1994; Pearson and Soares 1998). Foliar $NH_3$ uptake may affect plant metabolism in various ways and result in changes in parameters as metabolic compounds, enzyme activity, root uptake and plant growth (Pérez-Soba et al. 1994; Gessler and Rennenberg 1998; Pearson and Soares 1998). Metabolic changes related to the $NH_3$ assimilatory capacity of the plant generally lead to an increase in the pool of N-containing metabolites, such as amino acids and total N content (Van Dijk and Roelofs 1988; Pérez-Soba et al. 1994; Clement et al. 1997; Gessler and Rennenberg 1998). Visible symptoms, such as black spots and necrosis in the leaves, arise when $NH_3$ uptake by the shoot exceeds the assimilation capacity of the plant (Van der Eerden 1982; Fangmeier et al. 1994).

## 3. Impact of $NH_3$ on growth and N metabolism of *Brassica oleracea*

The present case study was aimed at investigating the impact of a range of $NH_3$ levels on growth and N metabolism of *Brassica oleracea* L. Plants were grown on a Hoagland nutrient solution containing 3.75 mM nitrate (for experimental details see Castro et al. 2004). *B. oleracea* was chosen because it is an economically important crop plant with a relatively high RGR, and it is a suitable species because of its preference for nitrate (Pearson and Stewart 1993) as well as its sensitivity to $NH_4^+$ (Britto and Kronzucker 2002). *Brassica* species originate from saline, sulfur-rich environments and are considered to have a high S requirement for growth (Westerman et al. 2000). Therefore, the impact of $NH_3$ on S compounds was measured as well.

Upon $NH_3$ exposure the shoot biomass production was slightly increased at levels up to 4 $\mu$l l$^{-1}$, whereas it was decreased at levels $\geq$ 6 $\mu$l l$^{-1}$ $NH_3$. Root biomass production was decreased significantly at 6 and 8 $\mu$l l$^{-1}$ $NH_3$, showing that exposure of the shoot to $NH_3$ had a negative effect on root growth (in the used experimental conditions, the formation of $NH_4^+$, by dissolution of atmospheric $NH_3$ into the nutrient solution, was prevented). Relative growth rate (RGR), calculated on a plant basis was only significantly decreased at 8 $\mu$l l$^{-1}$ $NH_3$. Exposure to 6 and 8 $\mu$l l$^{-1}$ $NH_3$ affected root biomass production relatively more than shoot biomass production, resulting in a higher shoot to root ratio (S/R,

**Table 1.** Impact of NH$_3$ on growth of *Brassica oleracea*. Seedlings (26 days old) were exposed for 7 days. Shoot and root growth (g FW) was calculated by subtracting the final fresh weight from the initial fresh weight. RGR, relative growth rate (g g$^{-1}$ day$^{-1}$) on a plant basis. S/R, shoot to root ratio on a fresh weight basis. DMC, dry matter content (%). Data represent the mean of 2 experiments, with 3 measurements per experiment with 3 plants in each (±SD). Means followed by different letters are statistically different at p< 0.01. Statistical analysis was performed by using an unpaired Student's t-test. For further experimental details see Castro et al. (2004).

| [NH$_3$] | 0 µl l$^{-1}$ | 2 µl l$^{-1}$ | 4 µl l$^{-1}$ | 6 µl l$^{-1}$ | 8 µl l$^{-1}$ |
|---|---|---|---|---|---|
| Shoot growth | 1.90±0.07$^c$ | 2.04±0.04$^b$ | 2.45±0.25$^b$ | 1.78±0.45$^a$ | 1.69±0.30$^a$ |
| Root growth | 0.55±0.20$^b$ | 0.36±0.06$^b$ | 0.48±0.08$^b$ | 0.10±0.07$^a$ | 0.20±0.11$^a$ |
| RGR | 0.20±0.01$^a$ | 0.20±0.01$^a$ | 0.20±0.01$^a$ | 0.16±0.04$^a$ | 0.15±0.03$^b$ |
| S/R | 3.3±0.6$^a$ | 4.2±0.3$^a$ | 4.1±0.4$^a$ | 5.8±1.1$^b$ | 5.9±1.5$^b$ |
| Shoot DMC | 14.1±1.2$^a$ | 14.2±1.5$^a$ | 13.1±1.0$^a$ | 13.0±1.2$^a$ | 14.0±0.9$^a$ |
| Root DMC | 6.4±1.2$^a$ | 6.1±0.8$^a$ | 7.2±0.5$^a$ | 11±0.4$^c$ | 9.1±0.9$^b$ |

Table 1). Shoot dry matter content (DMC) was not affected upon exposure to NH$_3$, whereas root dry matter content was decreased at 6 and 8 µl l$^{-1}$ NH$_3$ (Table 1).

Exposure to NH$_3$ resulted in a substantial increase in shoot total N content at all atmospheric levels (Fig. 1a). This was mainly due to an increase in the soluble N fraction (amino acids, amides and NH$_4^+$), viz. 1.5 fold and 5.6-fold at 4 µl l$^{-1}$ and at 8 µl l$^{-1}$, respectively, compared to that of the control (0 µl l$^{-1}$, results not shown). Root total N content was only increased at 2 µl l$^{-1}$ NH$_3$ (Fig. 1a). Shoot nitrate content was increased at all NH$_3$ levels, but most at 4 µl l$^{-1}$. Root nitrate content was increased at 2 µl l$^{-1}$, not affected at 4 µl l$^{-1}$, and decreased at 8 µl l$^{-1}$ (Fig. 1c). The free amino acid content in the shoot increased with increasing NH$_3$ levels (8% and 15% at 4 µl l$^{-1}$ and 8 µl l$^{-1}$, respectively), while no effect was observed in the roots (Fig. 1e).

Shoot sulfur content was not affected by exposure to 2 µl l$^{-1}$ NH$_3$, but decreased at higher levels. Root total sulfur was increased at 2 to 6 µl l$^{-1}$, and decreased at 8 µl l$^{-1}$ (Fig. 1b). Shoot sulfate content was increased at 4 µl l$^{-1}$, and decreased at 6 and 8 µl l$^{-1}$. Root sulfate content was increased at 2 µl l$^{-1}$, not changed at 4 µl l$^{-1}$, and decreased at 6 and 8 µl l$^{-1}$ (Fig. 1d).

The impact of atmospheric NH$_3$ on total S and sulfate (Fig. 1b,d) can be explained by changes in RGR (Table 1), rather than by a direct effect of NH$_3$ exposure on S compounds. Noteworthy is the relatively high sulfate content found in this species. Other experiments with *Brassica* seedlings also showed that a high percentage (90%) of total S is present as sulfate, and only 10% as organic S (Castro et al. 2003). Therefore, for this species the definition of "sulfur requirement for growth" may have to be redefined, as "organic sulfur need for growth" (Castro et al. 2003). In the shoot, the organic N/S ratio increased with increasing NH$_3$ levels, which correlates well with the increase in free amino acid content. Changes in the organic N/S ratio in the root were minor.

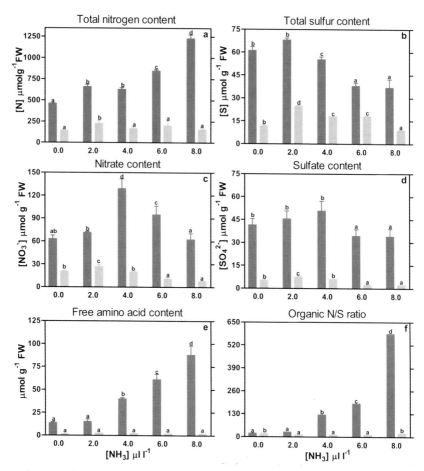

**Fig. 1.** Impact of $NH_3$ on N and S compounds in *Brassica oleracea*. Seedlings (26 days old) were exposed for 7 days. Shoot data is given in dark-grey bars, root data in light-grey bars. Data on total nitrogen, nitrate and free amino acids contents represent the mean of 2 experiments, with 3 measurements per experiment with 3 plants in each (±SD). Data on total S and sulfate content represent the mean of 3 measurements with 3 plants in each (±SD). The organic N/S ratio, a parameter was calculated by subtracting the nitrate and sulfate contents from total nitrogen and sulfur contents, respectively. Different letters indicate significant differences at $p< 0.01$. Statistical analysis was performed by using an unpaired Student's t-test. For experimental details see Castro et al. (2004).

## 4. Impact of $NH_3$ on nitrate uptake by *Brassica oleracea*

The net nitrate uptake rate (NNUR) was not affected at 2 µl l$^{-1}$ but was reduced by 25% upon exposure to ≥4 µl l$^{-1}$ $NH_3$ (Table 2). It has been suggested that a decrease in NNUR upon $NH_3$ may be due to a down-regulation of the nitrate transporters by reduced N

**Table 2.** Net nitrate uptake rate by the root and foliar atmospheric NH$_3$ uptake by *Brassica oleracea* seedlings. Net nitrate uptake rate (NNUR; $\mu$mol g$^{-1}$ FW plant day$^{-1}$) was measured according to Stuiver et al. (1997). Atmospheric NH$_3$-enhanced N flux ($\mu$mol g$^{-1}$ FW plant day$^{-1}$) was calculated from the data on RGR and total N presented in Table 1 and Fig. 1, respectively. Atmospheric NH$_3$ uptake ($\mu$mol g$^{-1}$ FW plant day$^{-1}$) was calculated by adding the increase in total plant N flux to the decrease in NNUR upon exposure NH$_3$. Values between brackets represent the atmospheric NH$_3$ uptake as a percentage of NNUR at 0 $\mu$l l$^{-1}$ NH$_3$. Means followed by different letters are statistically different at p< 0.01. Statistical analysis was performed by using an unpaired Student's t-test.

| [NH$_3$] | 0 $\mu$l l$^{-1}$ | 2 $\mu$l l$^{-1}$ | 4 $\mu$l l$^{-1}$ | 6 $\mu$l l$^{-1}$ | 8 $\mu$l l$^{-1}$ |
|---|---|---|---|---|---|
| NNUR | 103±6$^b$ | 102±8$^b$ | 74±13$^a$ | 70±5$^a$ | 76±7$^a$ |
| NH$_3$-enhanced N flux | --- | 28±4$^a$ | 38±7$^a$ | 43±6$^a$ | 83±7$^b$ |
| NH$_3$ uptake | --- | 29(28) | 67(65) | 76(73) | 110(106) |

compounds (for instance specific amino acids or its derivatives) transferred from the shoot to the root via the phloem (Stulen et al. 1998 and references therein). However, the present data provide no direct support for the involvement of amino acids in the decrease in NNUR, since its level in the root was not affected (Fig. 1e) upon NH$_3$ exposure, even not at high levels.

## 5. Atmospheric NH$_3$ – nutrient or toxin?

The requirement for growth of an element can be defined as "the minimum rate of uptake and utilization, which is sufficient to obtain the maximum yield, quality and fitness". Physiologically, the plant's N requirement for growth can be expressed as the rate of N uptake and its assimilation per gram plant biomass produced with time. Similar to observations with S, the N requirement for growth is dependent on the ontogeny and developmental stage of the plant (De Kok et al. 2000, 2002). In the vegetative stage, and at optimal nutrient supply, the plant's N requirement for growth can be estimated from data on relative growth rate (RGR) and plant N content as "$N_{requirement} = N_{content} * RGR$" (after Williams 1948). Based on model calculations, Stulen et al. (1998) estimated that in theory at an atmospheric level of 2 $\mu$l l$^{-1}$ NH$_3$, the foliar uptake might contribute up to 50% to the N requirement of plants growing at a RGR of 0.2 g g$^{-1}$ day$^{-1}$ (may vary between species differing in N requirement). From the current experimental data the foliar uptake of NH$_3$ at the various atmospheric levels and its possible contribution to the plant's N requirement could be estimated. NH$_3$ exposure resulted in enrichment of the total N content of the plant (Fig.1), which could analogue to the formula of William (1948) be defined as an atmospheric NH$_3$-enhanced N flux and calculated as follows:

Atmospheric NH$_3$-enhanced N flux at [NH$_3$] =
$$\{((RGR * N_{content}) \text{ at } [NH_3] - ((RGR * N_{content}) \text{ at } 0 \mu l \, l^{-1} NH_3)\}$$

[NH$_3$] represents the NH$_3$ level plants have been exposed to. The atmospheric NH$_3$-enhanced N flux is expressed as $\mu$mol N g$^{-1}$ FW plant day$^{-1}$. The net nitrate uptake rate

(NNUR) was decreased upon exposure to 4 to 8 µl l$^{-1}$ NH$_3$ (Table 2). If one would assume that the observed decrease in NNUR would be due to a partial transfer of the plant from pedospheric nitrate to foliarly absorbed NH$_3$ as nitrogen source for growth, then the sum of the atmospheric NH$_3$-enhanced N flux and the decrease in NNUR ($\mu$mol N g$^{-1}$ FW plant day$^{-1}$) would represent the total atmospheric NH$_3$ uptake by the plant. On basis of these estimations the possible contribution of the atmospheric NH$_3$ uptake by the foliage at 2, 4, 6 and 8 µl l$^{-1}$ NH$_3$ could have accounted for 28, 65, 73 and 106 % of the plant's N requirement, respectively. However, from the observations that there was already an enrichment in the total N content upon exposure to a level as low as 2 µl l$^{-1}$ NH$_3$ one may conclude that at a sufficient nitrate supply to the roots either the foliar absorbed NH$_3$ was hardly utilized for structural growth or there was a poor shoot to root signaling in tuning of the rate of metabolism of the absorbed reduced nitrogen in the shoot to the nitrate uptake by the root. This is in contrast with observations upon exposure of *B. oleracea* for instance to the sulfurous air pollutant H$_2$S, where there was a strong interaction between the rate of uptake and metabolism of H$_2$S in the shoot and the uptake of sulfate by the roots (Westerman et al. 2000). This might be explained by differences in the factors determining the internal resistance to NH$_3$ and H$_2$S. In contrast to NH$_3$, where the internal resistance is determined by its high solubility and dissociation in the aqueous phase of the mesophyll cells, that of H$_2$S is largely determined by its rate of metabolism (De Kok et al. 2002).

## 6. Conclusions

The present results demonstrated that at atmospheric levels up to 4 $\mu$l l$^{-1}$, NH$_3$ appeared not to be toxic to *B. oleracea*. Higher NH$_3$ levels had a negative effect on growth, likely due the negative effect of high intracellular NH$_4^+$ levels, or possible formed metabolites during cellular metabolism. The current NH$_3$ levels in field conditions range from 0.03 $\mu$l l$^{-1}$ to 1.2 $\mu$l l$^{-1}$, the latter as a peak level (Stulen et al. 1998). It is unlikely that these levels would negatively affect growth of *B. oleracea* under field conditions. This is in agreement with the observations of Van der Eerden (1982), who exposed various plant species, including *B. oleracea*, to 0.8 - 1.4 $\mu$l l$^{-1}$ NH$_3$ and found no differences in growth. Apparently *B. oleracea* is able to cope with even higher levels of NH$_3$, most likely because of its higher nitrogen requirement and high RGR.

It remains to be questioned to what extent foliarly absorbed NH$_3$ is used as nitrogen source for growth, if the nitrogen supply to the roots is sufficient. Even at atmospheric NH$_3$ levels (e.g. 2 $\mu$l l$^{-1}$), which only would cover a limited proportion of the N requirement if used as nitrogen source, there was an enhancement of the nitrogen content of the shoots and roots. Apparently there was no direct regulatory control of and/or interaction between atmospheric and pedospheric nitrogen utilization in *B. oleracea*.

## References

Bassirirad H (2000) Kinetics of nutrient uptake by the roots: responses to global changes. New Phytol 147:155-169

Britto DT, Kronzucker HJ (2002) NH$_4^+$ toxicity in higher plants: a critical review. J Plant Physiol 159:567-584

Castro A, Stulen I, De Kok LJ (2003) Nitrogen and sulfur requirement in *Brassica oleracea* cultivars. In: Davidian J-C, Grill D, De Kok LJ, Stulen I, Hawkesford MJ, Schnug E, Rennenberg H (Eds) Sulfur transport and assimilation in plants. Backhuys Publishers, Leiden, pp 181-183

Castro A, Aires A, Rosa E, Bloem E, Stulen I, De Kok LJ (2004) Distribution of glucosinolates in *Brassica oleracea* cultivars. Phyton 44:133-143

Clement JMAM, Loorbach J, Meijer J, Van Hasselt PR (1997) The impact of atmospheric ammonia and temperature on growth and nitrogen metabolism in winter wheat. Plant Physiol Biochem 34:159-164

De Kok LJ, Westerman S, Stuiver CEE, Stulen I (2000) Atmospheric H$_2$S as plant sulfur source: interaction with pedospheric sulfur nutrition - a case study with *Brassica oleracea* L. In: Brunold C, Rennenberg H, De Kok LJ, Stulen I, Davidian J-C (Eds) Sulfur nutrition and sulfur assimilation in higher plants: molecular, biochemical and physiological aspects. Paul Haupt Publishers, Bern, pp 41-56

De Kok LJ, Castro A, Durenkamp M, Stuiver CEE, Westerman S, Yang L, Stulen I (2002) Sulphur in plant physiology. Proceedings No. 500, International Fertiliser Society York, 1-26

Erisman JW, Schaap M (2004) The need for ammonia abatement with respect to secondary PM reductions in Europe. Environ Pollut 129:159-163

Fangmeier A, Hadwiger-Fangmeier A, Van der Eerden L, Jager H-J. (1994) Effects of atmospheric ammonia on vegetation - a review. Environ Pollut 86:43-82

Gessler A, Rennenberg H (1998) Atmospheric ammonia: mechanisms of uptake and impacts on N metabolism of plants. In: De Kok LJ, Stulen I (Eds) Responses of plant metabolism to air pollution and global change. Backhuys Publishers, Leiden, pp 81-94

Högberg P, Högbom L, Schinkel H (1998) Nitrogen-related root variables of trees along an N-deposition gradient in Europe. Tree Physiol 18:823-828

Husted S, Schjoerring JK (1996) Ammonia fluxes between oilseed rape plants and the atmosphere in response to changes in leaf temperature, light intensity and air humidity. Plant Physiol 112:67-74

Hutchinson GL, Millington RJ, Peters DB (1972) Atmospheric ammonia: absorption by plant leaves. Science 175:771-772

Krupa SV (2003) Effects of atmospheric ammonia (NH$_3$) on terrestrial vegetation: a review. Environ Pollut 124:179-221

Lea PJ, Miflin BJ (1974) An alternative route for nitrogen assimilation in higher plants. Nature 251:614-616

Leith ID, Pitcairn CER, Sheppard LJ, Hill PW, Cape JN, Fowler D, Tang S, Smith RI, Parrington JA (2002) A comparison of impacts of N deposition applied as NH$_3$ or as NH$_4$Cl on ombrotrophic mire vegetation. Phyton 42 (3):83-88

Pearson J, Stewart GR (1993) The deposition of atmospheric ammonia and its effects on plants. New Phytol 125:285-305

Pearson J, Soares S (1998) Physiological responses of plant leaves to atmospheric ammonia and ammonium. Atmos Environ 32:533-598

Pérez-Soba M and Van der Eerden LJM (1993) Nitrogen uptake in needles of Scots pine (*Pinus sylvestris* L.) when exposed to gaseous ammonia and ammonium fertilizer in the soil. Plant Soil 153:231-242

Pérez-Soba M, Stulen I, Van der Eerden LJM (1994) Effect of atmospheric ammonia on the nitrogen metabolism of Scots pine (*Pinus Sylvestris*) needles. Physiol Plant 90:629-636

Pitcairn CER, Fowler D, Leith ID, Sheppard LJ, Sutton MA, Kennedy V, Okello E (2003) Bioindicators of enhanced nitrogen deposition. Environ Pollut 126:353-361

Rennenberg H and Gessler A (1999) Consequences of N deposition to forest ecosystems – recent results and future research needs. Water Air Soil Pollut 116:47-64

Rogers HH, Aneja PA (1980) Uptake of atmospheric ammonia by selected plant species. Environ Exp Bot 20:251-257

Schjoerring JK, Husted S, Mattsson M (1998) Physiological parameters controlling plant-atmosphere ammonia exchange. Atmos Environ 32:491-498

Sheppard LJ, Leith ID (2002) Effects of $NH_3$ fumigation on the frost hardiness of *Calluna* – does N deposition increase winter damage by frost? Phyton 42(3)183-190

Stuiver CEE, De Kok LJ, Westerman S (1997) Sulfur deficiency in *Brassica oleracea* L.: development, biochemical characterization, and sulphur/nitrogen interactions. Russian J Plant Physiol 44:505-513

Stulen I, Pérez-Soba M, De Kok LJ, Van der Eerden L (1998) Impact of gaseous nitrogen deposition on plant functioning. New Phytol 139:61-70

Van der Eerden LJM (1982) Toxicity of ammonia to plants. Agric Environ 7:223-235

Van der Eerden LJM, Dueck ThA, Berdowski JJM, Greven H, Van Dobben HF (1991) Influence of $NH_3$ and $(NH_4)_2SO_4$ on heath land vegetation. Acta Bot Neerl 40:281-296

Van Dijk HFG, Roelofs JGM (1988) Effects of excessive ammonium deposition on the nutritional status and conditions of pine needles. Physiol Plant 73:494-501

Van Hove LWA, Koops AJ, Adema EH, Vredenberg WJ, Pieters GA (1987) Analysis of the uptake of atmospheric ammonia by leaves of *Phaseolus vulgaris* L. Atmos Environ 21:1759-1763

Westerman S, De Kok LJ, Stuiver CEE, Stulen I (2000) Interaction between metabolism of atmospheric $H_2S$ in the shoot and sulfate uptake by the roots of curly kale (*Brassica oleracea*). Physiol Plant 109:443-449

Williams RF (1948) The effects of phosphorus supply on the rate of intake of phosphorus and nitrogen and upon certain aspects of phosphorus metabolism in gramineous plants. Austr J Sci Res ser B1:333-359

Wollenweber B, Raven JA (1993) Implications of N acquisition from atmospheric $NH_3$ for acids base and cation–anion balance of *Lollium perenne*. Physiol Plant 89:519-523

# How sensitive are forest trees to ozone? - New research on an old issue

Rainer Matyssek[1], Gerhard Wieser[2], Angela J. Nunn[1], Markus Löw[1], Christiane Then[1,2], Karin Herbinger[3], Manuela Blumenröther[4], Sascha Jehnes[5], Ilja M. Reiter[1], Christian Heerdt[6], Nina Koch[1], Karl-Heinz Häberle[1], Kris Haberer[5], Herbert Werner[6], Michael Tausz[3,7], Peter Fabian[6], Heinz Rennenberg[5], Dieter Grill[3], and Wolfgang Oßwald[4]

[1] Ecophysiology of Plants, Technische Universität München, Am Hochanger 13, 85354 Freising, Germany
[2] Federal Office and Research Centre for Forests, Alpine Timberline Ecophysiology, Rennweg 1, 6020 Innsbruck, Austria
[3] Institut für Pflanzenwissenschaften, Karl-Franzens Universität Graz, Schubertstraße 51, A-8010 Graz, Austria
[4] Pathology of Woody Plants, Technische Universität München, Am Hochanger 13, 85354 Freising, Germany
[5] Forest Botany & Tree Physiology, Albert-Ludwigs Universität Freiburg i. Br., Georges-Köhler-Allee 53/54, 79110 Freiburg, Germany
[6] Ecoclimatology, Technische Universität München, Am Hochanger 13, 85354 Freising, Germany
[7] School of Forest and Ecosystem Science, University of Melbourne, Water Street, Creswick, Victoria 3363, Australia

**Summary.** Studies on juvenile individuals under artificial environments dominate knowledge about the sensitivity of trees to $O_3$. Field approaches based on free-air $O_3$ fumigations of adult forest trees are a novel choice. Such a case study on beech and spruce (at the Kranzberg Forest near Munich, Germany) is used to address four long-standing issues in $O_3$ research: (1) Can a "unifying theory" of $O_3$ sensitivity be verified? (2) Are responses to $O_3$ consistent at different scaling levels in trees? (3) Are branch-bag experiments relevant for $O_3$ risk assessment of crowns? (4) Are saplings surrogates of adult trees when both are assessed under the same field conditions? Preliminary evidence from the ongoing long-term study confirms (1) and (3) but negates (2) and (4). In the absence of acute risks for adult trees, responsiveness of leaves cannot rule out long-term constraints by chronic $O_3$ stress.

**Key words.** Ozone, Risk assessment, Adult forest trees, Scaling levels

## 1. Introduction

The question raised in the title may appear surprising after so many years of research into the impacts of ozone ($O_3$) on trees (Miller and McBride 1999; Matyssek and Sandermann 2003). However, the available knowledge derives predominantly from chamber experi-

*Plant Responses to Air Pollution and Global Change*
Edited by K. Omasa, I. Nouchi, and L. J. De Kok ( Springer-Verlag Tokyo 2005 )

ments with juvenile individuals (Sandermann et al. 1997). In contrast, information on adult field-grown trees is rather scarce (Kolb and Matyssek 2001) and hence, although such a basis is needed, ecologically meaningful risk assessment is not feasible (Skärby et al. 1998; Matyssek and Innes 1999). Furthermore, extrapolations of physiological performance and sensitivity from juvenile to adult trees are questionable (Kolb et al. 1997; Kolb and Matyssek 2001), and the debates continue on how to assess and express $O_3$ doses in physiologically and ecologically meaningful terms for trees (Karlsson et al. 2004).

Given the long history of $O_3$ research and the above-outlined limitations, can new experimental concepts promote the mechanistic understanding of adult-tree performance under chronic $O_3$ impact and provide a basis for site-relevant risk assessment? New perspectives have been opened by establishing free-air $O_3$ exposure of adult forest trees, as exemplified in Central Europe in a mixed stand of about 60-year-old *Fagus sylvatica* and *Picea abies* trees (Nunn et al. 2002; Werner and Fabian 2002).

The following account concentrates on an ongoing study at Kranzberg Forest, Freising/Germany. After providing a short overview of the conceptual approach, the following questions will be addressed: (1) Can the "unifying theory" of $O_3$ sensitivity linking responsiveness and total $O_3$ uptake as outlined by Reich (1987) be verified comparing adult beech and spruce trees? (2) Are responses to chronic $O_3$ impact consistent at different scaling levels in adult trees? (3) Are findings from branch-bag experiments relevant for whole-tree risk assessment? (4) Do adult and juvenile trees compare in $O_3$ sensitivity, when both ages are assessed under field conditions?

## 2. Conceptual approach

Studies were undertaken in a c. 60-year-old beech/spruce stand at Kranzberg Forest (Freising/Germany, 48°25' N, 11°39' E; for stand description see Pretzsch et al. 1998). Scaffolding and a research crane provided access to sun and shade crowns. A free-air $O_3$ fumigation system (Nunn et al. 2002; Werner and Fabian 2002) was installed within the forest canopy, allowing entire crowns of five neighbouring beech and spruce trees each to be exposed to an experimentally enhanced, twice-ambient $O_3$ regime ($2xO_3$). Comparisons were made with trees under ambient air ($1xO_3$ = control). Maximum $O_3$ levels at $2xO_3$ were restricted to 150 nl $O_3$ $l^{-1}$ to prevent risk of acute $O_3$ injury. Online $O_3$ analyses and 160 passive-samplers distributed across the fumigated and non-fumigated canopy were employed for continuous $O_3$ monitoring (Nunn et al. 2002).

Two-year-old beech saplings were planted into 30 l containers (6 seedlings per container) with natural forest soil. Nine containers each were transferred into the upper sun crown of adult beech trees of both $O_3$ regimes or to the forest floor. Containers were regularly watered to avoid soil drought. In addition, to address the still open question whether branch-bag experiments are representative for whole-crown $O_3$ sensitivity (Sandermann et al. 1997), transparent fumigation cuvettes, tracking ambient climatic conditions and $O_3$ levels (Havranek and Wieser 1994), were mounted around terminal sections of branches in the sun crown of beech under both $O_3$ regimes. Eight cuvettes received $1xO_3$ air in trees under $2xO_3$ and vice versa. Repeatedly, during and at the end of the growing season, gas exchange behaviour and biochemical characteristics were assessed in leaves of juvenile and adult trees.

## 3. Leaf longevity – and the unifying theory of $O_3$ sensitivity

Beside genotype and site conditions, $O_3$ sensitivity is also governed by species-specific differences in morphology, physiological behaviour, and longevity of foliage. Because conifers generally have a lower stomatal conductance ($g_s$) and hence on a seasonal basis take up less $O_3$ than deciduous trees, the latter have been suggested to be prone to $O_3$ injury (Reich 1987). Due to their lower $g_s$, conifers require a longer exposure period to match the same $O_3$ uptake of deciduous trees under the same $O_3$ exposure. Thus, although never experimentally evaluated, based on a "unifying theory" on $O_3$ sensitivity Reich (1987) argued that, when accounting for foliage longevity and prolonged $O_3$ impact, conifers might be similarly or more $O_3$-sensitive than deciduous trees.

$O_3$ flux ($FO_3$) into the foliage of spruce and beech as based on $g_s$ and $O_3$ concentration (Matyssek et al. 1995) was modelled according to (Nunn et al. 2005; cf. Emberson et al. 2000). Given a significantly lower $g_s$ in spruce as compared to beech (Table 1), mean daily $FO_3$ throughout the leaf-bearing time was 8- to 12-times higher in beech as compared to spruce. As a consequence, spruce needles require 4 to 5 years' (Table 1) exposure to gain a cumulative $O_3$ uptake similar to that in beech over one growing season (i.e. ≈167 days).

At Kranzberg Forest, the maximum life-span of spruce needles is 9 years on average, although only about 20 % of the foliage is older than five years (cf. Matyssek et al. 1995). Thus, in agreement with conclusions by Reich (1987), the risk of injury and premature needle loss is high when $O_3$ uptake into spruce needles, after four to five years of chronic stress, is equivalent to that absorbed by beech leaves within one growing season. Nevertheless, in Central Europe, spruce does not tend to develop the "mottling" symptoms associated with $O_3$ injury (Vollenweider, pers. comm.). One-year-old needles of spruce displayed a decline in photosynthesis, similar to beech, during the $2^{nd}$ and $3^{rd}$ year of whole-crown $O_3$ exposure (Fig. 1). Thus, spruce appears more sensitive to $O_3$ than beech, consistent with the conclusions of Reich (1987). However, up until now, after 4 years of free-air $O_3$ fumigation, neither beech nor spruce displayed $O_3$-caused reductions in stem growth at Kranzberg Forest.

Table 1. Mean maximum stomatal conductance for water vapour ($g_{max}$), mean cumulative $O_3$ uptake (CU) in sun and shade crowns of beech and spruce at Kranzberg Forest across 2000 throughout 2002 under the two $O_3$ regimes, and the ratio of CU in beech versus spruce ($R_{F/P}$). Data are means of 5 trees ± SE. **=P<0.01, ***=P<0.001 for differences between species. $g_{max}$-values determined at 25°C leaf temperature, 40 % air humidity, light saturation, and 360 ppm $CO_2$.

| | | Fagus sylvatica | | Picea abies | | |
|---|---|---|---|---|---|---|
| crown position | $O_3$ regime | $g_{max}$ [mmol m$^{-2}$ s$^{-1}$] | CU [mmol m$^{-2}$] | $g_{max}$ [mmol m$^{-2}$ s$^{-1}$] | CU [mmol m$^{-2}$] | $R_{F/P}$ |
| shade | 1x$O_3$ | 66.6± 5.6 | 5.5±0.6 | 9.2±1.1*** | 1.5±0.1** | 3.7 |
| | 2x$O_3$ | 57.3± 2.1 | 7.0±1.1 | 5.8±0.6*** | 1.3±0.1** | 5.3 |
| sun | 1x$O_3$ | 163.9±12.7 | 15.0±1.7 | 20.2±1.8*** | 3.7±0.4** | 4.0 |
| | 2x$O_3$ | 132.8±15.6 | 21.1±0.3 | 15.2±1.3*** | 3.9±0.2*** | 5.4 |

Note that due to the contrasting morphology of foliage and different length of growing seasons, $g_{max}$ and CU were expressed per unit projected leaf area during the foliated season in deciduous beech, and per unit total needle surface area throughout the year in spruce.

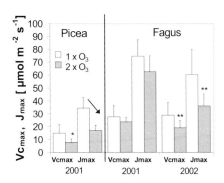

**Fig. 1.** $V_{cmax}$ and $J_{max}$, mirroring rubisco activity and electron transport capacity, respectively, for adult beech and spruce trees (shade foliage each) in Kranzberg Forest during 2001 and 2002 under 1x$O_3$ (open bars) and 2x$O_3$ (closed bars). Data are means of 5 trees ± SE. **=P<0.01; *=P< 0.05; arrow shows conspicuous, but non-significant 2x$O_3$ effect in spruce; assessments at 25°C leaf temperature, 40 % air humidity and under saturation of light ($V_{cmax}$, $J_{max}$) and $CO_2$ ($J_{max}$).

## 4. $O_3$ response of adult beech at different scaling levels

$O_3$ sensitivity of adult forest trees has rarely been examined experimentally by means of free-air $O_3$ fumigation. Given the genetic non-uniformity and, hence, scatter in the $O_3$ responses of field grown trees, analysing response patterns from the molecular to the stand level is the preferential approach and this should allow scaling towards a functional integration in tree performance (Sandermann and Matyssek 2004).

As compared to 1x$O_3$, the glucose level of foliage tended to be increased from June to September under 2x$O_3$, while the opposite was indicated by starch (Table 2). These signals support the view that $O_3$ enhances glycolytic activity (Landolt et al. 1997; Einig et al. 1997). Except for June, the redox state of ascorbate indicated reduction at 2x$O_3$ while salicylic acid concentration was slightly enhanced in the shade crown under 2x$O_3$ (Table 2). 2x$O_3$ neither caused significant reduction of $CO_2$ uptake in the sun crown (Table 2) nor in annual stem increment (not shown). Overall, trends in response to 2x$O_3$ were inconsistent across the scaling levels.

## 5. Are branch-bag experiments representative for crowns?

After exposure to the $O_3$ regimes employed in the branch cuvettes, leaves of enclosed twigs were compared in late summer with leaves on neighbouring twigs outside the cuvettes in terms of gas exchange and antioxidant levels. Few significant differences in foliar gas exchange and biochemical parameters were revealed (as shown for photosynthetic capacity, $A_{max}$, and α-tocopherol in each $O_3$ regime; Fig. 2).

Differences in foliar gas exchange and biochemical parameters were within the natural variation observed at the study site (Wieser et al. 2003). Most important, findings

**Table 2.** Levels of $CO_2$ uptake, glucose, starch, redox state of ascorbate (red/tot ascorbate) and content of salicylic acid in relation to unfumigated leaves [$1xO_3$ = 100%] in sun and shade leaves of adult beech (*Fagus sylvatica*) trees exposed to $2xO_3$ during 2003. Data represent the mean of measurements made on 5 trees ± SE. n.d. = not determined. Glucose and starch determined after Moore et al. (1997); ascorbate after Okamura (1980), modified by Knörzer et al. (1996); and salicylic acid modified after Meuwly and Métraux (1993). $CO_2$ uptake rate at light saturation and 360 $\mu l$ $CO_2$ $l^{-1}$; *** = $p < 0.001$ in relation to $1xO_3$.

| Crown position | Month | Glucose | Starch | Red/tot ascorbate | Salicylic acid | $CO_2$ uptake |
|---|---|---|---|---|---|---|
| sun | Jun | 105±5 | 70±16 | 176±69 | 132±51 | 55.4±11*** |
| | Jul | 165±27 | 37±9 | 74±13 | 148±54 | 78.2±24 |
| | Sep | 171±29 | 109±36 | 92±10 | 139±37 | 109.7±31 |
| shade | Jun | 79±10 | 68±31 | 138±35 | 105±53 | n. d. |
| | Jul | 97±20 | 29±13 | 78±10 | 163±38 | n. d. |
| | Sep | 130±4 | 57±9 | 78±22 | 161±52 | 77.5±12 |

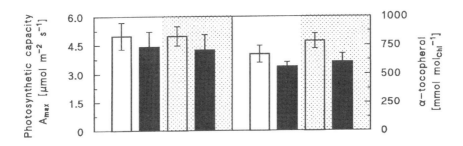

**Fig. 2.** Photosynthetic capacity ($A_{max}$) and α-tocopherol levels of leaves of adult beech trees (*Fagus sylvatica*) on twigs outside (open bars) and inside twig chambers (solid bars) after $O_3$ exposure to $1xO_3$ (open area) or $2xO_3$ (hatched area) throughout the growing season 2003. Data are means of 5 trees ± SE. Note that cuvettes receiving $2xO_3$ were mounted into branches of $1xO_3$ trees and *vice versa*. $A_{max}$ determined at 20°C leaf temperature, 310 Pa kPa⁻1 leaf to air vapor pressure difference, light saturation, and 360 ppm $CO_2$; α-tocopherol after Wildi and Lütz (1996).

demonstrate that branches are indeed representative regarding the $O_3$ responsiveness of crowns. Still, the extent to which branches are autonomous with respects to defence metabolism is undecided (cf. Sprugel et al. 1991). However, branch autonomy in defence is plausible, as ascorbate regeneration through the Asada/Halliwell pathway requires NADPH (Smirnoff 1996; Noctor and Foyer 1998) from the photosynthetic light reactions of the branch foliage. Also, the ascorbate pool correlates with the photosynthetic capacity and supply of soluble carbohydrates (cf. Smirnoff and Pallanca 1996), as ascorbate biosynthesis requires glucose-6-phosphate (Smirnoff et al. 2001).

## 6. Young trees as surrogates of adult trees in $O_3$ sensitivity?

Independent of the $O_3$ regime, the growing season extended, on average, throughout 167±1 and 186±1 days in saplings and adult beech trees, respectively. In late summer, $2xO_3$ caused a significant decline in $A_{max}$ of saplings and indicated some $O_3$ effect on antioxidant levels (Fig. 3). Discrepancies have also been found in antioxidant levels and gas exchange between juvenile and adult *Fagus sylvatica* and *Picea abies* trees at the same study site (Wieser et al 2002, 2003). These findings underline that even under same field conditions saplings are inadequate surrogates for estimating the $O_3$ susceptibility of adult trees. However, elevated $O_3$ did not affect photosynthesis of saplings grown in the shade at the forest floor (not shown) which is the natural habitat of beech saplings. Thus, ecologically meaningful comparisons have to take into account the typically contrasting growth conditions between young and old beech trees, when scaling $O_3$ sensitivity from saplings towards older ontogenetic stages (Kolb and Matyssek 2001).

## 7. Conclusions

In response to question 1 of the introduction, the results obtained in the present study so far support the "unifying theory" regarding $O_3$ uptake and sensitivity of foliage types (Reich 1987). Question 2 must be negated at this stage, because depending on the investigated parameter, $O_3$ responsiveness distinctly varies across scaling levels. Regarding question 3, the results suggest that branch-bags installed at individual branches are useful surrogates for examining $O_3$ sensitivity of entire crowns in adult forest trees. Rather different is the conclusion about question 4, whether juvenile trees are surrogates in judging $O_3$ sensitivity of adult trees. Even when comparing these two ontogenetic stages in the field under the same micro-climatic conditions of sun crowns, saplings responded

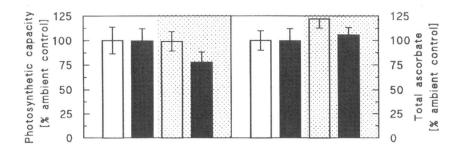

**Fig. 3.** Photosynthetic capacity and ascorbate level of beech leaves (*Fagus sylvatica*) in proportion to respective responses under $1xO_3$ in adult trees (open bars) and saplings (solid bars) after exposure to $1xO_3$ (open area) and $2xO_3$ (hatched area) throughout the growing season 2003. Data are means of 5 trees ± SE. Photosynthetic capacity determined as in Fig. 2; ascorbate after Tausz et al. (1996).

differently to ozone. Thus, it is uncertain to predict which tree parameter may respond at advanced ontogenetic stages in the field, and to which extent.

The presented study did not provide proof of an imminent risk imposed by site-relevant $O_3$ impact on the survival of adult trees, but leaf level responses cannot rule out long-term constraints by chronic $O_3$ stress. Nevertheless, approaches covered by issues 1 to 4 offer appropriate experimental means that can broaden evidence about adult-tree $O_3$ sensitivity at given forest sites beyond the current state of knowledge. The experiment will continue for another three years, before final conclusions on whole-tree $O_3$ sensitivity can be drawn. It is envisaged to integrate analyses of belowground processes and responses at the molecular level with data similar to those presented here.

Acknowledgements. The research presented in this chapter is supported by the European Commission (EU Project CASIROZ, EVK2-2002-00165), "Bayerisches Staatsministerium für Umwelt, Gesundheit und Verbraucherschutz" and the DFG through SFB 607.

## References

Einig W, Lauxmann U, Hauch B, Hampp R, Landolt W, Maurer S, Matyssek R (1997) Ozone-induced accumulation of carbohydrates changes enzyme activities of carbohydrate metabolism in birch leaves. New Phytol 137:673-680

Emberson LD, Ashmore MR, Cambridge HM, Simpson D, Tuovinen JP (2000) Modeling stomatal ozone flux across Europe. Environ Pollut 109:403-413

Havranek WM, Wieser G (1994) Design and testing of twig chambers for ozone fumigation and gas exchange measurements in mature trees. Proc Roy Soc Edinburgh Sect B 102:541-546

Karlsson PE, Uddling J, Braun S, Broadmeadow M, Elvira S, Gimeno BS, Le Thiec D, Oksanen, E, Vandermeiern K, Wilkinson M, Emberson L (2004) New critical levels for ozone effects on young trees based on AOT40 and simulated cumulative leaf uptake of ozone. Atmos Environm 38:2283-2294

Knörzer OC, Durner J, Böger P (1996) Alterations in the antioxidative system of suspension- cultured soybean cells (*Glycine max*) induced by oxidative stress. Physiol Plant 97:388-396

Kolb TE, Fredericksen TS, Steiner KC, Skelly JM (1997) Issues in scaling tree size and age responses to ozone: a review. Environ Pollut 98:195-208

Kolb TE, Matyssek R (2001) Limitations and perspectives about scaling ozone impacts in trees. Environ Pollut 115:373-393

Landolt W, Günthardt-Goerg MS, Pfenninger I, Einig W, Hampp R, Maurer S, Matyssek R (1997) Effect of fertilization on ozone-induced changes in the metabolism of birch (*Betula pendula*) leaves. New Phytol 137:389-397

Matyssek R, Sandermann H (2003) Impact of ozone on trees: an ecophysiological perspective. Prog Bot 64:349-404

Matyssek R, Reich PB, Oren R, Winner WE (1995) Response mechanisms of conifers to air pollutants. In: Smith WK., Hinckley TH. (Eds) Physiological ecology of coniferous forests. Academic Press, New York, pp 255-308

Matyssek R, Innes JL (1999) Ozone – a risk factor for forest trees and forests in Europe. Water Air Soil Pollut 116:199-226

Meuwly P, Metraux J-P (1993) Orthoanisic acid as internal standard for the simultaneous quantification of salicylic acid and its putative biosynthetic precursors in cucumber leaves. Anal Biochem 214:500-505

Miller PR, McBride JM (1999) Oxidant air pollution impacts in the montane forest of southern California. Ecological Studies Vol. 134, Springer, Berlin, Heidelberg, New York, pp 317-336

Moore BD, Palmquist, DE, Seemann JR (1997) Influence of plant growth at high $CO_2$ concentrations on leaf content of ribulose-1,5 -bisphosphate carboxylase/oxygenase and intracellular distribution soluble carbohydrates in tobacco, snapdragon and parsley. Plant Physiol 115:241-248

Noctor G, Foyer Ch (1998) Ascorbate and glutathione: keeping active oxygen under control. Ann Rev Plant Physiol Plant Mol Biol 49:249-279

Nunn AJ, Reiter IM, Häberle K-H, Werner H, Langebartels C, Sandermann H, Heerdt C, Fabian P, Matyssek R (2002) „Free-Air" ozone canopy fumigation in an old-growth mixed forest: concept and observations in beech. Phyton 42:105-119

Nunn AJ, Kozovits AR, Reiter IM, Heerdt C, Leuchner M, Lütz C, Liu X, Grams TEE, Häberle KH, Werner H, Matyssek R (2005) Comparison of ozone uptake and sensitivity between a phytotron study with young beech and a field experiment with adult beech (*Fagus sylvatica*). Environ Pollut (in press)

Okamura M (1980) An improved method for determination of L-ascorbic acid and L-dehydroascorbic acid in blood plasma. Clin Chim Acta 103:259-268

Pretzsch H, Kahn M, Grote R, (1998) Die Fichten-Buchen-Mischbestände des Sonderforschungsbereiches "Wachstum oder Parasitenabwehr?" im Kranzberger Forst. Forstw Cbl 117:241-257

Reich PB (1987) Quantifying plant response to ozone: a unifying theory. Tree Physiol 3:63-91

Sandermann H, Wellburn AR, Heath RL (1997) Forest decline and ozone: a comparison of controlled chamber and field experiments. Ecological Studies 127, Springer, Berlin Heidelberg New York, 400p

Skärby L, Ro-Poulsen H, Wellburn FAM, Sheppard LJ (1998) Impacts of ozone on forests: a European perspective. New Phytol 139:109-122

Smirnoff N (1996) The function and metabolism of ascorbic acid in plants. Ann Bot 78: 661-669

Smirnoff N, Pallanca JE (1996) Ascorbate metabolism in relation to oxidative stress. Biochem Soc Transact 24:472-478

Smirnoff N, Conklin PL, Loewus FA (2001) Biosynthesis of ascorbic acid in plants: a renaissance. Annu Rev Plant Physiol Plant Mol Biol 52:437-467

Sprugel DG, Hinckley TM, Schaap W (1991) The theory and practice of branch autonomy. Annu Rev Ecol Syst 22:309-334

Tausz M, Kranner I, Grill D (1996) Simultaneous determination of ascorbic acid and dehydroascorbic acid in plant materials by high-performance liquid chromatography. Phytochem Analysis 7:69-72

Werner H, Fabian P (2002) Free-air fumigation on mature trees: a novel system for controlled ozone enrichment in grown-up beech and spruce canopies. Environ Sci Pollut Res 9:117-121

Wieser G, Hecke K, Tausz M, Häberle K-H, Grams TEE, Matyssek R (2003) The influence of microclimate and tree age on the defense capacity of European beech (*Fagus sylvatica* L.) against oxidative stress. Ann For Sci 60:131-165

Wieser G, Tegischer K, Tausz M, Häberle K-H, Grams TEE, Matyssek R (2002) Age effects on Norway spruce (*Picea abies*) susceptibility to ozone uptake: a novel approach relating stress avoidance and defence. Tree Physiol 22:583-590

Wildi B, Lütz C (1996) Antioxidant comparison of selected high alpine plant species from different altitudes. Plant Cell Environ 19:138-146

# Northern conditions enhance the susceptibility of birch (*Betula pendula* Roth) to oxidative stress caused by ozone

Elina Oksanen

Department of Biology, University of Joensuu, POB 111, 80101 Joensuu, Finland

**Summary.** Impacts of increasing tropospheric ozone, together with the most important interactive stress factors (such as drought, soil N, and early frost) on European white birch (*Betula pendula* Roth) has been extensively investigated. In this paper, a summary of those studies, conducted both in chambers and realistic open-field conditions, is presented. Typical ozone responses in birch were found as impaired foliage, stem and root growth, altered shoot to root ratio, delayed bud burst, visible foliar injuries, enhanced leaf senescence, disturbed stomatal conductance, impaired net photosynthesis and water-use efficiency, reduced concentrations of Rubisco, pigments, starch, soluble protein and nitrogen, increased concentrations of phenolic compounds, increased transcription levels of genes encoding stress proteins PR-10 and PAL, increased stomatal density, reduced leaf thickness, ultrastructural injuries in chloroplasts and mitochondria, increased mesophyll cell wall thickness and reduced volume for intercellular space. Many of the negative responses were significantly promoted by simultaneous soil drought or early frost. Although elevated soil N counteracted ozone-caused growth reductions, delayed leaf senescence and increased shoot to root ratio may lead to disturbances in winter hardening processes and predispose the trees e.g. to soil drought due to a lower water uptake per higher water loss through transpiration.

**Key words.** Birch, Ozone, Drought, Nitrogen, Frost

## 1. Northern conditions enhance forest damage

During the last decade, impacts of increasing ozone concentrations on north-European forest trees have attracted considerable attention (e.g. Pääkkönen et al. 1993; Oksanen and Saleem 1998; Yamaji et al. 2003). There is increasing evidence that the prevailing ozone concentrations in Europe can cause visible leaf injury, growth and yield reductions in trees, as well as altered sensitivity to other biotic and abiotic stresses (e.g. Pääkkönen et al. 1998a; De Temmerman et al. 2002). The trees in northern environment are more susceptible to ozone, because the nights in summertime become too short to recover from ozone injury through repair processes driven by dark respiration (De Temmerman et al. 2002). In addition, cool and humid conditions in northern Scandinavia tend to promote ozone uptake by canopy cuticles (surface deposition) and stomatal conductance to result in high ozone flux into the leaf mesophyll (Karlsson et al. 2004; Uddling et al. 2004). Therefore, the growth of northern trees can be affected by present ozone concentrations in spite of lower ozone concentrations and shorter ozone episodes as compared to more southern latitudes.

Intensive monitoring of impacts of environmental stress factors on European forests has revealed that deposition of nitrogen, acidity and heavy metals exceeded critical loads over a large proportion of the monitoring plots (Monitoring programme of EU and ICP Forests, Executive Report 2002), indicating enhanced risks e.g. for tree root damage, storm damage, and crown damage by drought, frost and pests and changes in the plant diversity of ground vegetation. The results from monitoring suggest that forests in northern Europe are also more sensitive to excess nitrogen as compared to southern Europe. Therefore, active research is needed on interactive effects of ozone and other environmental stress factors on northern trees.

## 2. Birch is sensitive to ozone

Since early 1990's we have conducted numerous chamber and open-field experiments to elucidate the effects of increasing tropospheric ozone concentrations on European white birch (*Betula pendula* Roth), which is economically the most important hardwood species in Finland (e.g. Pääkkönen et al. 1993, 1995a,b, 1996, 1998a,b,c; Oksanen and Saleem 1999; Oksanen and Holopainen 2001; Tables 1-3). Already the first short-term studies conducted in laboratory chambers with young potted saplings revealed the high variability of ozone sensitivity among the birch clones (Pääkkönen et al. 1993): in the most sensitive genotypes ozone concentrations between 70 and 200 ppb resulted in 9-46% reductions in height growth (Pääkkönen et al. 1993; Oksanen and Holopainen 2001), 4-17% reduced dry mass of stem (Oksanen and Holopainen 2001), 58% reduced dry mass of roots (Pääkkönen et al. 1993), 41% smaller leaf area (Pääkkönen et al. 1993), and 47% smaller RGR for the whole plant (Pääkkönen et al. 1998c). The most tolerant genotypes, on the other hand, were practically unaffected by ozone enrichment (Pääkkönen et al. 1993, 1996, 1998b; Oksanen and Saleem 1999).

In more realistic open-field experiments (Kuopio FACE site, 62°13´ N 27°35´ E, central Finland) over one to five growing seasons with ozone concentrations of 1.2-1.8 times higher than the ambient level, dry mass of stem reduced by 23-30% (Pääkkönen et al. 1996; Saleem et al. 2001), dry weight of roots by 8-75% (Pääkkönen et al. 1996; Saleem et al. 2001), stem height growth by 17% (Pääkkönen et al. 1993), foliage growth by 11-45 % (Pääkkönen et al. 1993, 1996, 1998a,b; Oksanen 2001), and net assimilation rate by 9% (Oksanen 2001) in the most sensitive clones. Furthermore, activity of Rubisco declined by 18% (Pääkkönen et al. 1998a), net photosynthesis by 3-18% (Pääkkönen et al. 1996, 1998a), and starch content by 14% (Oksanen 2001). These changes were accompanied by increased visible foliar injuries, enhanced leaf senescence (Pääkkönen et al. 1995a,b; Oksanen 2003a), altered shoot to root ratio and a shift in carbon allocation towards defensive foliar phenolic compounds (Pääkkönen et al. 1998b; Saleem et al. 2001; Yamaji et al. 2003; Tables 1-2).

According to these experiments, good ozone tolerance of young birch saplings was a result of several physiological and anatomical characteristics: (1) investment in foliage growth (rather than stem biomass), leading to a higher leaf area ratio (LAR = foliage area per total dry mass of plant) and a more active photosynthetic compensation (Pääkkönen et al. 1996; Oksanen and Holopainen, 2001); (2) a more complete night-time closure of

**Table 1.** Summary of ozone responses on growth, productivity and visible injuries of birch (*Betula pendula*) originated from Finland, and interactions with soil N, drought and early frost. Responses are shown as significant increases (↑), significant decreases (↓), or non-significant effects (n.s.) as compared to trees grown under ambient air (field experiments) of close to zero ozone concentrations (chamber experiments). n.a. = not analysed.

| Response | $O_3$ | $O_3$ + high N | $O_3$ + soil drought | $O_3$ + frost |
|---|---|---|---|---|
| **Growth and productivity** | | | | |
| Height growth | ↓(↑)* | n.s | n.s. | n.s. |
| Leaf biomass | ↓(↑)* | ↑ | ↓ | n.s. |
| Stem biomass | ↓(↑)* | n.s. | ↓ | n.s. |
| Root biomass | ↓(↑)* | ↑ | ↓ | n.s. |
| Stem volume | ↓ | n.a. | n.a. | n.a. |
| Leaf area | ↓ | n.s. | ↓ | n.s. |
| Leaf size | ↓(↑)* | ↑ | n.s. | n.s. |
| Specific foliage mass (mg/cm$^2$) | ↑ | ↓ | ↑ | n.a. |
| Net assimilation rate | ↓ | n.a. | n.a. | n.a. |
| Shoot:root ratio | ↑↓ | ↑ | n.s. | n.a. |
| Number of over-wintering buds | ↓ | n.a. | n.a. | n.a. |
| Bud burst | ↓ | n.a. | n.a. | n.a. |
| **Visible foliar changes** | | | | |
| Necrotic/chlorotic injuries | ↑ | n.s. | ↑↓ | ↑ |
| Leaf senescence | ↑ | ↓ | ↑ | n.a. |

*) growth stimulations were found only in short-term studies with low ozone concentrations.

stomata as compared to sensitive clones, thus enabling the plant to avoid ozone flux during the dark period (Oksanen and Holopainen 2001); (3) and thicker leaves and mesophyll cell walls (improving the capacity for apoplastic ozone detoxification through ascorbate), a lower volume of intercellular air space in leaf mesophyll, and a thicker palisade mesophyll layer (Pääkkönen et al. 1995a, 1998b).

However, the last long-term experiment extending from May 1996 to May 2003 with soil-growing trees of a sensitive clone 5 and a tolerant clone 2 revealed that ozone sensitivity of birch increased with exposure time and tree size (Oksanen 2003a,b). After seven year's exposure to elevated ozone (1.4-1.7x ambient) ozone-induced biomass reduction for the total tree was 54% in the sensitive clone and 39% in the tolerant clone, whereas the stem height growth was reduced by 20% and 9%, and base diameter by 28% and 21%, respectively. Increasing ozone sensitivity during the exposure could be explained by several related reasons: (1) low net photosynthesis to stomatal conductance ratio in the late season leading to (2) lower carbohydrate gain (indicated by reduced starch content), (3) cumulative carry-over effects of the multi-year exposure mediated by impaired bud formation, affecting negatively the early growth of foliage in the next year, and

**Table 2.** Summary of ozone responses on gas exchange, biochemistry and foliar gene expression of birch (*Betula pendula*) originated from Finland, and interactions with soil N, drought and early frost. Responses are shown as significant increases (↑), significant decreases (↓), or non-significant effects (n.s.) as compared to trees grown under ambient air (field experiments) of close to zero ozone concentrations (chamber experiments). n.a. = not analysed.

| Response | $O_3$ | $O_3$ + high N | $O_3$ + soil drought | $O_3$ + frost |
|---|---|---|---|---|
| **Gas exchange** | | | | |
| Stomatal conductance | ↓↑ | n.a. | ↓ | ↓ |
| Net photosynthesis | ↓ | n.a. | n.s. | n.s. |
| WUE (water-use efficiency) | ↓ | n.a. | n.a. | n.a. |
| **Biochemistry of leaves** | | | | |
| Rubisco concentration | ↓ | n.a. | n.s. | n.s. |
| Chlorophyll concentration | ↓ | n.a. | ↑ | ↓ |
| Carotenoid concentration | ↓ | n.a. | n.a. | ↓ |
| Nitrogen concentration | ↓/n.s. | ↑ | ↑ | n.s. |
| Phenolic concentration | ↑ | n.a. | n.a. | n.a. |
| Starch concentration | ↓ | n.a. | n.a. | |
| Soluble protein concentration | ↓ | n.a. | n.s. | n.s. |
| **Foliar gene expression** | | | | |
| PR-10 transcripts[1] | ↑ | n.a. | ↑ | n.a. |
| PAL transcript[2] | ↑ | n.a. | ↑ | n.a. |

[1] Pathogenesis-related protein 10; [2] Phenylalanine-ammonialyase.

(4) reduced capacity for photosynthetic compensation and repair for ozone damage as indicated by onset of foliar injuries under high ozone uptake (Oksanen 2003a,b). In addition to reduced number of over-wintering buds, bud break was delayed significantly in both clones under elevated ozone, as found also in a chamber study with different birch genotypes (Prozherina et al. 2003; Table 1).

Ozone-caused enhancement of birch growth has been reported in some short-term open-field studies (≤ two growing seasons) using considerably low ozone concentrations (24-h mean of 32-40 ppb) (Oksanen and Rousi 2001; Yamaji et al. 2003). These growth stimulations were accompanied by altered resource allocation leading to changes in shoot to root biomass ratio, and induction of several ozone defence strategies (e.g. increase in antioxidative phenolic compounds in leaves, stomatal closure) (Oksanen and Rousi 2001; Yamaji et al. 2003). The most serious concern about ozone-induced disturbances in carbon allocation is for those trees, where shoot growth is favoured at the expense of roots, thereby predisposing the trees to drought stress due to higher water loss/uptake ratio. (Tables 1-2).

## 3. Combined action of ozone, soil drought, N and early frosts

During the warm periods of summer, elevated ozone and soil drought are stresses likely to simultaneously affect the trees, especially in southern parts of Finland. Our experiment with birch indicated that drought protected the plants from ozone injuries under high-stress conditions in the chamber experiment, whereas enhanced ozone damage appeared as reduced growth, lower starch content, and accelerated leaf senescence in a combined exposure to ozone and drought in a more realistic open-field experiment (Pääkkönen et al. 1998a,b). In addition to water deficit, more frequent early spring frosts are predicted to occur in Northern Europe due to lengthened growing seasons (Root et al. 2003). In our chamber experiment with six birch genotypes, co-occurring ozone enrichment disturbed the recovery processes from acute frost occurrence through structural chloroplast damage (Prozherina et al. 2003; Table 3).

In northern forests, availability of nutrients (nitrogen, in particular) has been basically low, and therefore there is an increasing concern about elevating nitrogen deposition due to traffic, industry, intensive agriculture, and other human actions. In birch, significant ozone x high soil N interactions were found as an increase in leaf and root biomass production, as increased shoot to root ratio, as a delayed autumnal leaf senescence, and as relatively thinner photosynthesizing palisade (Pääkkönen and Holopainen 1995). Therefore, our results suggest that increasing nitrogen supply may confer the birch trees with greater resistance to ozone. However, excess N may also lead to disturbances in processes of winter hardening and in shoot to root balance, affecting water and nutrient status (Table 3).

Table 3. Summary of ozone responses on leaf structure of birch (*Betula pendula*) originated from Finland, and interactions with soil N, drought and early frost. Responses are shown as significant increases (↑), significant decreases (↓), or non-significant effects (n.s.) as compared to trees grown under ambient air (field experiments) of close to zero ozone concentrations (chamber experiments). n.a. = not analysed.

| Response | $O_3$ | $O_3$ + high N | $O_3$ + soil drought | $O_3$ + frost |
|---|---|---|---|---|
| **Leaf structure** | | | | |
| Stomatal density | ↑ | n.a. | ↓ | n.s. |
| Leaf thickness | ↓ | n.s. | n.a. | ↓ |
| Palisade layer thickness | ↑ | ↓ | n.a. | ↓ |
| Intercellular air space | ↓ | ↑ | n.a. | n.s. |
| Mesophyll cell wall thickness | ↑ | n.a. | ↑ | n.a. |
| Chloroplast injuries | ↑ | n.a. | ↑ | ↑ |
| Starch grains | ↓ | n.a. | ↓ | ↓ |
| Mitochondrial injuries | ↑ | n.a. | n.a. | n.a. |
| Plastoglobuli | ↑ | n.a. | ↑ | n.s. |
| Tannin deposition | ↑ | n.a. | ↑ | n.a. |
| Phenolic deposition | ↑ | n.a. | ↑ | n.a. |

## 4. Conclusions and future needs

Taken together, we have strong evidence that increasing ozone together with other relevant environmental stresses pose a real risk factor for birch establishment, production, and sustainable forestry in Finland due to deteriorating above- and belowground processes, and tree vitality in long term (Table 1-3). Although forest production is very important for Finnish economy, the view that any response must be related to economic loss reflects an out-dated view of forestry that is especially inappropriate in Europe (where the timber production is no longer seen as the primary function of forests). Many other European and US ozone studies with trees have shown that the regulatory capacity of resource allocation rather than productivity may be the most significant, in the long term, for the individual fitness and survival of trees. Chronic ozone stress can eventually lead to losses in species and genetic diversity, and therefore it needs to be examined to what extent forest decline through diseases or changes in competitiveness between the tree species may arise from predispositions induced by ozone. In addition, there are still several gaps of fundamental knowledge on mechanisms of ozone responses of trees. For example, incomplete senescence program leading to carbon and resource retention in leaves, the interplay among ozone-induced formation of reactive oxygen species (ROS), localized cell death, early senescence, and carbon allocation to roots (responses of fine roots, in particular), changes in cell wall permeability due to lignification and a possible cross-linking of some structural proteins, and protective role of photorespiration to avoid photo-oxidation in senescing leaves should be unravelled.

## References

De Temmerman L, Vandermeiren K, D'Haese D, Bortier K, Asard H, Ceulemans R (2002) Ozone effects on trees, where uptake and detoxification meet. Dendrobiology 47:9-19

Executive Report 2002. The Condition of Forests in Europe. United Nations Economic Commission for Europe, European Commission, Germany, ISSN 1020-587X, 35 p

Karlsson PE, Uddling J, Braun S, Broadmeadow M, Elvira S, Gimeno BS, Le Thiec D, Oksanen E, Vandermeiren K, Wilkinson M, Emberson L (2004) New critical levels for ozone effects on young trees based on AOT40 and simulated cumulative leaf uptake of ozone. Atm Env 38:2283-2294

Oksanen E (2001) Increasing tropospheric ozone level reduced birch (*Betula pendula*) dry mass within five years period. Water Air Soil Pollut 130:947-952

Oksanen E (2003a) Physiological ozone responses of birch (*Betula pendula* Roth) differ between soil-grown trees in a multi-year exposure and potted saplings in a single-season exposure. Tree Physiol 23:603-614

Oksanen E (2003b) Responses of selected birch (*Betula pendula*) clones to ozone change over time. Plant Cell Environ 26:875-886

Oksanen E, Holopainen T (2001) Responses of two birch (*Betula pendula* Roth) clones to different ozone profiles with similar AOT40 exposure. Atmos Environ 35:5245-5254

Oksanen E, Rousi M (2001) Differences in *Betula* origins in ozone sensitivity based open-field fumigation experiment over two growing seasons. Can J For Res 31:804-811

Oksanen E, Saleem A (1999) Ozone exposure results in various carry-over effects and prolonged reduction in biomass in birch (*Betula pendula* Roth). Plant Cell Environ 22:1401-1411

Pääkkönen E, Holopainen T (1995) Influence of nitrogen supply on the response of birch (*Betula pendula* Roth.) clones to ozone. New Phytol 129:595-603

Pääkkönen E, Paasisalo S, Holopainen T, Kärenlampi L (1993) Growth and stomatal responses of birch (*Betula pendula* Roth.) clones to ozone in open-air and chamber fumigations. New Phytol 125:615-623

Pääkkönen E, Holopainen T, Kärenlampi L (1995a) Ageing-related anatomical and ultra-structural changes in leaves of birch (*Betula pendula* Roth.) clones as affected by low ozone exposure. Ann Bot 75:285-294

Pääkkönen E, Metsärinne S, Holopainen T, Kärenlampi L (1995b) The ozone sensitivity of birch (*Betula pendula*) in relation to the developmental stage of leaves. New Phytol 132:145-154

Pääkkönen E, Vahala J, Holopainen T, Karjalainen R, Kärenlampi L (1996) Growth responses and related biochemical and ultrastructural changes of the photosynthetic machinery in birch (*Betula pendula* Roth.) exposed to low level ozone fumigation. Tree Physiol 16:597-605

Pääkkönen E, Vahala J, Pohjola M, Holopainen T, Kärenlampi L (1998a) Physiological, stomatal and ultrastructural ozone responses in birch (*Betula pendula* Roth) are modified by water stress. Plant Cell Environ 21:671-684

Pääkkönen E, Günthardt-Goerg M, Holopainen T (1998b) Responses of leaf processes in a sensitive birch (*Betula pendula* Roth) clone to ozone combined with drought. Ann Bot 82:49-59

Pääkkönen E, Seppänen S, Holopainen T, Kärenlampi S, Kärenlampi L, Kangasjärvi J (1998c) Induction of genes for the stress proteins PR-10 and PAL in relation to growth, visible injuries and stomatal responses in ozone and drought exposed birch (*Betula pendula*) clones. New Phytol 138:295-305

Prozherina N, Freiwald V, Rousi M, Oksanen E (2003) Effect of spring-time frost and elevated ozone on early growth, foliar injuries and leaf structure of birch (*Betula pendula* Roth) genotypes. New Phytol 159:623-636

Root TJ, Price JT, Hall KR, Schneider SH, Rosenzweig C, Pounds JA (2003) Fingerprints of global warming on wild animals and plants. Nature 421: 57-60

Saleem A, Loponen J, Pihlaja K, Oksanen E (2001) Effects of long-term open-field ozone exposure on leaf phenolics of European silver birch (*Betula pendula* Roth). J Chem Ecol 27:1049-1062

Uddling J, Günthard-Goerg MS, Matyssek R, Oksanen E, Pleijel H, Selldén G, Karlsson PE (2004) Biomass reductions of juvenile birch were more strongly related to stomatal uptake of ozone than to ozone indices based on external exposure. Atmos Environ 38:4709-4719

Yamaji K, Julkunen-Tiitto R, Rousi M, Freiwald V, Oksanen E (2003) Ozone exposure over two growing seasons alters root to shoot ratio and chemical composition of birch (*Betula pendula* Roth). Glob Change Biol 9:1363-1377

# Physiological responses of trees to air pollutants at high elevation sites

Dieter Grill[1], Hardy Pfanz[2], Bohumir Lomsky[3], Andrzej Bytnerowicz[4], Nancy E. Grulke[4], and Michael Tausz[1,5]

[1]Institut für Pflanzenwissenschaften, Karl-Franzens-Universität Graz, Schubertstraße 51, A-8010 Graz, Austria
[2]Institut für Angewandte Botanik, Universität Duisburg-Essen, Campus Essen, 45117 Essen, Germany
[3]VULHM, Jiloviste, Strnady, 15604 Zbraslav-Praha, Czechia
[4]USDA Forest Service, Pacific Southwest Research Station, 4955 Canyon Crest Drive, Riverside CA 92507-6090, USA
[5]School of Forest and Ecosystem Science, University of Melbourne, Water Street, Creswick, Victoria 3363, Australia

**Summary.** At high elevations a combination of environmental factors restricts the distribution of forest ecosystems. In addition to these natural limitations, high mountains are particularly prone to the deposition of air pollutants, which can lead to detrimental effects on the already struggling ecosystems. In the present chapter we review two typical examples of pollution impact on mountain forests. (1) The effects of high concentrations of $SO_2$ on spruce forests of the Ore Mountains in Central Europe and (2) the effects of photo-oxidants (mainly $O_3$) on mixed conifer forests in the San Bernardino Mountains in Southern California. Particular attention is paid to the potential interaction between natural stress factors and anthropogenic pollution impact. The development of oxidative stress and antioxidative defence systems play a key role in plant responses to adverse environmental conditions. Components of these systems have been used as stress markers, a task that is complicated due to their involvement in plant responses to both natural factors and pollution. We present a multivariate approach, which has been evaluated under various different field conditions as a step towards the distinction of the effects of different stress factors on forest trees in the field.

**Key words.** Ozone, Oxidative stress, High elevation, Sulphur dioxide, Antioxidants

## 1. Introduction

High elevation environments put considerable restraints to tree growth and hence forest ecosystems reach the limits of their distribution. Low temperatures, short growing season, high irradiance with a higher UV proportion, low nutrient turnover rates in the soils, and high atmospheric concentrations of ozone contribute to this complex "high elevation stress". As in most environmental stress situations, "high elevation stress" becomes manifest as oxidative stress in plant cells. Photo-oxidative stress – caused by high irradiance in

*Plant Responses to Air Pollution and Global Change*
Edited by K. Omasa, I. Nouchi, and L. J. De Kok ( Springer-Verlag Tokyo 2005 )

combination with factors limiting $CO_2$ fixation in the Calvin cycle (e. g. low temperatures, stomatal closure) - but also direct action of oxidative compounds ($O_3$), or mechanical injuries (e. g. by the abrasive effects of ice crystals) all lead to the generation of reactive oxygen species (ROS) in plant tissues. Tree species reaching the alpine timberline must be adapted to these conditions. All plants have evolved antioxidative defence systems – consisting of low molecular weight antioxidants in aqueous and lipid phases, enzymes, and photoprotective pigments – to detoxify ROS or repair their adverse effects. If the capacity of these systems is too low to keep up with the generation rates of ROS, cell structures are destroyed and tissue damages emerge. Given the high natural stress levels at high elevations, any additional factors (e. g. anthropogenic air pollutants) can have detrimental effects.

Unfortunately, mountains are particularly prone to the deposition of air pollutants. Slopes exposed to predominant winds often form a natural barrier to the movement of air masses, which carry pollutants. The present paper reviews two "classical" examples to study the effects of air pollution on high elevation forests. (1) A typical situation in Central Europe caused by the impact of $SO_2$ from industrial sources. While $SO_2$ emissions have been decreasing due to technological progress in exhaust treatment and due to the collapse of old-fashioned heavy industries in many regions, areas such as the Ore Mountains in eastern Germany are still exposed to the effects of this gaseous pollutant. (2) Photochemical smog formation in the huge urban region of the Los Angeles basin leads to high concentrations of photochemical oxidants (with the most abundant compound $O_3$) being transported to the pine forests at the San Bernardino Mountains, where they lead to significant typical damages in these ecosystems.

## 2. Sulphur dioxide stress in European forest systems

Starting with the early 1950's of the last century the increasing $SO_2$ pollution caused an extremely devastating effect on European forests. The most famous and best examined sites are located in the Czech-German Ore Mountains (Lomsky et al. 2001, 2002). Due to the intensive combustion of sulphur-rich brown coal and lignite, unfiltered $SO_2$ was emitted in high amounts leading to extremely elevated concentrations in the plateau regions during inversion periods (Table 1).

Compared to the so-called clean air regions mean monthly $SO_2$ concentrations were 7-8 times higher and peak concentrations reached 2000 $\mu g$ $SO_2$ $m^{-3}$ and even higher values. Sulphur dioxide uptake by leaves can be traced and quantified by the total sulphur content (Materna 1981, Pfanz and Beyschlag 1993, Pfanz et al. 1993). Under the high emission load 4-year-old spruce needles accumulated 3.5 - 4.5 mg S $g^{-1}$dry weight (dw) whereas control needles had values around 1.2 mg S $g^{-1}$dw. In some sites of the Ore Mountain plateau sulphur pollution was so extreme that some trees had only one needle age class left. After bud burst the young needles developed but were soon increasingly damaged due to the presence of sulphur dioxide. Even though the young needles were still developing their photosynthetic capacities decreased dramatically to cease as early as mid autumn of the same year. During the winter the needles solely respired and were shed after reddening within the next spring season (Pfanz et al. 1993). The about up to 30-year-old but only 1-2 m high trees were then still alive, but had an appearance of the needle-less skeletons. Nevertheless in the following June the new needles emerged from

**Table 1.** Sulphur dioxide concentrations in the air and total sulphur content of spruce needles in a heavily SO$_2$ affected location in the Czech Ore Mountains (Maly Haj) and a control site in Franken, Germany (Würzburg). SO$_2$-concentrations are presented as maximum hourly peak values, or as annual mean values. Values for SO$_2$ are presented in $\mu$g SO$_2$ m$^{-3}$, those for the total S-content of the needles in mg S g$^{-1}$ dw. (Data from Pfanz and Beyschlag 1983).

| Forest site | SO$_2$ concentration | | Approx. S content of needles | |
|---|---|---|---|---|
| | Annual mean value | max. ½ hour peak value | 1-year old | 4-years old |
| Control site Würzburg | 15-20 | 400 | 0.9 | 1.3 |
| Maly Haj, Ore Mountains | 130 | 2000 and more | 1.2 | 3.5 – 4.5 |

the buds and a new one-year-needle-cycle started.

Since several years, mainly since the reunion of Germany and the opening of the iron curtain between Eastern and Western Europe brown coal emissions were largely desulphurized and the emission loads of S to the forest ecosystems significantly decreased (Lomsky et al. 2001, 2002). Within years the amount of sulphur gases taken up by leaves and needles decreased in parallel. However yet, acidification effects within the soils did not disappear as quickly as the above-ground effects. Depending on soil properties (buffering capacities, original pH, base saturation) soil pH could still be low due to the acidifying properties of the acidic rain and mist in these regions. Acidification led to a leaching of cations in the soil thus creating a nutritional imbalance for the vegetation. In the end of the 1990's the symptoms of the so-called "novel forest decline" were found in forests exposed to high acid input. Prominent Mg deficiency in the conifer needles developed in large areas of the Ore Mountain (up to 8000 ha) and other forested mountains of the Sudetic Arch. However, since the time of atmospheric sulphur emission reduction, ozone and nitrogen oxides have simultaneously increased in parallel in the Sudetic Mountain range exerting their deleterious influence on the still imbalanced forest ecosystems.

## 3. Oxidative stress in plants at high elevations

Studies at altitudinal gradients from valley sites to the alpine timberline revealed that plants at high elevation sites are exposed to higher levels of oxidative stress. Because of the prevailing economic (timber harvest) and environmental importance (protection from erosion and avalanches, water catchment management) most research has been done on forest trees, but corresponding effects were also found in herbaceous plants. Among trees, evergreen (conifer) species have been studied in more detail, because their foliage has to withstand conditions for longer periods. Along such an altitudinal gradient, changes in needle colour from dark green to yellow green can often be detected visually. A corresponding decrease in needle chlorophyll concentrations has been reported repeatedly (Polle and Rennenberg 1992; Tausz et al. 1997, 1999; Hecke et al. 2003). This effect was interpreted as an adaptation to lower requirements in energy absorption and would decrease the amount of energy being absorbed in excess of what is needed in carbon fixation. Photo-oxidative stress is a common situation under adverse environmental conditions, where an imbalance between light driven electron transport and electron consumption promotes the generation of ROS (Fig. 1).

Alternatively, photo-oxidation processes are directly responsible for the decreased chlorophyll concentrations at high elevations, because chlorotic symptoms can be observed in sun needles (and sometimes only at the upper needle surface), but not in shade needles. Possibly, the natural defence capacity against photo-oxidative stress is exceeded, although concentrations of protective substances increase with increasing elevation. These protection systems comprise low molecular weight antioxidative compounds directly in the lipid phase of membranes – such as α-tocopherol or carotenoids (Munné-Bosch and Alegre 2002) – or in the aqueous phase – for example glutathione or ascorbate (Smirnoff and Wheeler 2000; Grill et al. 2001). Detoxifying enzymes – such as superoxide-dismutase (SOD), peroxidases, and reductases – remove ROS directly or regenerate antioxidants. The action of these enzymes becomes manifest in the redox states of antioxidant pools (reduced ascorbate/dehydroascorbate, reduced glutathione/oxidised glutathione). The regeneration of antioxidants is necessary to keep them in their protective, reduced form.

Photo-oxidative stress due to abiotic factors originates in the chloroplast (Fig. 1), whereas oxidative air pollutants (*e.g.* $O_3$) enter leaves through stomata and initially react in the apoplastic aqueous phase of the cell walls. Despite their low solubility and high reactivity in water they can reach the plasmalemma. Apoplastic antioxidative systems (mainly ascorbate and cell wall peroxidases) guarantee fast breakdown of oxidants. The crucial regeneration of apoplastic dehydroascorbate (the oxidised form) occurs in the symplast, *i.e.* it is dependent on translocation processes across the membrane and on the capacity of aqueous systems in the cytoplasm (similar to those in the chloroplast as depicted in Fig. 1). Interaction of photo-oxidative processes in the chloroplast and oxidative reactions originating in the apoplast lead to typical response patterns of the antioxidative systems.

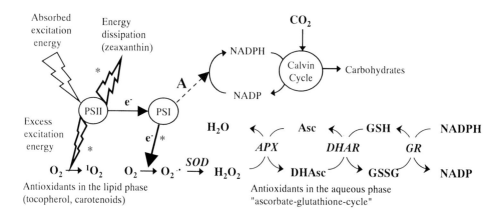

**Fig. 1.** Simplified scheme of photo-oxidative stress in the chloroplast. Environmental factors, such as low temperature or drought limit the electron flow to the Calvin cycle at A. This leads to an excess of absorbed excitation energy. In this case, alternative pathways of electron flow and energy transfer come into play (bold arrows marked with asterisks). APX ascorbate peroxidase, Asc reduced ascorbate, DHAsc oxidised ascorbate, DHAR dehydroascorbate reductase, GR glutathione reductase, GSH reduced glutathione, GSSG oxidised glutathione, SOD superoxide dismutase.

At high elevations, decreasing chlorophyll concentrations are accompanied by increasing concentrations of antioxidative compounds, such as ascorbate and glutathione. These compounds are linked in the ascorbate-glutathione cycle (Fig. 1), where they are regenerated from the oxidised to the reduced form employing NADPH as the reductant. Reduced ascorbate is also necessary for the regeneration of α-tocopherol and the xanthophyll conversions (see below, Smirnoff and Wheeler 2000). Increasing concentrations of α-tocopherol mark increased protection of the photosynthetic membranes from oxidation and counteracts detrimental lipid peroxidations. Compared to chlorophylls, carotenoid concentrations decrease to a lesser extent with increasing elevation, which results in more carotenoid molecules per unit chlorophyll at high elevation sites. The xanthophyll cycle – acting through reversible conversions of three xanthophylls violaxanthin, antheraxanthin, and zeaxanthin – provides flexible protection against excess excitation energy. At low light, violaxanthin functions as a light harvesting pigment and increases the quantum efficiency of photosynthesis, whereas under excess light violaxanthin is de-epoxidised via the intermediate antheraxanthin to zeaxanthin, which supports the dissipation of excitation energy as heat (Demmig-Adams 2003). High irradiance in combination with factors that decrease carboxylation rates (*e. g.* stomatal closure, low temperatures, pollution effects) makes it very likely that trees at high elevations frequently experience situations of excess excitation energy. Therefore, zeaxanthin concentrations are often higher at high elevations than at valley sites. Despite more energy dissipation, higher production of ROS may lead to a shift in the redox state of antioxidants towards their oxidised form.

Such patterns in the antioxidative systems have been observed in *Pinus ponderosa* in California (Tausz et al. 1999, 2001), *Pinus canariensis* in Tenerife (Tausz et al. 1998), or *Picea abies* in the Alps (Polle and Rennenberg 1992; Tausz et al. 1998). Recent studies confirmed comparable responses of *Pinus cembra* (Wieser et al. 2001), but slightly different changes have been observed in the antioxidative systems in the deciduous conifer *Larix decidua:* Whereas evergreen conifers seem to reinforce their photoprotection systems (more zeaxanthin), in *Larix* needles mainly glutathione concentrations responded to elevation. A possible explanation is that *Larix* generally maintains higher carboxylation rates (for example it keeps stomata open for longer, Wieser 1999) thus minimising imbalance between energy capture and electron consumption, but maximising possible pollutant uptake (Hecke et al., unpublished results). This underlines the necessity of studies on the species of interest and cautions against generalisations across different ecological types. Among others, such studies are presently performed on European beech (*Fagus sylvatica*) e.g. in the CASIROZ project (see Matyssek et al., this issue).

## 4. Natural stress *versus* ozone impact – lessons from studies in California

In the Alps and other European regions, atmospheric $O_3$ concentrations are highest at highest elevations, where natural "high elevation stress" is also maximal. This prevents the easy separation of ozone effects from natural high elevation effects at these field sites. The San Bernardino Mountains at the eastern border of the Los Angeles basin in Southern California have been subjected to high ozone loads for over 40 years. A long-term ozone gradient from high to moderate concentrations from the western to the eastern side of the mountain range is well documented. In contrast to European mountains, this gradient is reverse to the gradient of natural stresses, because the plots with highest ambient $O_3$ levels are located at lower elevations (and receive more precipitation) than comparably cleaner air plots.

Detailed studies on *Pinus ponderosa* growing in the mixed conifer forests of the San Bernardino Mountains east of Los Angeles, California revealed differences in patterns of protective compounds in the foliage between sites exposed to high pollution (but low natural stress) and those exposed to high natural stress levels (but only moderate pollution) (Tausz et al. 1999). In agreement with studies in European mountains, glutathione concentrations and ratios of some carotenoids over chlorophyll increased and total chlorophyll concentrations decreased with increasing elevation and increasing exposure to natural stress, but lower pollutant exposure (Tausz et al. 1999). Trees at higher elevated and high natural stress sites had more zeaxanthin, which is related to a higher requirement for energy dissipation. Pollutant exposure induced changes in the redox state of antioxidants (their regeneration rates), but not in their concentrations. While the proportion of oxidised glutathione and ascorbate in full sunlight was comparable among sites, only in needles from clean (but higher elevation) sites the antioxidant redox state changed towards more reduced overnight, possibly indicating a higher regeneration capacity of the system. Although the detailed mechanisms responsible for this phenomenon remain unclear, it becomes evident that pollution impact and photo-oxidative stress related to natural impacts may both affect the antioxidative protection systems, but elicit distinguishable patterns of responses (see below).

If antioxidative defence capacity is involved in resistance against natural stress as well as in protection from pollution effects, then interaction effects between pollution and natural stress are more complex then previously anticipated. For example, drought is commonly thought to protect plants from pollutants, because closed stomata largely reduce uptake. On the other hand, drought itself causes photo-oxidative stress and thereby potentially weakens the antioxidative defence. This question was addressed in a series of studies in the Sequoia National Park in California (Grulke et al. 2003a,b). There, montane mixed conifer forests experience considerable levels of ozone transported from the metropolitan region of San Francisco as well as the agricultural lands of the San Joaquin Valley to the west, and "chlorotic mottling" – the definitive ozone symptom – can be observed on *Pinus jeffreyi* needles. Due to shallow soils and inhomogeneous stand conditions, trees with good access to water (mesic) are growing near trees, which suffer considerable drought in the summer season (xeric). It has been demonstrated that due to stomatal closure, ozone uptake of xeric trees is 20% lower than that of mesic trees. Nevertheless, chlorotic mottling was even higher in xeric trees suggesting that it may be a more general symptom of ROS action rather than a selective ozone response (Grulke et al. 2003a).

## 5. The evaluation of multivariate response patterns

From the examples shown above it becomes evident that the use of stress-physiological responses as biomarkers for specific environmental impacts (*e. g.* bioindication of ozone) is not straightforward, because multivariate patterns are to be evaluated under field conditions. This question was addressed by the application of multivariate explorative statistics (such as principal component analysis or cluster analysis) to sets of biochemical stress

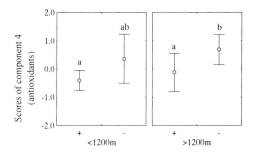

**Fig. 2.** *Picea abies* at mountain forest sites in Tyrol, Austria. Scores of the accumulated variable "antioxidant response" in relation to investigated sites grouped according to elevation and a "site quality factor" derived from vegetation ecological analysis. Ozone exposure was lower at low elevated sites and higher at high elevated sites. "Site quality factor" (+ or -) mainly includes water availability being more intermittent in "-" sites. Significant differences indicated by different letters (Modified after Wonisch et al. 1999).

variables. Figure 2 shows such an approach to a specific case study in an Austrian mountain forest. The original hypothesis was that ozone is a major stress factor for spruce trees in this area, and an array of biochemical stress markers was measured in needles to test it. Multivariate ordination extracted major axes of variation (accumulated variables = principal components), none of them being related to ozone exposure. In contrast, the accumulated variable representing oxidative stress response was related to a small-scale site quality factor (as determined by independent vegetation ecological analysis), which represented water availability. This result could not be found using single variables (Wonisch et al. 1999). Comparable approaches were successfully performed in several field studies (Tausz et al. 1998a,b, 2001, 2002).

Multivariate analysis also helped to clarify the relation between physiological stress responses and crown condition evaluation (including chlorotic mottling) in *P. jeffreyi* in the Sequoia National Park. Canopy morphology is used in two widely accepted assessment methods based on visible injury - the forest pest management score (FPM) and in the ozone injury index (OII). The appropriate statistical analysis revealed two major axes of relationships between visual assessment and biochemical responses, and only one was potentially attributable to higher ozone uptake. The other was related to drought and even less ozone uptake due to stomatal closure (Grulke et al. 2003b).

Acknowledgements. M. Tausz and D. Grill acknowledge support by the EU project CASIROZ (EVK2-2002-00165).

# References

Demmig-Adams B (2003) Linking the xanthophyll cycle with thermal energy dissipation. Photosynth Res 76:73-80

Elstner EF, Oßwald W (1994) Mechanisms of oxygen activation during plant stress. Proc Roy Soc Edinburgh B Biol 102:131-154

Grill D, Tausz M, De Kok LJ (2001) Significance of glutathione in plant adaptation to the environment. Kluwer Handbook Series of Plant Ecophysiology. Kluwer Publishers, Amsterdam, 262 p

Grulke NE, Johnson R, Esperanza A, Jones D, Nguyen T, Posch S, Tausz M (2003a) Canopy transpiration of Jeffrey pine in mesic and xeric microsites: $O_3$ uptake and injury response. Trees 17:292-298

Grulke NE, Johnson R, Monschein S, Nikolova P, Tausz M (2003b) Variation in morphological and biochemical $O_3$ injury attributes of mature Jeffrey pine within canopies and between microsites. Tree Physiol 23:923-929

Hecke K, Tausz M, Gigele T, Havranek WM, Anfodillo T, Grill D (2003) Foliar antioxidants and protective pigments in *Larix decidua* Mill from contrasting elevations in the Northern and Southern Tyrolean Limestone Alps. Forstwiss Centralbl 122:368-375

Lomsky B, Sramek V, Pfanz H (2001) An introduction to the development of tree and forest damage by air pollution in the Ore Mountains. J For Sci 47:3-7

Lomsky B, Materna J, Pfanz H (2002) $SO_2$-pollution and forest decline in the Ore Mountains. VULHM, Praha

Munné-Bosch S, Alegre L (2002) The function of tocopherols and tocotrienols in plants. Crit Rev Plant Sci 21:31-57

Pfanz H, Beyschlag W (1993) Photosynthetic performance and nutrient status of Norway spruce (*Picea abies* (L.) Karst.) at forest sites in the Ore Mountains (Erzgebirge). Trees 7:115-120

Pfanz H, Vollrath B, Lomsky B, Oppmann B, Hynek V, Beyschlag W, Bilger W, White MV, Materna J (1993) Life expectancy of spruce needles under extremely high air pollution stress. Performance of trees in the Ore Mountains. Trees 8:213-222

Polle A, Rennenberg H (1992) Field studies on Norway spruce trees at high altitudes: II. Defence systems against oxidative stress in needles. New Phytol 121:635-642

Smirnoff N, Wheeler GL (2000) Ascorbic acid in plants: biosynthesis and function. Crit Rev Plant Sci 19:267-290

Tausz M, Jiménez MS, Grill D (1998) Antioxidative defence and photoprotection in pine needles under field conditions – a multivariate approach to evaluate patterns of physiological responses at natural sites. Physiol Plant 104:760-764

Tausz M, Stabentheiner E, Wonisch A, Grill D (1998) Classification of biochemical response patterns for the assessment of environmental stress to Norway spruce. ESPR - Environmental Science & Pollution Research Special Issue No 1:96-100

Tausz M, Bytnerowicz A, Arbaugh MJ, Weidner W, Grill D (1999) Antioxidants and protective pigments of *Pinus ponderosa* needles at gradients of natural stresses and ozone in the San Bernardino Mountains in California. Free Rad Res 31:113-120

Tausz M, Bytnerowicz A, Arbaugh MJ, Wonisch A, Grill D (2001) Biochemical response patterns in *Pinus ponderosa* trees at field plots in the San Bernardino Mountains (Southern California). Tree Physiol 21:329-336

Tausz M, Wonisch A, Riberič-Lasnik C, Batič F, Grill D (2002) Multivariate analyses of tree physiological attributes – application in field studies. Phyton Ann Rei Bot 42(3):215-221

Wieser G (1999) Evaluation of impact of ozone on conifers in the Alps: a case study in spruce, pine and larch in the Austrian Alps. Phyton Ann Rei Bot 39:241-252

Wieser G, Tausz M, Wonisch A, Havranek WM (2001) Free radical scavengers and photosynthetic pigments in *Pinus cembra* L. needles as affected by ozone exposure. Biol Plant 44:225-232

Wonisch A, Tausz M, Haupolter M, Kikuta S, Grill D (1999) Stress-physiological response patterns in spruce needles relate to site factors in a mountain forest. Phyton Ann Rei Bot 39:269-274

# Complex assessment of forest condition under air pollution impacts

Tatiana A. Mikhailova[1], Nadezhda S. Berezhnaya[1], Olga V. Ignatieva[1], and Larisa V. Afanasieva[2]

[1]Siberian Institute of Plant Physiology and Biochemistry, Siberian Branch of the Russian Academy of Sciences, Lermontova, 132, Irkutsk, 664033, Russia
[2]Institute of General and Experimental Biology Siberian Branch of the Russian Academy of Sciences, Sahjanova, 6, Ulan-Ude, 670042, Russia


**Summary.** Studies of coniferous forests have been carried out in the Baikal region (East Siberia) – one of the greatest regions in the boreal zone. The region is characterized by presence of large industrial centers emitting considerable amounts of pollutants. The Scots pine (*Pinus sylvestris* L.) as a predominant region's species is the main subject of the investigations. Air pollution effects on pine treestands have been assessed using a set of morphostructural and physiological parameters as well as needle elements content. According to the results obtained, within the polluted areas the following parameters such as trees crown defoliation, length of shoots, needles mass, intensity of photosynthesis and respiration, content of elements in the needles are found to have changed to a great extent. A high level of a reverse correlation has been found between accumulation of elements-pollutants in the needles and most of the morphostructural parameters of pine trees polluted by industrial emissions.

**Key words.** Air pollution, Pine treestands, Needles elements content

## 1. Introduction

Air pollution is presently one of the key factors damaging forest ecosystems boreal zone included. Evaluation of its impact on ecosystems is complicated by a large number of various complex changes resulted from air pollution effects on the ecosystems, and by insufficiently elaborated methodical and methodological basis. While heavy forest damages are rather easily detected by remote methods, latent changes (physiological and biochemical) in the treestands condition as well as determination of a territorial scale of damages require direct natural observations. The region we investigate is characterized by both considerable differences in treestands weakening and wide industrial emissions spreading within the region territory. In this connection the objective of the study is to adequately reflect the boreal forests condition on large territories subjected to industrial emissions. Before, to achieve the objective, we worked out an original complex approach (Pleshanov et al. 2000). It is based on integration of methods of natural field observations, chemical analytical research, mathematical formalization and cartographical methods (mapping). And this approach implies analyzing the data obtained, taking into account interconnections between the parameters measured.

## 1.1 Study area

The approach mentioned above has been used to assess the forest condition of a large region in East Siberia, in particular, the condition of Baikal region forest ecosystems. The total area of the region is 800 thousand square km. It is directly connected with the world known Baikal Lake and is located in the center of Asia on the territories of two countries – Russia and Mongolia. Our research embraced the Russian part of the region entitled "Baikal Natural Territory"(BNT). We have examined the forests of southern, south-western, south-eastern and eastern parts of BNT (Mikhailova 2003), the remaining parts of the territory are still being examined. The environmental role of Baikal forests is very important, suffice to say, the forest ecosystems form up to 80% of water inflow to Lake Baikal. Coniferous forests are the principal type of vegetation in the region. The Scots pine (*Pinus sylvestris* L.) is the main subject of the investigation because it is a widely distributed species and the most sensitive one to air pollution. This species is assumed as an adequate forest condition indicator.

## 1.2 Parameters analyzed

We have studied parameters reflecting the treestand structure (age, height, diameter of trees, tree crown structure, understory vegetation); parameters reflecting structural-functional changes of the trees (length and mass of needles and shoots; nitrogen fractions in the needles, total and organic phosphorus, content of pigments, intensity of photosynthesis and respiration); parameters showing needles elemental composition alterations (elements-pollutants accumulation in the needles, nutrient elements content changes in the needles, disturbance of proportions between elements in the needles). In general we analyzed a large set of inter-related indicators. That is why, in our view, the subjectivity of the forest condition assessment is minimized.

The investigation was carried out by setting up sample plots following procedures developed within an International Cooperative Program (ICP Forests). A total 58 sample plots were set up within the studying area of about 500 thousand hectares (ha). The parameters were determined on each of the sample plots. Elemental composition, fractions of nitrogen and phosphorus, were examined in the 2-year-old needles from 40-year-old trees. In the pine needles the following elements were determined: sulfur, fluorine, mercury, lead, iron, zinc, cadmium, copper, manganese, aluminium, silicon, potassium, calcium, magnesium, sodium, as well as nitrogen and phosphorus.

## 1.3 Main industrial centers

A distinctive feature of the surveyed area is presence of a large number of industrial centers. Many of them differ significantly both in the amount of emissions and in their compositions. Pine treestands affected by emissions of the three largest industrial centers: Usolie-Angarsk (I), Shelekhov (II), and Irkutsk (III) have been surveyed. The total annual emissions of pollutants into the atmosphere from the first center amount to about 170,000 tons, from the second - about 30,000 tons, and from the third - about 52,000 tons (State Report 2004). The emissions of the Usolie-Angarsk industrial center contain the highest concentration of sulfur dioxide and mercury, the emissions of the Shelekhov industrial center contain the highest fluorine content, the amount of sulfur dioxide is far less than that from the first center, the emissions of the Irkutsk industrial center contain the highest content of lead, the content of sulfur dioxide three time less than from the first center, the amount of fluorine is far less than from the second center.

## 2. Results

### 2.1 Needles elemental composition

According to the prevalence of sulfur, mercury, fluorine and lead in the emissions of the three industrial centers, these elements accumulate in great quantities in the pine needles of trees growing in the vicinity of the centers, in comparison with the pine needles from the background areas (Fig. 1). But the polluted areas significantly differ not only in accumulation of prevailing pollutants, but also in the contents of other elements in the needles. There is a distinct disbalance of the elemental composition when considering quantitative proportions between the elements calculated for dry needles mass (Table 1). To illustrate, N:P:K proportion under any type of pollution changes due to increase of nitrogen level and reduction of potassium level; proportions of sulfur with nitrogen, phosphorus, potassium and microelements show pronounced disbalance, it is conditioned by the increase of sulfur share with simultaneous reduction of other elements shares; Mn:Fe proportion alters due to the pronounced decrease of manganese level and increase of iron level. On the whole, it should be noted that proportions between the elements allow one to evaluate the disbalance of needles elemental composition with higher reliability than by absolute values of elements concentrations.

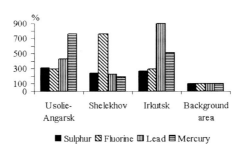

Fig. 1. Percental difference in the pine needle elemental contents between the territories polluted by the three industrial centers and the background areas.

**Table 1.** Proportions* of the elements in the needles of background and polluted pine trees.

| Center | N:P:K | Ca:K | N:S | P:S | K:S | Mn:S | Zn:S | Mn:Fe |
|---|---|---|---|---|---|---|---|---|
| Usolie-Angarsk | 72:10:18 | 64:36 | 89:11 | 53:47 | 68:32 | 1:99 | 3:97 | 49:51 |
| Shelekhov | 72:9:19 | 66:34 | 93:7 | 63:37 | 78:22 | 11:89 | 5:95 | 46:54 |
| Irkutsk | 71:9 19 | 60:40 | 95:5 | 71:29 | 86:14 | 15:85 | 5:95 | 65:35 |
| Bakground area | 66:9:25 | 57:43 | 97:3 | 82:18 | 92:8 | 48:52 | 11:89 | 83:17 |

*Proportions were calculated as a percentage share of the element from the sum of concentration of two (or three) elements in needles dry mass.

## 2.2 Vital state of treestands

Vital state of the treestands weakened by industrial emissions was determined via a set of representative parameters (Table 2). In accordance with the quantitative values of the parameters there were identified the following classes of treestand condition: low, moderately and heavily weakened, as well as relatively healthy (background). Treestands of different weakening classes are characterized by certain features (Mikhailova 2000). In heavily weakened treestands changes of morphophysiological parameters point to a very low activity of growth processes. Moderately weakened treestands are characterized by the state of pronounced chronic suppression, these treestands, however, still have fairly active growth processes. Low weakened treestands are physiologically characterized by emergence of minor metabolic disorders, which, to a certain extent, negatively influence the growth processes of the trees.

## 2.3 Connection between elements balance and morphostructural parameters

Since weakened treestands reveal significant changes both in the element accumulation and morphostructural parameters, it seemed important to investigate if there is connection between disturbance of the elements balance in the needles and morphostructural parameters in pine trees polluted by emissions from the three industrial centers under examination. Correlation analysis showed that the level of trees crown defoliation correlated with the content of most elements examined, for example, high correlation coefficients were revealed between the defoliation level and the content of S, F, Hg, Pb in the needles (Table 3). Reliable reverse correlations were also found between accumulation of elements-pollutants in the needles and the needles age, length, shoot length, needles mass on the shoots (Table 4).

## 2.4 Carbon assimilation activity of the trees

The data on the change of morphostructural parameters of the trees witness reduction of the quantity of assimilating phytomass in the weakened treestands. To assess assimilation

**Table 2.** Representative parameters for assessment of vital state of weakened pine treestands.

| Parameters | Background treestands | Weakened treestands | | |
|---|---|---|---|---|
| | | Low | Moderately | Heavily |
| Defoliation, % | 0-25 | >25-40 | >40-55 | >55 |
| Radial increment reduction % | <10 | >10 | >25 | >35 |
| Needles age, year | 5-7 | 3-4 | 2-3 | 1-3 |
| Needles length, cm | 6.8 ± 0.6 | 6.5 ± 0.6 | 5.9 ± 0.5 | 5.2 ± 0.7 |
| Needle mass, mg | 54.4 ± 0.8 | 38.9 ± 0.7 | 32.3 ± 0.8 | 27.8 ± 0.6 |
| Shoot length, cm | 25.9 ± 2.2 | 18.5 ± 2.1 | 15.2 ± 1.8 | 12.3 ± 1.7 |
| Shoot mass, g | 14.8 ± 2.7 | 8.2 ± 1.4 | 6.7 ± 1.6 | 4.2 ± 1.3 |
| Protein N/ nonprotein N | 6.3 - 7.0 | 4.7 - 6.2 | 3.0 - 4.6 | <3.0 |
| Acid soluble P $\times 10^{-3}$, % DW | 44.0 ± 2.3 | 37.5 ± 3.4 | 25.1 ± 1.8 | 13.5 ± 4.6 |
| Manganese $\times 10^{-1}$, %DW | 3.6 ± 0.2 | 2.9 ± 0.3 | 2.0 ± 0.2 | 1.3 ± 0.1 |
| Potassium $\times 10^{-1}$, % DW | 3.9 ± 0.10 | 4.1 ± 0.10 | 3.7 ± 0.07 | 3.0 ± 0.10 |
| Calcium $\times 10^{-1}$, % DW | 5.1 ± 0.1 | 5.9 ± 0.3 | 6.5 ± 0.2 | 7.2 ± 0.4 |

**Table 3.** Correlation coefficients between the level of trees crown defoliation and the elements content in the needles of pine trees polluted by emissions of the three industrial centers (the confidence level is 0.05, and the number of sampling is 15 values for each parameters, the table gives only significant levels of the coefficients).

| Center | S | F | Si | Cu | Hg | Pb | K | Mn |
|---|---|---|---|---|---|---|---|---|
| Usolie-Angarsk | 0.75 | - | 0.63 | 0.81 | 0.59 | 0.75 | -0.80 | -0.59 |
| Shelekhov | 0.70 | 0.82 | 0.80 | - | - | 0.58 | -0.64 | -0.86 |
| Irkutsk | 0.58 | 0.52 | 0.57 | 0.73 | - | 0.74 | -0.55 | -0.53 |

**Table 4.** Correlation coefficients between the elements-pollutants content in the needles and the morphostructural parameters of the pine trees polluted by emissions from the three industrial centers (the confidence level is 0.05, and the number of sampling is 15 values for each parameters).

| Parameters | Usolie-Angarsk center | | Shelekhov center | | Irkutsk center | |
|---|---|---|---|---|---|---|
| | Mercury | Sulfur | Fluorine | Sulfur | Lead | Sulfur |
| Needles age | -0.51 | -0.60 | -0.76 | -0.51 | -0.76 | -0.62 |
| Needles length | -0.69 | -0.58 | -0.62 | -0.55 | -0.72 | -0.51 |
| Shoot length | -0.58 | -0.62 | -0.71 | -0.61 | -0.60 | -0.55 |
| Mass of shoot needles | -0.67 | -0.60 | -0.72 | -0.51 | -0.53 | -0.58 |

activity in the trees we studied the content of pigments in the needles of the second year, intensity of photosynthesis and respiration of second year shoots. These parameters were calculated per 1 g (dry or wet) needles mass and per needles mass of a whole shoot. Specific parameters better reflect functional changes, and the calculation results for the whole shoot needles mass better reflect how structural changes influence a total assimilation crown activity (Mikhailova et al. 2004).

The pigments content calculated per 1 g needles mass shows even some increase of chlorophyll *a* and carotenoides (Fig. 2a). This seems to demonstrate compensatory protective response of trees to a negative factor weakening photosynthetic function. Evaluation of the pigments content per the whole shoot needles mass reveals significant reduction of the chlorophylls and carotenoides content (Fig. 2b). This confirms large

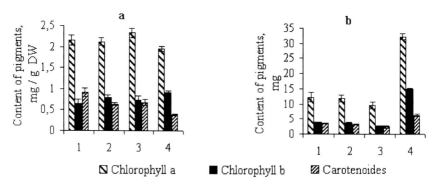

**Fig. 2a,b.** Content of the pigments in the needles of pine trees polluted by the three industrial centers: 1-Usolie-Angarsk, 2-Shelekhov, 3-Irkutsk. 4-Background area. **a** The data calculated per 1 g needles mass. **b** The data calculated per whole shoot needles mass.

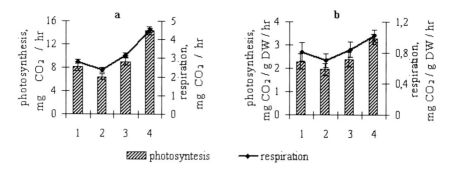

**Fig. 3a,b.** Intensity of photosynthesis and respiration of the needles of pine trees polluted by the three industrial centers: 1-Usolie-Angarsk, 2-Shelekhov, 3-Irkutsk. 4-Background area. **a** The data calculated per whole shoot mass. **b** The data calculated per 1 g needles mass.

morphostructural differences between the shoots of the weakened trees and the shoots of background trees. Alike, the study results of intensity of photosynthesis and respiration show that the differences between background and weakened trees are more pronounced with the calculation for the needles mass of the whole shoot (Fig. 3a), than with the calculation per 1 g needles mass (Fig. 3b).

These results prove decrease of carbon assimilation to be caused by functional and, to a greater extent, structural disturbances of the assimilation organs of the pine trees weakened by air pollution.

## 3. Conclusions

Impact of industrial emissions on coniferous treestands leads to disbalance of the needles elemental composition, which is confirmed by changes in the elements accumulation and violations of proportions between the elements. A high level of correlations between the disturbance of needles elements composition and the change of most morphostructural parameters has been found. The data obtained witness reduction of the quantity of assimilating phytomass in the weakened treestands, which is supported by the following phenomena: decrease of the needles age; reduction of length and mass of needles as well as of shoots; increase of the trees crown defoliation level. Changes in the pigments content, in the intensity of photosynthesis and respiration of the weakened trees are more pronounced with the calculation for the whole shoot needles mass. Structural and functional violations of the assimilating organs bring about reduction of carbon assimilation by forest ecosystems of the boreal zone.

Acknowledgements. The authors are grateful to the Russian Fund Fundamental Researches for supporting this study (the Grant number is 03-04-49565).

## References

Mikhailova T (2000) The physiological condition of pine trees in the Prebaikalia (East Siberia). Forest Pathology 30:345-359

Mikhailova T (2003) The effect of industrial emissions on forests of the Baikal Natural Territory. Geography and Natural Resources 1:51-59 (in Russian)

Mikhailova T, Suvorova G, Berezhnaya N, Ignatieva O, Yankova L (2004) Morphophysiological indices for assessment of changes of carbon sink in pine treestands polluted by industrial emissions. In: Vaganov E (Ed) Structural-Functional Organization and Dynamics of Forests. Sukachev Forest Institute of the Russian Academy of Sciences Siberian Branch, Krasnojarsk, pp 180-182 (in Russian)

Pleshanov A, Mikhailova T, Berezhnaya N, Toshakov S (2000) Methodological approach for complex cartographying ecosystems disturbed by industrial impacts. In: Vorobjev V (Ed) Problems of the regional ecology. Russian Academy of Sciences Siberian Branch, Novosibirsk, pp 44-45 (in Russian)

State Report (2004) About the environmental conditions of Irkutsk region in 2002. In: Korzun N (Ed) State Committee for Environmental Protection, Irkutsk, 328p (in Russian)

# Evaluation of the ozone-related risk for Austrian forests

Friedl Herman[1], Stefan Smidt[1], Wolfgang Loibl[2], and Harald R. Bolhar-Nordenkampf[3]

[1]Federal Office and Research Centre for Forests, Seckendorff-Gudent Weg 8, A-1130 Vienna, Austria
[2]ARC Systems Research, Austrian Research Centers, A-2444 Seibersdorf, Austria
[3]Institute of Ecology and Conservation Biology, University of Vienna, Althanstraße 14, A-1090 Vienna, Austria

**Summary.** Forest ecosystems are particularly affected by the impact of stress factors. Apart from natural stressors, there are also anthropogenic ones such as air pollutants. Ozone is considered to be the most phytotoxic air pollutant in Austria on a regional scale. In the past decade, concentrations in forested areas have increased significantly by up to 1.6 ppb per year. In order to assess the ozone impact on forests, a nationwide mapping of the AOT 40 for the forested area of Austria was done with the objective to take the ozone Critical Level as defined by the UN-ECE as the basic concept and to develop a threshold value with the help of three approaches. The adaptation of trees to the pre-industrial ozone levels, the parameters affecting stomatal uptake such as light intensity and water vapour saturation deficit, and the hemeroby (altered and natural) of the forest stands were taken into consideration. 61 % of the forest area showed a level of more than 10,000 ppb.h. Lower altitude areas with predominantely altered stands were affected more heavily.

**Key words.** Ozone risk modelling, Norway spruce forests, AOT40 modification, Adaptation, Hemeroby

## 1. Introduction

47 % (= 43.88 million hectares) of the Austrian territory is covered by forests and the need to preserve the forest area through sustainable management has utmost priority. This is due to the fact that, according to the Forest Development Plan (Bundesministerium für Land- und Forstwirtschaft 1991), 31 % of the forest areas have above all a protective function. In order to protect forest health and to survey the condition of forests, comprehensive networks and research activities have been designed (Bundesministerium für Land- und Forstwirtschaft 2000). Norway spruce is the main tree species with 61 % of the growing stock. This share increases up to 1000 m a.s.l. and then it decreases again (Schieler et al. 1995).

Ozone is considered to be the most phytotoxic air pollutant on a regional scale (Krupa and Kickert 1989). In Europe, the long-term ozone concentrations have been increasing since the eighties (Brönnimann et al. 2000; Guicherit and Roemer 2000; Bytnerowicz et

*Plant Responses to Air Pollution and Global Change*
*Edited by K. Omasa, I. Nouchi, and L. J. De Kok (Springer-Verlag Tokyo 2005)*

al. 2004a,b; Ministerium für Umwelt und Forsten 2004) particularly in remote areas (0.5 to 2.0 % per year; Vingarzan 2004), whereas the peak concentrations are decreasing (de Leeuw 2000). In Austria, annual mean values have increased by up to 0.2 ppb per year in forest areas in the last two decades (Smidt and Herman 2004); they also increase with altitude (Smidt 1998).

Negative ozone impacts on forest trees are well documented (Sandermann et al. 1997; Herman et al. 1998; Fuhrer 2000; Chappelka et al. 2001; Herman et al. 2001; Smidt and Herman 2004). In the course of various interdisciplinary research projects in the field it has been proved that, given the present ozone level, old trees show measurable reductions of $CO_2$ uptake (Bolhar-Nordenkampf 1989; Lütz et al. 1998). Based on the EU Council Directive 92/72/EEC (1992), the Austrian Ozone Law (BGBl. 34/2003) defines goals for 2010 and 2020, respectively. With the European Critical Level AOT40 ("accumulated exposure over a threshold of 40 ppb") the ozone risk for the vegetation was assessed. The AOT40 (limit value: 10,000 ppb.h for 6 months of one growing season) was exceeded by up to five times (Loibl and Smidt 1996). However, those patterns of high Critical Level exceedances do not correspond with the forest health status, and it was obvious that the AOT40 concept had overestimated the ozone induced stress. Therefore, various approaches were outlined including the potential uptake of ozone and the probable adaptation to the ozone impact on trees.

## 2. Material and methods

### 2.1 European Critical Level (AOT40) and Austrian Ozone Law

The Critical Level AOT40 served as the basis for assessing the risk of ozone for forest trees in the first approach. The goals of the Austrian Ozone Law, which will be the legal basis in 2010 and 2020, respectively (Table 1), serve as the basis for the evaluation of ozone data of two investigation areas in the Austrian Alps.

### 2.2 Data base for trend analyses and calculations

The ozone levels are shown for 30 monitoring sites in forest areas of Austria as annual mean values from 1993-2000. The altitudinal dependence is illustrated by the AOT40 values of these stations (Spangl 2003). The trend analysis (NEUMANN-Test; Sachs

Table 1. AOT40 and Austrian Ozone Law.

| AOT40 (prov.) | 10,000 ppb.h, April - September | UN-ECE (1994) |
|---|---|---|
| Goal 2010 | 9,000 ppb.h *) | Austrian Ozone Law (BGBl. 34/2003) |
| Goal 2020 | 3,000 ppb.h *) | Austrian Ozone Law (BGBl. 34/2003) |

*) May – July, 8.00 – 20.00 MEZ; mean value over 5 years.

1997) for the years 1990-2002 is based on annual mean values, provided that at least 8 continuous observation years were available. The exceedances of the goals of the Austrian Ozone Law were calculated for two investigation areas in Achenkirch/Tyrol (Northern Alps, 920 m a.s.l.) and in Bodental/Carinthia (Southern Alps, 1010 m a.s.l.), based on half-hourly mean values (http://bfw.ac.at/600/2023.html).

## 2.3 Data base for ozone risk modelling

For the evaluation of risk maps for the forest area with regard to Norway spruce, the following comprehensive data sets have been applied:

- Ozone data: Hourly ozone mean values from 120 monitoring sites for the year 1994 (Spangl 2002).
- Austrian Forest Inventory: A grid of 5,600 sample plots (Schieler et al. 1995).
- Meteorological data: Air humidity, solar radiation, temperature from 78 monitoring sites over 10 years (1981–1990) measured by the Central Institute of Meteorology and Geodynamics (Vienna).
- Digital elevation model: Map with a grid-cell size of 1x1 km (Loibl et al. 1994).
- Forest hemeroby map: Hemeroby is a measure of the anthropogenic influence on a biotic community. Thus, the hemerobic degree is based on the effect of anthropogenic interference with the development into a climax community (final natural state). According to Grabherr et al. (1998), nine degrees of hemeroby were established to describe the forest stands ranging from natural (9) to artificial (1). For the modelling approach two superior classes were defined: natural forests (hemeroby classes 6-9) and altered forests (hemeroby classes 1-5).

### 2.3.1 Ozone interpolation model

The ozone interpolation model served as the basis for all further approaches. The diurnal variation of ozone concentration depending on daytime and altitude is described by an analytical function that reflects the dependence of ozone concentration at the relative altitude (= elevation above valley floor) and the daytime. The equation generates hourly standard ozone-elevation curves showing the increase of ozone concentration with altitude (Loibl et al. 1994).

In order to integrate hourly influences, hourly ozone-elevation gradient curves were devised considering the specific hourly monitoring data. The application of a digital elevation model and the modified ozone-elevation dependence functions allow to model day-specific hourly "ozone concentration surfaces". The ozone concentration is calculated cell by cell by integrating the local elevation and the respective day-time into the ozone-elevation dependence equation. After considering the remaining residuals, the ozone concentration surfaces reflect the hourly day-specific local ozone distribution.

### 2.3.2 Modelling of the AOT40 (basis approach)

The AOT40 has been calculated based on the ozone measurements of 1994, using the above-mentioned data sets and the ozone-elevation dependence function. The calculation refers to 1,700 hourly ozone concentration maps for all daylight hours during the vegetation period. The exceedances above 40 ppb of all hours have been gathered in one single AOT40 map. The exceedances were distinct, especially at higher altitudes of the alpine regions, the reason being mostly the increased ozone levels with altitude which persist also during the night.

The basis approach represents the AOT40 concept of the UN-ECE (1994). Since the timber growth (Schadauer 1996) and forest health status (Federal Research Centre 1998) did not reflect the high Critical Level exceedances, it became clear that further approaches had to be developed.

## 2.4 Development of new approaches

The 1$^{st}$ approach modifies the AOT40 threshold value based on the assumption that forests have adapted themselves to pre-industrial levels of ozone, which increase with altitude. Therefore, an elevation-dependent AOT gradient was applied varying between 32 ppb at sea level and 56 ppb at the timberline. (Historical ozone data sets show an average "pre-industrial" ozone concentration of 13 ppb at 200 m and 29 ppb at 1800 m a.s.l.; Bolhar-Nordenkampf et al. 1999.) The AOT40 threshold shows a difference of 27 ppb from the 13 ppb average concentration at 200 m (the typical altitude for ozone hazard tests leading to the AOT40 rule), which results in a 56 ppb threshold at the timberline (29 ppb + 27 ppb).

The 2$^{nd}$ approach modifies the AOT40 threshold value according to the potential for ozone uptake related to the stomatal conductance of mature Norway spruce trees, based on field measurements and data from literature. Stomatal conductance is heavily influenced by light intensity and the water vapour saturation deficit (Grünhage et al. 1999, 2001; Emberson et al. 2000; Fuhrer 2000; Karlsson et al. 2000; Larcher 2000; Wieser and Emberson 2004). Hourly maps of relative humidity, air temperature and light intensity were applied to estimate AOT correction factor maps, which lends new weight to the AOT40 exceedances, considering the $O_3$ : $H_2O$ molecule size ratio and taking into account the ozone concentration variation inside intercellular spaces (Bolhar-Nordenkampf et al. 1999).

The 3$^{rd}$ approach combines the two approaches and applies them to areas with either natural or altered forests. Natural forest stands are supposed to have adapted themselves to the pre-industrial ozone level. Thus, the results of the 1$^{st}$ approach were related to these areas. The results of the 2$^{nd}$ approach were related to areas with altered forests, for which the ozone uptake triggered by stomata opening has to be applied (Loibl et al. 1999).

# 3. Results

## 3.1 Trend analysis and calculations

Annual mean values (1993-2000) of ozone ranged from 11.5 ppb at lower altitudes up to 49 ppb at the altitudes of the timberline (2000 m a.s.l.; Smidt 2004). The AOT40 values increased significantly with the altitude (Fig. 1); the 10,000 ppb.h level was exceeded at almost all altitudes.

Fig. 2 outlines the ozone trends (mean increase of the ozone mean values, ppb p.a.). Significant positive trends were predominant, whereas the negative trends were rare. There was a clear increase in ozone concentrations up to 1000 m a.s.l.

The results obtained from two long-term measuring stations served as examples to demonstrate that measures to reduce emissions must be taken in order to reach the goals set for 2010 and 2020. The measuring station in Achenkirch/Tyrol delivered a value of 14,848 ppb.h in 2003; the value measured at the station in Bodental/Carinthia was 13,835 ppb.h in the same year, the target values described in the ozone law were decisively exceeded in both areas.

## 3.2 Modelling of the AOT40

Fig.3 illustrates the ozone exposure of the forested areas. The AOT40 was exceeded in the whole forested area and the most frequent AOT40 values were calculated between 30,000 and 40,000 ppb.h. Taking into consideration the adaptation of Norway spruce to the pre-industrial ozone concentrations and the ozone uptake, the AOT 10,000 ppb.h was exceeded on 61 % of the forested grid cells (Loibl et al. 2004; Fig. 4).

The 61 % of the forest area that showed a level of more than 10,000 ppb.h were mainly located between altitudes of 400 and 1400 m a.s.l., and the maximum was reached at an elevation of 600-700 m a.s.l. (Fig. 5). This is the altitude where most of the altered forests are found (Fig. 6).

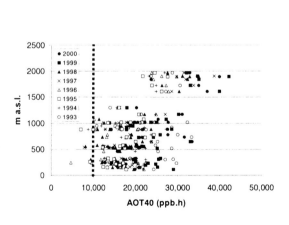

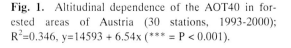

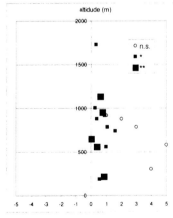

**Fig. 1.** Altitudinal dependence of the AOT40 in forested areas of Austria (30 stations, 1993-2000); $R^2=0.346$, $y=14593 + 6.54x$ (*** = $P < 0.001$).

**Fig. 2.** Ozone trend (ppb per year; 30 stations, 1990-2002). *= $P < 0.05$, ** = $P < 0.01$.

**Fig. 3.** Ozone exposure of forested areas of Austria (AOT40 basis approach; ppb.h; 1994).

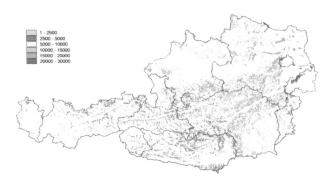

**Fig. 4.** AOT elevation gradient applied for natural forests and climate influence on stomata opening for altered forests (ppb.h; 1994).

## 4. Conclusions

The results obtained from long-term measurements at sites of relevance to forests show that ozone continues to be an important atmogenic risk factor for Austrian forests. The calculation of trends during the previous decade showed a significant increase in ozone levels (referring to the EU critical level concept) up to an altitude of 1000 m a.s.l. Modelling of AOT40 values for the entire forest area delivered unrealistic ozone risk patterns with values exceeding 50,000 ppb.h. This called for a modification of risk assessment

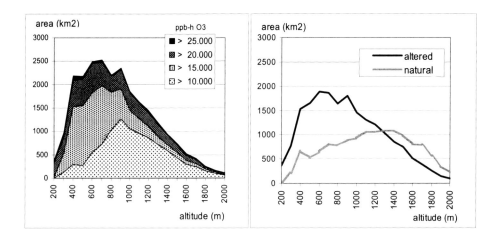

**Fig. 5.** Forest areas with ozone levels of > 10,000 ppb.h at different altitudes (1994).

**Fig. 6.** Distribution of areas of natural and altered forests at different altitudes.

considering the adaptation to a higher ozone concentration, the factors influencing ozone uptake, and hemeroby. The modification of the AOT40 concept demonstrated that primarily forests at altitudes between 400 m and 1000 m a.s.l. showed the highest ozone levels (<= 25,000 ppb.h). In addition, the integration of the hemeroby approach illustrated that more than half of the forest area at the affected altitudes were covered with altered forests, which are consequently exposed to a higher risk.

# References

Bolhar-Nordenkampf HR (1989) Stressphysiologische Ökosystemforschung Höhenprofil Zillertal. Phyton (Horn, Austria) 29(3) 302pp

Bolhar-Nordenkampf HR, Loibl W, Gatscher B (1999) Adaptation to pre-industrial ozone concentrations - the cause for a fundamental change in the Austrian ozone risk map. Proceedings UNECE -Meeting Suisse, Bern - Gerzensee

Brönnimann S, Schuepbach E, Zanis P, Buchmann B, Wanner H (2000) A climatology of regional background ozone at different elevations in Switzerland (1992-1998). Atmos Environ 34:5191-5198

Bundesministerium für Land- und Forstwirtschaft (1991) Waldentwicklungsplan. Vienna

Bundesministerium für Land- und Forstwirtschaft (2000) The sustainable future of mountain forests in Europe, 3rd International Workshop Igls – Tyrol (Austria), May 3-5, 2000. Vienna

Bytnerowicz A, Godzik B, Grodzinska K, Fracek W, Musselman R, Manning W, Badea O, Popescu F, Fleischer P (2004a) Ambient ozone in forests in the Central and Eastern European mountains. Environ Pollut 130: 5-16

Bytnerowicz A, Musselman R, Szaro R (2004b) Effects of air pollution on the Central and Eastern

European mountain forests. Environ Pollut 130:5-134
Chappelka A, Karnosky D, Percy K (2001) Impacts of air pollution on forest ecosystems. Environ Pollut 115:319-481
De Leeuw FA (2000) Trends in ground level concentrations in the European Union. Environ Sci Policy 3:189-199
Emberson LD, Wieser G, Ashmore MR (2000) Modelling of stomatal conductance and ozone flux of Norway spruce: comparison with field data. Environ Pollut 109:393-402
Federal Research Centre (1998) Forest condition in Europe. 1998 Executive Report Geneva
Fuhrer J (2000) Risk assessment for ozone, effects on vegetation in Europe. Environ Pollut 109:361-542
Grabherr G, Koch G, Kirchmeir H, Reiter K (1998) Hemerobie österreichischer Wälder. Österreichische Akademie der Wissenschaften, Veröffentlichungen des Österreichischen MaB Programmes, Bd. 17. Universitätsverlag Wagner, Innsbruck
Grünhage L, Jäger HJ, Haenel HD, Löpmeier FJ, Hanewald K (1999) The European critical levels for ozone: improving their usage. Environ Poll 105:163-173
Guicherit R, Roemer M (2000) Tropospheric ozone trends. Chemosphere – Global Change Sci 2:167-183
Herman F, Lütz C, Smidt S (1998) Description and evaluation of stress to mountain forest ecosystems – Resuls of long-term field experiments. Env Sci Pollut Res, Special Issue 1, 104pp
Herman F, Smidt S, Huber S, Englisch M, Knoflacher M (2001) Evaluation of pollution-related stress factors for forest ecosystems in Central Europe. Env Sci Pollut Res 8:231-242
Karlsson PE, Pleijel H, Karlsson GP, Meddin EL, Skärby L (2000) Simulations of stomatal conductance and ozone uptake to Norway spruce saplings in open-top chambers. Environ Pollut 109:443-452
Krupa SV, Kickert RN (1989) The greenhouse effect: Impacts of ultraviolet-B (UV-B) radiation, carbon dioxide ($CO_2$), and ozone ($O_3$) on vegetation. Environ Pollut 61:263-393
Larcher W (2000) Physiological plant ecology. Ecophysiology and stress physiology of functional groups. 4$^{th}$ Edition. Springer Berlin
Loibl W, Winiwarter W, Kopcsa A, Züger J, Baumann R (1994) Estimating the spatial distribution of ozone concentrations in complex terrain using a function of elevation and day time and Kriging techniques. Atmos Environ 28:2257-2566
Loibl W, Smidt S (1996) Areas of ozone risk for selected tree species. Env Sci Pollut Res 3:213-217
Loibl W, Winiwarter W, Kopcsa A, Züger J (1999) Critical Levels-Karten für Ozon für ausgewählte Waldgebiete: Berechnung eines modifizierten AOT40 Level II – Ozonaufnahme abhängig von Witterungsbedingungen. Endbericht Bundesministerium für Land- und Forstwirtschaft, Seibersdorf
Loibl W, Bolhar-Nordenkampf HR, Herman F, Smidt S (2004) Modelling critical levels of ozone for the forested area of Austria – Modifications of the AOT40 concept. Env Sci Pollut Res 11:171-180
Lütz C, Kuhnke-Toss R, Thiel S (1998) Natural and anthropogenic influences on photosynthesis in trees of Alpine forests. Environ. Sci Pollut Res, Special Issue 1:88-95
Ministerium für Umwelt und Forsten (2004) Jahresbericht 2003. Mitteilungen der Forschungsanstalt für Waldökologie und Forstwirtschaft 53 (04), 43pp
Sachs (1997) Angewandte Statistik. Springer Berlin
Sandermann H, Wellburn AR, Heath RL (1997) Forest decline and ozone. Ecological Studies 127. Springer Berlin
Schadauer (1996) http://fbva.forvie.ac.at/700/700.html
Schieler K, Büchsenmeister R, Schadauer K (1995) Österreichische Forstinventur. FBVA-Berichte

92 (Vienna)

Smidt S (1998) Risk assessment for air pollutants for forested areas in Austria, Bavaria and Switzerland. Env Sci Pollut Res, Special Issue 1:25-31

Smidt S (2004) Trends von Luftschadstoffen in österreichischen Waldgebieten. http://bfw.ac.at/600/2166.html

Smidt S, Herman F (2004) Evaluation of air pollution-related risks for Austrian mountain forests. Environ Pollut 130:99-112

Spangl W (2002) Data sets. Umweltbundesamt Vienna

Spangl W (2003) Luftgütemessstellen in Österreich. Umweltbundesamt, Berichte BE-231

UN-ECE (1994) Critical Levels for ozone. In: Fuhrer J, Achermann B (Eds) Eidgen. Forschungsanstalt für Agrikulturchemie und Umwelthygiene, Bern

Vingarzan R (2004) A review of surface ozone background levels and trends. Atmos Environ 38:3431-3442

Wieser G, Emberson LD (2004) Evaluation of the stomatal conductance formulation in the EMEP ozone deposition model for *Picea abies*. Atmos Environ 38:2339-2348

# Causes of differences in response of plant species to nitrogen supply and the ecological consequences

David W. Lawlor

Crop Performance and Improvement, Rothamsted Research, Harpenden, Herts., AL5 2JQ, UK

**Summary.** Plant species differ in response to nitrogen (N) supply, depending on environment. Mechanisms of $NH_4^+$ and $NO_3^-$ uptake, reduction of $NO_3^-$ to $NH_4^+$, and $NH_4^+$ incorporation into amino acids, depend on light energy which drives photosynthesis, providing reductant, ATP and carbohydrates. Amino acids are used in protein synthesis, which is the basic process determining development and growth. The mechanisms are similar, but differ quantitatively, between species. Genetic factors (the 'genetic potential') regulate protein synthesis and metabolism and control organ growth and composition, but rates of processes depend on temperature and supply of N and other nutrients, water etc. Species differ in biochemical composition, size and growth rates: these determine requirements for, and responses to, N etc, under different conditions. With N supply below that required to meet the genetic potential, maximum growth is not achieved: with N in excess, growth is maximal. Interaction between genetic potential and N supply results in differences in growth, production, survival and reproductive success of species and thus their occurrence ('richness') in complex ecosystems. Slow growing, low capacity species of nutrient-depleted soils are adapted to limited N. Rapidly growing, large capacity plants requiring large N-supply for maximal growth are associated with nutrient-rich habitats. With increased N-deposition from pollution, plants able to accumulate N and grow and reproduce rapidly, out-compete slow growing species requiring little N, and expand their range. However, there is considerable interaction with other nutrients, pH etc. Thus, there is a biochemical explanation for changes in species composition and production of ecosystems with increasing N-pollution. Examples of N-effects on moorland and pasture ecosystems are given.

**Key words.** Nitrogen pollution, Plants, Species-richness, N-metabolism, Ecosystem competition

## 1. Nitrogen and ecosystems

With long-term, relatively uniform climatic and soil conditions, ecosystems change slowly, with rather constant long-term dry matter production and species diversity. Altering the environmental conditions, e.g. the supply of nutrients, affects individual species within the ecosystem in different ways because they have different requirements for, and responses to, nutrients: This affects numbers and proportions of plant species in ecosystems (species-richness), and their productivity. With industrialisation, involving massive burning of fossil fuels and fixation of $N_2$ for N-fertilisers (currently of similar magnitude to global biological N-fixation) N-supply, in different forms, to ecosystems has increased substantially and widely: it is now ca. 50 kg N ha$^{-1}$ year$^{-1}$ over much of western Europe. Consequently, substantial changes are occurring in ecosystems, but mechanisms are not well understood or quantified. Here biochemical mechanisms underlying differences in species responses to N supply, and the consequences for changes in ecosystems, are briefly discussed.

## 1.1 Nitrogen use by plants & sources

Nitrate ($NO_3^-$) and ammonium ($NH_4^+$) ions are used by different species to different extents, depending on conditions: species associated with microorganisms, e.g. legumes with Rhizobia, assimilate gaseous $N_2$. Nitrate results from microbial mineralization of soil organic matter, rates depending on the amount and N-content of the organic matter, and on soil bacterial populations and their activity. This depends greatly on temperature, water content, pH, availability of other nutrients (P, K) etc. In base- and nutrient-rich, well drained, warm low-land soils with large organic matter content and pH close to neutrality, mineralization is fast. Inputs of N from pollution may be a small proportion of the total N supply and turn-over. Also, other nutrients, and conditions (e.g. temperature) are favourable for growth. In more extreme conditions such as heaths and moors on base- and nutrient-poor silicate rocks, in cool, upland areas, soils are often very wet, and pH and nutrient content - including N - low. N from pollution may be a large proportion of total N circulation, whilst other conditions required for growth are severely limiting. However, both types of ecosystem are affected by N-input: two cases are discussed after considering the plant-factors responsible for the responses.

### 1.1.1 Nitrogen, plant composition and growth

Genetic information, acquired by plants during their evolution, determines the types and amounts of proteins produced by plants, and establishes the potential for development and growth (Lawlor 2001). Synthesis of proteins requires nitrogen (here other requirements are assumed not to be limiting) so there is a very strong, linear relation between N absorbed (less with N supplied) and protein content. Active proteins are responsible for synthesis of all cellular components. Thus, with increasing N, more pigments e.g. chlorophyll, and proteins, e.g. of electron transport, ATP synthesis and $CO_2$ assimilation (including the most abundant - Rubisco) of chloroplasts are formed, as are respiratory components. More, generally larger cells are formed. Consequently with increasing N, leaves are bigger, with more metabolic components, have greater photosynthetic capacity, and capture and use light energy more efficiently. Growth of other organs is stimulated: more and larger branches (tillers in grasses) and leaves increase light interception; this together with greater photosynthetic capacity stimulates vegetative growth, which often leads to more reproductive organs, greater seed production and better germination and survival of offspring. The result is greater ecological competitiveness, such as shading out species of smaller stature. However, the response of species which also require large amounts of other nutrients, or particular conditions, will be small. Thus, species metabolically adapted to very low pH or limited P- or K-supply but able to exploit large N-supply to grow larger will have an advantage under such conditions. However, if P and K are abundant, they will be disadvantaged by species tolerant of low pH, but better able to exploit large N-supply if P and K are available. Of course, linear increase in metabolic capacity and number and size of organs with increasing N-supply is finite, because the mechanisms from gene expression, via protein synthesis to formation of products attains maximum rate and the genetic capacity is fully used. Production then slows and growth reaches a plateau as the N-supply continues to rise: the system has reached its 'genetic potential'. However, N may continue to accumulate, with $NH_4^+$ and $NO_3^-$ used by different species to different extents, depending on conditions. $NH_4^+$ is toxic to cells, so must be immediately assimilated; $NO_3^-$ may be accumulated in vacuoles and assimilated later. For ecological success, balanced C and N metabolism is crucial. Obviously photosynthesis and respiration must be optimized to N-supply to avoid toxicity and achieve greatest efficiency. Large N content may increase susceptibility to pests, herbivores, diseases and extreme conditions, e.g. frost and drought.

## 1.1.2 Genetic potential

This is an important concept, emphasizing that the total activity of a species - all biochemical, metabolic, physiological, growth and reproductive processes - as well as responses to environment - are genetically determined. However, the potential includes differing flexibility in response to environment. There are short- and long-term changes (adaptations) in structure and physiological function, in gene expression, and in population genetics (Linhart and Grant 1996). These increase or decrease development, growth, survival and reproduction, depending on conditions. Species differ greatly in their potential. Thus, the whole gene → protein → biochemistry → physiology → species → ecology → environment continuum must be considered to understand the effects of N-deposition on individual species and total ecosystem production and species-richness. A species' genetic potential is the result of an extremely long evolutionary process, with selection tending to optimize to specific conditions; thus some species are very specialized, best adapted to growing within a narrow range of conditions (often extreme, e.g. with very deficient nutrient supply, or in very acid soils), whilst others are able to grow well over a wider range of conditions. Genetic potential thus sets the plant's requirements for the environmental conditions, for example the temperature required for growth and development, or the amount, proportion and rate of N-supply. Species differ in their genetic potential; species A may grow slowly but mature and reproduce at small size with low N supply and be unable to respond to more N. Species B is much more productive than A but requires more N. However, species C can grow even more vigorously than B, but requires a very large amount of N. If the environment, including the supply of nutrients, water, etc, does not match the potential demand, then performance - development, growth, seed production etc – is below potential. Processes are maintained relatively balanced over a wide range of conditions, allowing some growth and production, but resources below the potential are 'stresses' and reduce ecological fitness. Species respond to many environmental factors- temperature, light, water, nitrogen, other nutrients etc with some dominant - in complex combinations, analysis based on one or a few factors, may fail to show the conditions determining the relative genetic potential of species. Variation in populations provide the basis for selection: thus species may adapt rapidly genetically, with functional changes but few morphological changes (Snaydon and Davies 1976; Silvertown et al 2002) Evolution may tend to optimise species' survival in particular environments in the long-term, with small differences in response to conditions having large ecological effects. Understanding of genetic potential, and the biochemical mechanisms, related to ecological success is limited but necessary to interpret ecological changes resulting from changing environment.

## 1.2 Effects of N-supply on species and ecosystems

### 1.2.1 Heaths and moors

Wet, acid heaths and moors in north-western Europe are characterized by often very slow growing species, of small stature and tolerant of low fertility, e.g. lichens, mosses, and vascular plants including monocots, e.g. *Narthecium ossifragum* and *Nardus stricta*, and dicots, e.g. *Erica* and *Calluna* spp, *Gentiana pneumonanthe*, and also *Drosera rotundifolia* and *Pinguicula* which obtain N and other nutrients through insectivory. Slow growth rates are probably an evolutionary, genetic adaptation to such conditions, including limited N supply (Pons et al. 1994). Large changes in species composition have occurred in many such areas, related to increased N-deposition (see Aerts and Bobbink 1999; NEGTAP 2001). Lichens and bryophytes, despite (or perhaps because of) their greatly and progressively increased N-content over the last 120 years, and *Drosera, Erica* and *Calluna*, have decreased in frequency: they are type A species considered earlier. There may be little evidence of direct competition (e.g. shading) and N-toxicity or greater susceptibility to biotic and abiotic factors (herbivory, frost etc) may be important. Often grass species such as *Molinia caerulea* and *Holcus lanatus* (and *Brachypodium pinnatum* on calcareous sites) increase: they represent type B species which greatly increase growth (particularly tillering) with more N. They may out-compete and eventually eliminate type A, by shading, removal of other nutrients and water (when scarce). Where N-supply increases and other resources are plentiful, rapidly growing species with large nutrient demand e.g. dicots such as *Urtica doica* and monocots e.g. grasses *Arrhenatherum elatius* and *Holcus*, become dominant, so B-types are out-competed. Where A and B or B and C overlap the species composition of the vegetation will depend on the relative sensitivity of species to small change in N, and on the interaction with other factors. Ultimately, woody perennial shrubs and tree invade and dominate.

### 1.2.2 Park grass experiment at Rothamsted, UK

Since 1856 the effects of applications of fertilizers and lime (to regulate pH of the clay-loam soil with no free calcium carbonate) have been tested on what was old grazed pasture. The original aim was to assess the nutritional requirements for optimal hay yields, but it has evolved into a major ecological experiments, of relevance to understanding the effects of N and acid pollutants on vegetation. Details are given by Thurston et al (1976; Rothamsted Experimental Station 1991). A zero-input (control) treatment provides a base line. Different amounts and combinations of inorganic fertilizers have been applied yearly. The N treatments are sodium nitrate, at 48 and 96 kg N ha$^{-1}$ year$^{-1}$ (N*1 and N*2) and ammonium sulphate at 48, 96 and 144 kg N ha$^{-1}$ year$^{-1}$ (N1, N2, N3) applied in spring. These are combined (not factorially) with 35 kg P ha$^{-1}$ year$^{-1}$ (as superphosphate), and 225 kg K ha$^{-1}$ year$^{-1}$ (as potassium sulphate). The vegetation is cut and removed, providing annual total above ground dry matter yield (TDM: biomass). Species composition and contribution of individual species to TDM are measured but less frequently. Measurement of species-richness – by presence in a relatively large area - over-represents infrequent species compared to their proportion in TDM. For the control soils, initial pH was ca. 5.7 and is now 5.2 due to cation removal and acid pollutant deposition. N1 –N3 treatments greatly acidify the soil (to pH 3.5) whereas N*1, N*2 did not (pH 5-6). Lime has been applied more recently to some parts of the treatments to increased pH to 7; despite pH increasing relatively slowly there are clear effects.

During the experiment yield and species composition of control plots have changed, perhaps related to depletion of P, K etc with removal of biomass, to increased acid and N deposition (probably from less than 5 to 50 kg N ha$^{-1}$), and to changes in the weather. However, the great differences in species composition and yield related to treatments are impressive. Initially, (averaged 1856-65) TDM was 2.8 t ha$^{-1}$, with some 50 vascular plant species. Grasses dominated, forming ca. 76% of TDM: *Lolium* and *Holcus* were most abundant, with *Arrhenatherum*, *Agrostis tenuis* and *Festuca rubra*. Legumes and other species contributed, respectively, ca. 5 and 16% of TDM. Diversity of control plots decreased over time; in 1975-76 there were 35-40 species and similar TDM (1986-90 average 2.5 t ha$^{-1}$). Without lime, grasses, legumes, and other dicots formed respectively ca. 62, 5 and 33% TDM: the dominant grass was *F. rubra*, followed by *Agrostis*, and *Trifolium pratense* was the main legume, with rosette and low-growing dicots important. With lime, TDM increased to 3.1 t ha$^{-1}$, but grasses decreased to 38%, particularly *Agrostis* and also *F. rubra*, and legumes increased to 10% and other dicots to 52%. Application of N*1 and N*2 (without P, K and little change in pH) doubled TDM to ca. 4 t ha$^{-1}$ with 60% as grasses (mainly *Alopecurus pratensis, Anthoxanthum odoratum*, and *Arrhenatherum*) and 40% as dicots but almost no legumes. Species decreased to ca. 20. Liming increased TDM to 4.8 t ha$^{-1}$, with 72% grasses (*Helictotrichon* becoming as important as *F. rubra* and *Dactylis* with much less *Alopecurus, Anthoxanthum* and *Agrostis*), increasing legumes (*T. pratense*) to 3% and slightly decreasing other dicots (to 24%). There were about 18 species. Thus N increased TDM, decreased species-richness, and dominant grasses changed. There were also large effects of altering soil pH. With N1 alone pH was ca. 3.7 which probably decreased availability of P but increasing aluminium: TDM was much decreased to 1.1 t ha$^{-1}$, with 98% grasses (*Agrostis capillaris* dominant) and only 2% other species and no legumes. Surface litter (due to absence of earthworms and very few bacteria) accumulated but increasing its pH decreased grasses (to 60% of the 4.3 t ha$^{-1}$ TDM), especially *Agrostis* but increased *Dactylis glomerata, Helictotrichon pubescens* and *Holcus*, and increased legumes (5%) and other dicots (30%). Species-richness increased from about 6 before liming to ca. 15 currently, so the rate of change is slow. Thus N, together with pH (probably via nutritional and toxicity factors) has altered composition drastically.

Application of P and K also changed TDM and species-richness. Without N, TDM increased to 4.7 and 6.2 t ha$^{-1}$ without and with lime respectively, but resulted in fewer species, about 25 and 23 respectively, with grasses (particularly *Agrosis, F. rubra* and *Anthoxanthum*) more important at low than high pH (which greatly increased *Arrhenatherum*), and legumes and other dicots increasing at higher pH. Application of P and K with N*1 and N*2 further increased maximum TDM to 4.8 and 6.6 t ha$^{-1}$ respectively without lime, and to 7.0 and 7.8 t ha$^{-1}$ with it. With N1 without lime, grasses (mainly *Agrostis, Alopecurus, F. rubra* and *Holcus*) again formed about 88% of TDM, with few legumes (2%) and other dicots (10%). Liming decreased grasses to 60% (*Agrostis* and *Alopecurus* replaced by *Arrhenatherum*), increased legumes to 8% and other species to 33%. At N*2 without lime, grasses dominated (85%) mainly *Alopecurus* and *Arrhenatherum*, with few legumes (2%), or other species (10%). With lime the proportion of grasses decreased slightly (*Alopecurus* replaced by *Arrhenatherum*). Application of P and K with N2, gave TDM of 3.9 without and 6.2 t ha$^{-1}$ with lime; with N3 TDM was 6.1 and 7.8 t ha$^{-1}$ without and with lime respectively. Only grasses occurred on unlimed plots at N1 and N2, with *Anthoxanthum* most abundant, some *Agrostis* but little *Holcus*, whilst at N3 *Holcus* was almost completely dominant. Liming N2, decreased grasses to ca 75% of DM (*Arrhenatherum, Alopecurus*) with 15% legumes and 10% other species although at N3 grasses (*Arrhenatherum*, followed by *Anthoxanthum* or *Alopecurus*) were again dominant (95%): legumes were absent and other species only 5% of TDM.

Explanation of the effects of N and its interactions in such a complex system over a long period, with changing weather and soil conditions, can only be attempted tentatively (at best) here. Most species (ca 30-40) occur where no nutrients have been applied, with pH >6, and TDM small: production is determined by deposition of N from pollution, animal excreta etc and removal of nutrients from the soil matrix. Species which do not require much nutrient because they are biochemically and physiologically adapted (*Leontodon hispidus, Lotus corniculatus, Briza media, Agrostis, Festuca rubra*; type A) can establish and survive (but have low productivity) because of limited competition from more dominant forms (type B), especially grasses such as *Arrehnatherum, Dactylis* and *Holcus*. The canopy is short and open, with patches at the soil surface brightly lit, encouraging establishment, survival and production of seed of short species (rosette plants and some legumes). The decrease in species in the early years of the experiment may reflect cessation of grazing, which opened the sward, removed grasses, introduced seeds of uncommon or immobile species and supplied manure (particularly P, K and perhaps micronutrients), rather than a direct effect of N *per se*. The experiment shows clearly that increasing N stimulated TDM production by more vigorous, N-requiring grasses (*Arrehnatherum, Dactylis* and *Holcus*). Competition for light probably decreased growth of shorter, less vigorous species which were eventually eliminated and excluded but measurements have not been made. However, there is considerable interaction with other factors. At pH <5, increase in TDM with added N was small and from acid-tolerant grasses (*Agrostis*); many other species could not survive or compete probably because of their requirement for P and K, or sensitivity to increasing aluminium. Increasing pH decreased *Agrostis* and increased other less acid-tolerant grasses and dicots, possibly by increasing P availability. Adding P only, especially at low pH, increased TDM substantially, despite decreasing grasses (especially *Agrostis*, which may be inhibited by large P supply) by encouraging legumes and other species. Adding P and K further increased TDM production but altered grass species. However, adding K and P with lime substantially increased yields (ca 90%) but decreased grasses, with *Arrhenatherum* and *Alopecurus/Dactylis* displacing *Agrostis/Festuca rubra* of the unlimed plots, increasing the legumes substantially and changing other species. An interpretation is that production of control plots is limited by both N and P, and can respond to either P or N application because of the shift in species composition especially at low pH. Adding N alone stimulates acid tolerant, low P requiring grasses (*Agrostis, F. rubra*) which do not form dense canopies, so allowing other species to grow. Increasing the pH greatly improves P, and *Agrostis* decreases, *F. rubra* increases and *Helictotrichon* enters. However increasing P without altering the pH had a very large effect, decreasing *Agrostis* but stimulating *F. rubra* and *Holcus*, legumes and other species, whereas increasing pH did not improve production, but composition shifted to *Arrhenatherum* and *Anthoxanthum*, possibly caused by increased P and imbalance with K or other factors resulting from the changing pH. One of the most spectacular effects is the almost complete dominance of acid-tolerant *Holcus* with N2 and N3 plus P and K: obviously this species requires large amounts of N, P and K to grow well and thus dominate and produce large TDM. Park Grass shows that low TDM, *per se*, is not required for species-richness, nor does high productivity cause species-poverty. Much N stimulates vigorous grasses but how particular species respond depends on P, K and pH: some are very tolerant of low pH and can exploit N (*Agrostis*), others (particularly *Holcus*) require N, P and K but tolerate low pH. Dicots, including legumes, generally require more neutral pH and substantial P and K: those on Park Grass do not exploit N as effectively as grasses.

Species may occur in many treatments, although varying widely in contribution to dry matter production. Populations of *Anthoxanthum odoratum* (an out-breeding grass with intra-specific variation, reproducing mainly by seed) on different plots differ in nutrient requirements, size etc related to treatment and sward structure etc. Such genetic differences are adaptive, rapidly selected, and specific (Snaydon and Davies 1976; Silvertown et al. 2002). In contrast, *Taraxacum officinale* is apomictic, with many 'species'; of the twelve on Park Grass two or three occur in any one treatment, where their characteristics are also adaptive. It range, however, is limited compared to *Anthoxanthum*. Thus the genetic potential of a population is flexible and provides the basis for species adaptation to conditions.

## 2. Conclusions

Understanding effects of N-pollution on ecosystems requires understanding of species' genetically determined biochemical requirements for N, and the interaction with many other environmental factors. Increased N-input to ecosystems with slow-growing, small species adapted to very small N-content or N-turn-over, will stimulate more competitive, rapidly growing, larger species, i.e. those with the genetic potential to exploit N, depending on their other requirements. However, species-richness is determined by many factors, with N-pollution a major - but not the only - one.

Acknowledgement. Thanks to Paul Poulton for help with understanding the Park Grass Experiment.

## References

Aerts R, Bobbink R (1999) The impact of atmospheric nitrogen deposition on vegetation processes in terrestrial, non-forest ecosystems. In : Langan SJ (Ed) The impact of nitrogen deposition on natural and semi-natural ecosystems. Kluwer Academic Publishers, Dordrecht, pp 85-122

Lawlor DW, Lemaire G, Gastal F (2001) Nitrogen, plant growth and crop yield. In: Lea PJ, Morot-Gaudry J-F (Eds) Plant nitrogen. Springer Verlag, Berlin, pp 343-367

Linhart YB, Grant MC (1996) Evolutionary significance of local genetic differentiation in plants. Annu Rev Ecol Syst 27:237-277

NEGTAP 2001 (2001) Transboundary air pollution: acidification, eutrophication and ground-level ozone in the UK. UK Department for Environment, Food and Rural Affairs (DEFRA)

Pons TL, van der Werf A, Lambers H (1994) Photosynthetic nitrogen use efficiency of inherently slow- and fast-growing species: possible explanation for observed differences. In: Roy J, Garnier E (Eds) A whole plant perspective on carbon-nitrogen interactions. SPB Academic Publishing, The Hague, pp 61-77

Rothamsted Experimental Station (1991). Guide to the classical experiments. AFRC Institute of Arable Crops Research

Silvertown J, McConway KJ, Hughes Z, Bliss P, Macnair M, Lutman P (2002) Ecological and genetic correlates of long-term population trends in the Park Grass Experiment. The American Naturalist 160: 409-420

Snaydon RW, Davies MS (1976) Rapid population differentiation in a mosaic environment. IV. Populations of *Anthoxanthum odoratum* L. at sharp boundaries. Heredity 37:9-25

Thurston JM, Williams ED, Johnston AE (1976) Modern developments in an experiment on permanent grassland started in 1856: effects of fertilizers and lime on botanical composition and crop and soil analyses. Ann Agron 27:1043-1082

# II. Plant Responses to Climate Change

# Long-term effects of elevated $CO_2$ on sour orange trees

Bruce A. Kimball[1] and Sherwood B. Idso[2]

[1]U.S. Water Conservation Laboratory, USDA, Agricultural Research Service 4331 East Broadway Road, Phoenix, Arizona 85040, USA
[2]Center for the Study of Carbon Dioxide and Global Change, Tempe, Arizona 85285, USA

**Summary.** The long-term responses of trees to elevated $CO_2$ are especially crucial (1) to mitigating the rate of atmospheric $CO_2$ increase, (2) to determining the character of future forested natural ecosystems and their spread across the landscape, and (3) to determining the productivity of future agricultural tree crops. Therefore, we initiated a long-term $CO_2$-enrichment experiment on sour orange trees in 1987. Four sour orange trees (*Citrus aurantium* L.) have been grown from seedling stage at 300 μmol mol-1 $CO_2$ above ambient in open-top, clear-plastic-wall chambers at Phoenix, Arizona. Four control trees have been similarly grown at ambient $CO_2$. All trees have been given ample water and nutrients.

The ratios of wood plus fruit annual biomass increments of the elevated-$CO_2$ trees to those of the control trees reached a peak of about 3.0 two years into the experiment, declined from about year 2 to year 8, and has plateaued at about 1.75 for the past 9 years. The enhancement ratio for net photosynthesis was about 2.8 in year 2 but had declined to 1.3 by the 14th year, indicating some acclimation to the elevated $CO_2$. Initial reductions in leaf N concentrations disappeared by year 7, while carbohydrate concentrations remained higher. Intrinsic water use efficiency determined from carbon isotope ratios was increased under elevated-$CO_2$ by the same amount as biomass, which implies water use was unchanged. Storage proteins have been detected in the leaves of the enriched trees, which might be a mechanism that enables a six-fold increase in the rate of bud burst in the spring.

**Key words.** Global change, $CO_2$, Orange, Citrus, Tree, Growth, Yield, Productivity

## 1. Introduction

The $CO_2$ concentration of Earth's atmosphere continues to rise, and general circulation models predict a consequent global warming and changes in precipitation patterns (IPCC, 2001). Plants in general are responsive to changing $CO_2$ concentrations, which portends changes in agricultural productivity around the world. At the same time, the ability of plants to absorb $CO_2$ during photosynthesis and then store the carbon in their bodies and/or sequester it in the soil has potential for mitigating the rate of rise of the atmospheric $CO_2$ concentration. The long-term responses of trees to elevated $CO_2$ are especially crucial (1) to mitigating the rate of atmospheric $CO_2$ increase, (2) to determining the character of future forested natural ecosystems and their spread across the landscape, and (3) to determining the productivity of future agricultural tree crops. This important nexus between trees and climate and future natural ecosystems and tree crop productivity led us to initiate a long-term $CO_2$-enrichment experiment on sour orange trees in 1987 (e.g., Idso et al. 1991a).

## 2. Methodology

Briefly, eight sour orange trees (*Citrus aurantium* L.) have been grown from seedling stage in four identically-vented, open-top, clear-plastic-wall chambers at Phoenix, Arizona (Idso et al. 1991a). Sour orange is an ornamental tree often used for root stocks in commercial citrus orchards because of its disease and frost resistance. The trees were planted directly into the ground (Avondale loam; Kimball et al. 1992) in July, 1987. The four chambers were constructed around pairs of trees. Initially, the chambers were 5.3 m long x 2.6 m wide x 2.0 m high. As the plants grew, the chambers were periodically enlarged until now they are 6.3 m long x 5.1 m wide x 9.0 m high. The target $CO_2$ concentration of the enriched chambers has been 300 $\mu$mol mol$^{-1}$ above that of the ambient chambers. The automatic sampling/control system was described by Kimball et al. (1992). Except for short periods of chamber enlarging and very infrequent mechanical problems, enrichment has been continuous 24 hr per day every day since November, 1987. The trees have been fertilized and flood irrigated similar to practice in commercial orchards so as to maintain ample nutrients and soil moisture.

## 3. Results and discussion

### 3.1 Biomass and fruit yield

Since 1987, the growth and productivity of the sour orange trees have been highly responsive to elevated $CO_2$ (Fig. 1; Idso et al. 1991a, 2002; Idso and Kimball 1991b, 1992a, 1992b, 1997, 2001). After two years, the above-ground biomass of the treated trees was about triple that of the ambient trees due to the 300 $\mu$mol mol$^{-1}$ increase in $CO_2$ concentration (Fig. 1). Similarly, the biomass of below-ground fine roots was 2-3 times greater after 3 years (Idso and Kimball 1991b; 1992c). Fruit production started about a year sooner in the enriched trees. Following the biomass peaks, there was a steady decline until 8 or 9 years into the experiment, and for the last nine years the ratios have been relatively steady at about a 75% increase in wood plus fruit (Fig. 1).

The growth response of the sour orange trees to a 300 $\mu$mol mol$^{-1}$ increase in $CO_2$ (Fig. 1) has been larger than generally observed for most plants, including woody species (Kimball 1983; Poorter 1993; Idso and Idso 1994; Ceulemans and Mousseau 1994; Wullschleger et al. 1997; Curtis and Wang 1998; Norby et al. 1999; Janssens et al. 2000 Kimball et al. 2002). However, for trees planted in unrestricted soil, there are

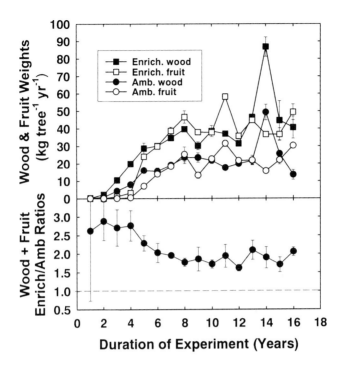

**Fig. 1.** Annual increments of sour orange wood and fruit biomass versus duration of exposure to $CO_2$ enrichment. Also the ratios of the $CO_2$-enriched wood plus fruit increments to those of the ambient trees. The wood data are based on monthly trunk circumference measurements converted to bio-volume using an allometric relationship established during years 2 and 3 of the experiment (Idso and Kimball 1992a). The conversion from volume to biomass was done using density measurements from pruned branches (unpublished data).

several instances of growth responses approaching that of our trees (e.g., Janssens et al. 2000). Focusing on citrus, Koch et al. (1986, 1987) obtained seedling growth increases of about 80% for a doubling of $CO_2$. Downton et al. (1987) observed about a 70% increase in productivity of 3-year old Valencia oranges enriched with $CO_2$ only during the 3$^{rd}$ year. Martin et al. (1995) observed a 87% increase in the growth of lemon at elevated $CO_2$ at supra-optimal temperatures, but the increase was only 21% at optimum temperatures.

### 3.2 Photosynthesis and respiration

Slightly less than two years into the experiment, leaf net photosynthetic rates of the enriched trees were about 11 μmol m$^{-2}$ s$^{-1}$ in the early morning and decreased to about 4 by mid-afternoon, whereas the leaves of the ambient trees were about 9 in the morning and decreased to 1 μmol m$^{-2}$ s$^{-1}$ by mid-afternoon (Idso et al. 1991b). Over years 2 and 3 of the study, the ratio of enriched to ambient was fairly steady at about 2.2, thus indicating no significant down regulation or acclimation (Idso and Kimball 1991a, 1992b). Another operative factor was that the enhancement at low light within the canopy more than compensated for self-shading produced by the $CO_2$-induced proliferation of leaf area (Idso et al. 1993c). Undoubtedly, another important aspect for the large growth response (Fig. 1) in our hot climate is that the elevated $CO_2$ raised the upper-limiting leaf temperature for positive net photosynthesis by approximately 7°C, which resulted in a 75% enhancement at a leaf temperature of 31°C, 100% enhancement at 35°C, and 200% at 42°C (Idso et al. 1995).

The degree of photosynthetic acclimation to elevated $CO_2$ is an important factor in determining the likely long-term productivity of tree crops. Gunderson and Wullschleger (1994) report that the average leaf photosynthesis rate of trees grown at elevated $CO_2$ and measured at ambient was only 0.79 of that of ambient-grown trees, although it was 44% greater when the rates were measured at the growth concentrations. Focusing on citrus trees, sweet orange has responded similarly to the overall average (Vu et al. 2002). Likewise, some acclimation has appeared in our sour orange trees, as indicated by a decline in the enhancement ratio of net photosynthesis from 2.8 in the second year (Idso et al. 1991b) to 1.28 in the 14$^{th}$ year (Adam et al. 2004). The decline was mostly due to a decrease in the maximum rate of $CO_2$ fixation (Vcmax), and to a lower activity and content of the primary $CO_2$-fixing enzyme, ribulose-1,5,-bisphosphate carboxylase/oxygenase (Rubisco).

From cuvette measurements taken in the second year of the experiment, it appeared that elevated $CO_2$ decreased dark leaf respiration by 20% (Idso and Kimball 1992b), but the cuvette technique is suspect (Amthor et al. 2001). Whether elevated $CO_2$ directly affects dark respiration remains controversial, yet other "dark" processes can be affected (e.g. Bunce 2002).

### 3.3 Stomatal conductance, water use, and intrinsic water use efficiency

Stomatal conductance of the orange trees was reduced about 10% by the elevated $CO_2$ (Idso et al. 1993a), a reduction that is smaller than most plant species but consistent with woody species (Curtis and Wang 1998; Kimball et al. 2002). No whole-tree measurements of water use have been done; however, we can speculate the water use response might be like that of cotton whose stomata are also relatively insensitive to $CO_2$ but have a large compensatory growth response resulting in little change in water use per unit of land area. In contrast, Wullschleger and Norby (2001) observed significant reductions in canopy transpiration of sweetgum exposed to free-air $CO_2$ enrichment. However, their treatments were imposed on 10-year-old trees. On the other hand, consistent with our cotton similarity hypothesis, the intrinsic water use efficiency (iWUE) of the sour orange trees determined from ratios of $^{13}C/^{12}C$ in leaf, wood core, and chamber air samples (Leavitt et al. 2003) increased by about exactly the same ratio as the biomass (Fig. 1) which implies no change in water use of the enriched trees compared to the ambient. Interestingly, the increase in iWUE of the sour orange trees per increase in $CO_2$ corresponded exactly to the increase in iWUE reported for 23 groups of naturally occurring trees scattered across the western U.S. that can be attributed to the increase in atmospheric $CO_2$ concentration that has occurred between 1800 and 1985 (Feng, 1999).

## 3.4 Elemental and chemical concentrations

There were no significant changes in N, P, K, Ca, Mg, Na, S, B, Cu, Fe, Mn, and Zn in the soil or roots of the well-fertilized sour orange trees three years into the experiment (Gries et al. 1993). However, concentrations of N, K, Ca, and Mn were slightly reduced in the leaves of enriched trees. Leaves from enriched trees sampled at bimonthly intervals from years 4-7 of the experiment had 4.8% less N (as well as chlorophyll a) than those from the ambient trees (Idso et al. 1996). Similarly working with bimonthly leaf samples, Peñuelas et al. (1997) reported there were clear seasonal trends in the concentration of most elements. There were initial decreases in the leaf concentration of N and the xylem-mobile and phloem-immobile Mn, Ca, and Mg, as well as a sustained increase in B. The initial reductions of N, Ca, Mn, and Mg gradually disappeared with time.

Not surprisingly, leaf carbohydrates have also been affected by elevated $CO_2$. During the $3^{rd}$ and $4^{th}$ years of the experiment, starch content per unit of leaf area was doubled while specific leaf mass increased 10-20% (Idso et al. 1993b). Interestingly, the area of each leaf also increased an average of about 10%. These changes were strongly positively correlated with mean air temperature. Soluble sugars in sun-acclimated leaves were doubled due to elevated $CO_2$ at 7.5 years into the experiment, whereas those in shade were unaffected (Schwanz et al. 1996).

Whether leaves were sun- or shade-acclimated made big differences in their ascorbate and glutathione antioxidant contents and activities 7.5 years into the experiment (Schwanz et al. 1996), but $CO_2$ treatment effects were not significantly different. The activities of superoxide dismutases were similar in the sun- and shade-acclimated leaves, but they decreased in response to elevated $CO_2$. In contrast, elevated $CO_2$ caused increases in ascorbate content of the sun-acclimated leaves. Similarly, the vitamin C content of the fruit was increased 7% based on samples taken from the $4^{th}$ through the $12^{th}$ years of the experiment (Idso et al. 2002).

One of the more interesting mechanisms that may help explain why the orange trees have a strong response to elevated $CO_2$ is that they have produced three putative storage proteins in their leaves with molecular masses of 33, 31, and 21 kDa (Nie and Long 1992; Idso et al. 2001). The evergreen sour orange trees generally have 2 years' worth of leaves at any given time. In the spring, there is bud burst that produces a new cohort of branches and leaves. The new branch growth following bud burst of the enriched trees was enormous compared to that of the ambient trees, reaching a peak 6 times greater (Idso et al. 2000). Amounts of the three proteins were generally lower in the $CO_2$-enriched leaves during the central part of the year, but they were higher in late fall, winter, and early spring (Idso et al. 2001). The decrease from their high wintertime levels in the $CO_2$-enriched trees possibly provided a source of nitrogen needed to sustain the rapid springtime branch growth. Leaves of an age greater than two years fall throughout the year, and during most of the year, the ratio of leaf fall from the enriched to ambient trees was steady at about 1.3. Surprisingly, around mid-October there was a sharp peak with the ratio reaching 2.7, indicating a significant qualitative difference in the behavior of the enriched and ambient trees. The enriched trees appeared to be reabsorbing N from second-year leaves during the process of accelerated senescence. This N was stored in the storage proteins of the first-year leaves, from which it was removed in the spring to sustain the enormous burst of new branch growth in the enriched trees.

# References

Adam NR, Wall GW, Kimball BA, Idso SB, Webber AN 2004 Acclimation of photosynthesis in leaves of sour orange trees grown at elevated $CO_2$ for 14 years. New Phytol 163:341-347

Amthor JS, Koch GW, Willms JR, Layzell DB (2001) Leaf $O_2$ uptake in the dark is independent of coincident $CO_2$ partial pressure. J Exp Bot 52:2235-2238

Bunce JA (2002) Carbon dioxide concentration at night affects translocation from soybean leaves. Ann Bot 90:399-403

Ceulemans R, Mousseau M (1994) Effects of elevated atmospheric $CO_2$ on woody plants. New Phytol 127:425-446

Curtis PS, Wang X (1998) A meta-analysis of elevated $CO_2$ effects on woody plant mass, form, and physiology. Oecologia 113:299-313

Downton WJS, Grant WJR, Loveys BR (1987) Carbon dioxide enrichment increases yield of Valencia orange. Aust J Plant Physiol 14:493-501

Feng X (1999) Trends in intrinsic water-use efficiency of natural trees for the past 100-200 years: a response to atmospheric $CO_2$ concentration. Geochimica et Cosmochimica Acta 63:1891-1903

Gries C, Idso SB, Kimball BA (1993) Nutrient uptake during the course of a year by sour orange trees growing in ambient and elevated atmospheric $CO_2$ concentrations. J Plant Nutr 16(1):129-147

Gunderson CA, Wullschleger SD (1994) Photosynthetic acclimation in trees to rising atmospheric $CO_2$: a broader perspective. Photosynth Res 39:369-388

Idso CD, Idso SB, Kimball BA, Park HS, Hoober JK, Balling Jr. RC (2000) Ultra-enhanced spring branch growth in $CO_2$-enriched trees: can it alter the phase of the atmosphere's seasonal $CO_2$ cycle? Environ Exp Bot 43:91-100

Idso KE, Idso SB (1994) Plant responses to atmospheric $CO_2$ enrichment in the face of environmental constraints: a review of the past 10 years' research. Agr Forest Meteorol 69:153-203

Idso KE, Hoober JK, Idso SB, Wall GW, Kimball BA (2001) Atmospheric $CO_2$ enrichment influences the synthesis and mobilization of putative vacuolar storage proteins in sour orange tree leaves. Environ Exp Bot 48:199-211

Idso SB, Kimball BA (1991a) Downward regulation of photosynthesis and growth at high $CO_2$ levels. Plant Physiol 96:990-992

Idso SB, Kimball BA (1991b) Effects of two and a half years of atmospheric $CO_2$ enrichment on the root density distribution of three-year-old sour orange trees. Agr Forest Meteorol 55:345-349

Idso SB, Kimball BA (1992a) Aboveground inventory of sour orange trees exposed to different atmospheric $CO_2$ concentrations for 3 full years. Agr Ecosyst Environ 60:145-151

Idso SB, Kimball BA (1992b) Effects of atmospheric $CO_2$ enrichment on photosynthesis, respiration and growth of sour orange trees. Plant Physiol 99:341-343

Idso SB, Kimball BA (1992c) Seasonal fine-root biomass development of sour orange trees grown in atmospheres of ambient and elevated $CO_2$ concentration. Plant Cell Environ 15:337-341

Idso SB, Kimball BA (1997) Effects of long-term atmospheric $CO_2$ enrichment on the growth and fruit production of sour orange trees. Glob Change Biol 3:89-96

Idso SB, Kimball BA (2001) $CO_2$ enrichment of sour orange trees: 13 years and counting. Environ Exp Bot 46(2):147-153

Idso SB, Kimball BA, Allen SG (1991a) $CO_2$ enrichment of sour orange trees: two-and-a-half years into a long-term experiment. Plant Cell Environ 14:351-353

Idso SB, Kimball BA, Allen SG (1991b) Net photosynthesis of sour orange trees maintained in atmospheres of ambient and elevated $CO_2$ concentration. Agr Forest Meteorol 54:95-101

Idso SB, Kimball BA, Akin DE, Krindler J (1993a) A general relationship between $CO_2$-induced reductions in stomatal conductance and concomitant increases in foliage temperatures. Environ Exp Bot 33(3):443-446

Idso SB, Kimball BA, Hendrix DL (1993b) Air temperature modifies the size-enhancing effects of atmospheric $CO_2$ enrichment on sour orange tree leaves. Environ Exp Bot 33:293-299

Idso SB, Wall GW, Kimball BA (1993c) Interactive effects of atmospheric $CO_2$ enrichment and light intensity reductions on net photosynthesis of sour orange tree leaves. Environ Exp Bot 33(3):367-375

Idso SB, Idso KE, Garcia RL, Kimball BA, Hoober JK (1995) Effects of atmospheric $CO_2$ enrichment and foliar methanol application on net photosynthesis of sour orange tree (*Citrus aurantium*; Rutaceae) leaves. Amer J Bot 82(1):26-30

Idso SB, Kimball BA, Hendrix DL (1996) Effects of atmospheric $CO_2$ enrichment on chlorophyll and nitrogen concentrations of sour orange tree leaves. Environ Exp Bot 36:323-331

Idso SB, Kimball BA, Shaw PE, Widmer W, Vanderslice JT, Higgs DJ, Montanari A, Clark WD (2002) The effect of elevated atmospheric $CO_2$ on the vitamin C concentration of (sour) orange juice. Agr Ecosyst Environ 90(1):1-7

IPCC (2001) Climate change 2001: the scientific basis: contribution of working group I to the third assessment report of the intergovernmental panel on climate change. Cambridge University Press, New York, USA

Janssens IA, Mousseau M, Ceulemans R (2000) Crop ecosystem responses to climatic change: tree crops. In: Reddy KR, Hodges HF (Eds) Climate change and global crop productivity. CABI Publishing, New York, pp 245-270

Kimball BA (1983) Carbon dioxide and agricultural yield: an assemblage and analysis of 430 prior observations. Agron J 75:779-788

Kimball BA, Mauney JR, LaMorte RL, Guinn G, Nakayama FS, Radin JW, Lakatos EA, Mitchell ST, Parker LL, Peresta GJ, Nixon II PE, Savoy B, Harris SM, MacDonald R, Pros H, Martinez J (1992) Carbon dioxide enrichment: data on the response of cotton to varying $CO_2$ irrigation and nitrogen. Report ORNL/CDIAC-44-NDP-037. Carbon Dioxide Information Analysis Center, Oak Ridge National Laboratory, US Dept of Energy, Oak Ridge, TN

Kimball BA, Kobayashi K, Bindi M (2002) Responses of agricultural crops to free-air $CO_2$ enrichment. Advances in Agronomy 77:293-368

Koch KE, Jones PH, Avigne WT, Allen LHJr (1986) Growth, dry matter partitioning, and diurnal activities of RuBP carboxylase in citrus seedlings maintained at two levels of $CO_2$. Physiol Plant 67:477-484

Koch KE, Allen LHJr, Jones P, Avigne WT (1987) Growth of citrus rootstock (Carrizo Citrange) seedlings during and after long-term $CO_2$ enrichment. J Amer Soc Hort Sci 112:77-82

Leavitt SW, Idso SB, Kimball BA, Burns JM, Sinha A, Stott L (2003) The effect of long-term atmospheric $CO_2$ enrichment on the intrinsic water-use efficiency of sour orange trees. Chemosphere 50:217-222

Martin CA, Stutz JC, Kimball BA, Idso SB, Akey DA (1995) Growth and topological changes of *Citrus limon* (L) Burm. F. 'Eureka' in response to high temperatures and elevated atmospheric carbon dioxide. J Amer Soc Hort Sci 120:1025-1031

Nie GY, Long SP (1992) The effect of prolonged growth in elevated $CO_2$ concentrations in the field on the amounts of different leaf proteins. In: Murata N (Ed) Research in photosynthesis, Vol. IV, Kluwer Academic Press, Dordrecht, The Netherlands, pp 855-858

Norby RJ, Wullschleger SD, Gunderson CA, Johnson DW, Ceulemans R (1999) Tree responses to rising $CO_2$ in field experiments: implications for the future forest. Plant Cell Environ 22:683-714

Peñuelas J, Idso SB, Ribas A, Kimball BA (1997) Effects of long-term atmospheric $CO_2$ enrichment on the mineral concentration of sour orange tree leaves. New Phytol 135:439-444

Poorter H (1993) Interspecific variation in the growth response of plants to an elevated ambient $CO_2$ concentration. In: Rozema J, Lambers H, Van de Geijn SC, Cambridge ML (Eds) $CO_2$ and biosphere. Kluwer Acacemic Publishers, Dordrecht, Netherlands, pp 77-97

Schwanz P, Kimball BA, Idso SB, Hendrix DL, Polle A (1996) Antioxidants in sun and shade leaves of sour orange trees (*Citrus aurantium*) after long-term acclimation to elevated $CO_2$. J Exp Bot 47(305):1941-1950

Vu JCV, Newman YC, Allen LHJr, Gallo-Meagher M, Zhang M-U (2002) Photosynthetic acclimation of young sweet orange trees to elevated $CO_2$ and temperature. J Plant Physiol 159:147-157

Wullschleger SD, Norby RJ (2001) Sap velocity and canopy transpiration in a sweetgum stand exposed to free-air $CO_2$ enrichment (FACE). New Phytol 150:489-498

Wullschleger SD, Norby RJ, Gunderson CA (1997) Forest trees and their response to atmospheric carbon dioxide enrichment: a compilation of results. In: Allen Jr. LH, Kirkham MB, Olszyk DM, Whitman CE (Eds) Advances in carbon dioxide research. American Society of Agronomy, Crop Science Society of America, and Soil Science Society of America, Madison, Wisconsin, USA, pp 79-100

# Plant responses to climate change: impacts and adaptation

David W Lawlor

Crop Performance and Improvement Division, Rothamsted Research, Harpenden, Herts., AL5 2JQ, UK

**Summary.** Pre-industrial global atmospheric carbon dioxide ($CO_2$) concentration (Ca), was ca. 280 $\mu$l l$^{-1}$, is now 376 $\mu$l l$^{-1}$, and may be 700 $\mu$l l$^{-1}$ by 2100. Temperature is now ca. 0.6°C greater than pre-industrially, and may be ca. 4°C greater by 2100. Warming is decreasing frost, snow and ice cover. Rain may increase in some areas, particularly high latitudes, but decrease in others. Also cloud, and therefore solar radiation reaching vegetation, will change. Agriculture must adapt to such changes, which affect crop growth, yield, and pests and diseases. Stimulation of yields of cereals, sugar beet, potatoes, pastures and forests by ca. 20-30% is expected with Ca of 700 $\mu$l l$^{-1}$. This results from increased photosynthesis. However, heat speeds respiration and crop development, shortening the growing season and thus limiting yield; a 4°C rise decreases dry matter by ca 20-30% in cereals and 20% in sugar beet. Quality of products (N-content, types of proteins in grain, and sugar beet storage tissue) are also affected. The biochemical and physiological mechanisms responsible for these effects are discussed, emphasising crop interaction with environment, including nutrition and water. Rates of environmental change are very rapid, however technology including plant breeding – but probably not current genetic engineering methods - may allow adjustment. Socio-economic pressures (consumer demand, world trade etc) may affect agriculture more than climate change *per se*. Advanced simulation modelling, linked closely to agronomic and plant research, is required to assess the consequences of putative changes on such complex systems.

**Key words.** Carbon dioxide, Temperature, Climate change, Plants, Crops

## 1. CO₂ and temperature: the primary environmental changes

Carbon dioxide ($CO_2$) in the atmosphere (Ca) has increased over 150 years from *ca* 270 to 380 $\mu$l l$^{-1}$ and will continue to increase to possibly *ca* 700 $\mu$l l$^{-1}$ by 2100. Temperatures globally (Tg) have also risen by *ca* 0.6°C over the period, with the 1900s and 1990s, respectively, the warmest century and decade in the last 1000 year, and they may increase further by *ca* 2-4°C by 2100 (IPCC 2001). These effects will intensify as global population and energy use increase. The consequences for natural and managed (crop) ecosystems is most important to the biosphere and human food supply. Ranges (extremes) and intensities of many other environmental factors will change; rainfall will be very important with amounts and patterns changing globally. Snowfall will decrease with warmer winters. Water supply to, and loss from ecosystems, will become more extreme, with

*Plant Responses to Air Pollution and Global Change*
Edited by K. Omasa, I. Nouchi, and L. J. De Kok ( Springer-Verlag Tokyo 2005 )

very wet and very dry periods. Such changes are already apparent. Also the solar energy incident on vegetation may change: increasing radiation will stimulate water loss but may not increase total photosynthesis (Ps) due to the non-linearity of Ps of leaves and canopies to radiation. Less radiation may improve water balance and thus increase Ps. Here the effects of some changes on plants and production are briefly analysed. Changes in ground-level and tropospheric ozone, and in UV radiation, are not considered, although important.

## 1.1 Experimental analysis, modelling and modifying plant responses to environmental change

Analysis of the effects of environment change (EC) on plant processes and production is a developing science, based in a long tradition of environmental physiology and agriculture, with mathematical simulation modelling combining the many, disparate aspects determining system responses. Understanding of the effects of increasing Ca on plants is based on many experiments in growth chambers, open-top chambers and with free-air $CO_2$ enrichment (see Kimball et al. 2002): a consensus about the main effects has developed but details of mechanisms are still unclear. Fewer studies of crop temperature ($T_c$) and the interaction with Ca have been done, and then mainly in chambers; many more are required as it is a very important, damaging consequence of EC. To integrate and apply experimental studies in the agricultural and environmental context, detailed mechanistic analysis and modelling are required to clarify plant genotype and environmental (so-called GxE) interactions. Ideally, a detailed understanding of the response of species to a very wide range of conditions is required – $CO_2$ and Tc change is only one facet of the general problem. Better understanding of the relative importance of EC factors and their influence on plant responses is needed, to allow appropriate action to be taken, e.g. changing agronomy and crop varieties. Also, plant mechanisms determining the responses must be better known. However analysis cannot supply the best adapted species. Selection breeding has been very successful in altering plants, predominantly crops for increased yield and quality, including those for food, fibre, construction and fuel required by man. It may be able to provide varieties better suited to future conditions. Molecular biology may provide methods for altering plants adapted to newer conditions but is untested in providing plants with specifically altered metabolism well adapted to complex environments. Currently, fewer resources are devoted to analysis and developing models of GxE, and to selection breeding, than are provided to molecular biology: also the disciplines are still rather distant: integration would be healthier.

## 1.2 Basic processes determining species' responses to environment

All species require energy and nutrients to grow and reproduce. Autotrophic plants obtain energy from solar radiation which is used in Ps to assimilate $CO_2$ and form carbohydrates, and to convert nitrate, ammonium and sulphate ions to amino acids. Growth and development then consume these basic products. However, particular temperatures are required for the processes to operate efficiently. Also, water and the radiation (photosynthetically active, as well as UV and IR) environment must be suitable, i.e. within the range required by a particular species. This is determined by the genotype of the plant, which is the product of evolutionary selection. If resources and conditions are optimal with respect to this genetic potential for growth, reproduction etc then species survival is more likely. Species generally have a wider physiological tolerance to conditions without than with competition, which limits the range of less efficient species. Thus, EC may alter conditions in particular geographical locations to be less favourable for some species and more so for others. As Ca is rather uniform (although within vegetation, and near natural $CO_2$ springs, urban areas etc, it may differ significantly from open areas), its effects globally are relatively predictable (because the mechanisms are now established), depending on interactions with other factors. Temperature depends on location, in many places varying over short distances and changing and fluctuating greatly diurnally, seasonally and annually. Species distribution with temperature is generally understood, but it has not been analysed or modelled in detail because the plant mechanisms are not well established. Similar, the role of nutrients and water in production are understood, but poorly quantified, especially differences between species. General requirements are known, but detailed understanding of mechanisms is lacking (even in important crops), limiting the ability to model, and to modify species responses by selection breeding or genetic engineering. Therefore, much is uncertain, and empirical data is inexact, so clouding interpretation.

### 1.2.1 Species response to $CO_2$

Experimental studies of the effects of $CO_2$ are numerous and frequently analysed and the references should be consulted for detailed analysis (see Lawlor and Mitchell 1991; Morison and Lawlor 1999; Kimball et al. 2002).

**Photosynthesis and respiration.** For C3 species, about $600 \mu l\ l^{-1}$ Ca saturates Ps, with other conditions optimal, so current Ca is insufficient for maximum Ps or dry matter production (DMP) and growth of species with the developmental and structural features allowing response to increased assimilate supply. The reason is known: the enzyme ribulose bisphosphate carboxylase-oxygenase (Rubisco), responsible for combining $CO_2$ with ribulose bisphosphate (RuBP) generated in the Calvin cycle of Ps, requires larger $CO_2$ concentration than current Ca. Rubisco also catalyses the reaction of oxygen with RuBP, leading to photorespiration and loss of $CO_2$. This inherent inefficiency is over-come by larger Ca. Although C4 plants were once considered insensitive to elevated Ca, because their $CO_2$ concentrating mechanism maintains a large $CO_2$ concentration in the chloroplasts and thus by-passes the inefficiency of Rubisco, they do respond similarly to C3's but less so (see Kimball et al. 2002). Decreased stomatal conductance (gs) is a very characteristic response of species to increased Ca, however, it decreases less than the rise in Ca, so the $CO_2$ concentration within the leaf and chloroplast increase thus stimulating Ps rate (Morison and Lawlor 1999).

Respiration converts assimilates into the required forms of energy, and provides the materials for growth and development of the plant and its maintenance. Decreased respiration rate with elevated $CO_2$ has been reported but the mechanism are unclear, but is now considered unlikely (Amthor 1997). This is important for assessing carbon balance of vegetation.

**Dry matter and growth.** When cultivated species such wheat, rice, sugar beet, and woody perennials are grown with Ca of 700 cf 380 $\mu$l l$^{-1}$, DMP, shoot, leaf, root and reproductive organ growth are stimulated by ca 20-30%, an average of numerous experiments with considerable variation which probably depends on conditions. C4 species are generally positively stimulated but less so. Increased production results from faster Ps (with decreased photorespiration and possibly dark respiration), producing more carbohydrates which stimulate formation, growth and survival of meristems, so more organs - branches, tillers, roots - result and these are generally larger. This, in turn, increases resource-capture, both light interception and nutrient acquisition, so increasing production further. Also more reproductive organs are formed and survive to maturity, thus increasing crop yields and reproductive potential and likely reproductive success. This is established for rice (Ziska et al. 1997), wheat (Lawlor and Mitchell 2000) and indeed is the main mechanism of increased yield in a range of species (Kimball and Kobayashi 2002). Species differ widely in capacity to form organs and the size to which they grow: those with large capacity may respond strongly to large Ca whereas those with small or no extra capacity may not. However, even so-called determinate species respond: sugar beet which cannot produce more shoots, forms more growth rings and thus increases DMP (Demmers-Derks et al. 1998). Increasing Ca will increase dry matter and yield of most C3 crops but this will only occur providing other resources, such as nutrients and water are available.

### *1.2.2 Interaction between $CO_2$ and nutrients*

Increased DMP requires more nutrients (N, P, K etc) which must come from the soil, air or, in agriculture, from fertilizers. Where N is available growth is stimulated, so N-fixing legumes benefit from greater Ca even when soil N is limiting. This is explained by the "Law of Diminishing Factors": if any nutrient is below the rate of supply required for growth at the maximum rate set by the genetic potential and temperature, then production will be smaller, and the stimulation by elevated $CO_2$ will be proportionally less. This is generally observed. Thus, ecosystems will generally respond with increased production but it is unlikely to be by 30%, especially over a long period as nutrition etc become limiting. Increased Ca may, and N-deficiency usually does, decreases Ps capacity, by mechanisms which are probably related. Abundant N increases Rubisco in leaves proportionally more than it increases ATP-synthase (a component of the energy transduction mechanism) suggesting a large capacity to synthesize Rubisco (which also has an N-storage function). Large Ca does decrease the N-content of leaves per unit dry matter, but Rubisco and ATP synthase change in proportion, suggesting that it is related to decreased N-availability (Theobald et al. 1998; Makino and Mae 1999). Experiments where $CO_2$ decreases N-content and Rubisco are generally N-limited (e.g. in small pots): under such conditions carbohydrates accumulate substantially and glucose particularly may inhibit gene expression but it is not a large effect where growth is active. However, the increased carbohydrate supply with elevated Ca does decrease the N-content of DM, including that of grain, with implications for human and animal nutrition and species (including herbivorous insects) survival.

## 1.2.3 Response of species to temperature

Differences in response to Tc are considerable, between species even within genera, suggesting that selection and optimization occur relatively quickly in evolution. The range of long-term average temperature and short-term fluctuations to which species are adapted may be narrow or broad. How species adapt, genetically and physiologically, is very poorly understood. Because the many enzymatic reactions in metabolism often respond somewhat differently to T, the over-all efficiency of the plant system must be optimized to a particular range by interaction of different pathways or by altering the total activity by changing the amount of particular components. If the genetic capacity to change the system (adaptation) is exceeded, then efficiency decreases, and growth, production, yield and competitiveness fall. Under extreme temperatures damage occurs, e.g. frost is a major determinant of species distribution, negatively by damage and positively by vernalisation, so warming climate will be detrimental to those requiring cold and allow others to migrate to higher latitudes and elevations. Temperature of organs greater than the optimum may decrease Ps but stimulate respiration and rate of development and growth, but shorten the duration of development. Experimental studies of the effects of warming expected from EC, compared with current normal conditions, show that these processes are crucial for production and yield of annual and perennial species. Less time is available for acquisition of nutrients and interception of solar energy, and for metabolic processes, so smaller, less productive plants results. Cooler temperatures may slow processes and extend their duration, thus enhancing DMP and yield, but at risk of encountering unfavourable conditions, e.g. low temperatures with onset of winter. For both wheat and sugar beet, a 4°C increase above current ambient temperatures decreased production by ca 20-30% (Lawlor and Mitchell, 2000; Demmers-Derks, 1998). Rice responds similarly with a 4°C warming eliminating the increase in grain yield which occurred with increased Ca (Ziska et al 1997). Grain set is greatly diminished above an upper threshold (the absolute value of which depends on the species) in part because of impaired pollen production and growth, so that higher Tc may increase the probability of poor seed set and thus decrease yields. Also the quality of plant components is sensitive to T, with cereal grains having a smaller protein content, and altered types and proportions of proteins and lipids.

## 1.2.4 Water, temperature and production

Decreased stomatal conductance (gs) slows the rate of evaporation from leaves (transpiration, Et): this raise their temperature (Tl) and Tc and, importantly, results in faster development of organs, more rapid maturation and a shorter growing period. This is additional to the generally warmer environment with EC. Often this Tc-effect is erroneously ascribed to $CO_2$ per se. Also decreased Et, from small gs, may be offset by the larger Tl so that total water loss of vegetation may not be greatly diminished. As DMP is increased by larger Ca, but Et either unaffected or somewhat decreased, the water use efficiency (WUE = DMP/ Et)) increases - often substantially. However, this should not be interpreted as a saving of the absolute amount of water used, which depends entirely on Et. Increased DMP will be more important: WUE per se is not significant and statements that increasing it (e.g. by genetic modification) will reduce water requirement of crops are wrong. Where water is limiting, even marginally, growth slows and fewer, smaller leaves, reproductive organs etc are made; this decreases production. Also gs decreases which slows Ps rate, and Ps metabolism is progressively inhibited. Consequently, production decreases, often substantially, with small decrease in supply. Increased Ca tends to slow Et and the onset of water stress, and increase Ps, so in future there may be marginal increases in production in drier areas. Where water supply is very marginal, decreased Et may substantially affect ecosystems, as more water is available, including the composition of vegetation, such as the balance of C3 and C4 grasses (Lecain et al. 2003). However, warmer conditions may offset any reduction in Et resulting from smaller gs, and resulting stress will decrease DMP. The outcome will depend on the balance of these factors. The converse – over supply of water- if drainage is inadequate to maintain optimal soil water for particular crops, will also decrease production via different mechanisms. If climate alters rain fall patterns in time and space, and changes amounts, agronomy and the projected global human population of 10 billion dependent on it, will have to adapt to altered yields within the context of a global economy, where limited areas are responsible for most production of staple foods. It is likely that altered water supply will be the most immediate consequence of EC and have the biggest direct impact. Effects on 'natural' ecosystems will be complex, with some areas becoming drier, others wetter. Consequently, changes in species composition of ecosystems is inevitable if T and Ca increase. Migration of 'weedy' species will occur. This will affect all other biota, possibly rather rapidly, with loss of species highly adapted to particular conditions. Such short- and long-term effects are very difficult to forecast, as are consequences for human society.

## 2. Relative importance of environmental change and technology on crops

It is important to put the magnitudes and rates of change into perspective: Amthor (1998) estimated that the greatly increased yields of crops over the last century have been caused by technological advances (e.g. fertilizers, pest and disease control, new varieties), with relatively little effect of increased Ca which is difficult to assess. The effect might have been larger in the early period of increasing Ca. The relative insensitivity of crops to increasing Ca is expected: assuming the change in Ca since 1950 has been 75 µl l$^{-1}$ and that yield increases by 30% with an 350 µl l$^{-1}$ increase in Ca compared to current ambient, then with the simplest linear analysis suggests a 6.4% increase in yield from additional $CO_2$ over 50 years. Wheat yields in much of western Europe have increased from an average of ca 2 to 8 t ha$^{-1}$ (400%) over 50 years (see Amthor 1998). A temperature increase of 4°C in atmosphere and soil decreases wheat yields by 30%, again assuming an 0.5°C increase in temperature since 1950 yields could have decreased by ca 3.8%, thus almost negating any positive effect of elevated Ca. Clearly, as Amthor's analysis (1998) shows, grain yields have increased with improved agronomy- nutrition, pest and disease control - and varieties much better able to respond to them. Future crop production will depend on the relative increase in Ca and $T_c$, on the nutritional and water status of the crops, and on secondary but very important factors such as disease and pests. Largely, $CO_2$ will have a positive effect on crops, but where other required resources are limited there will be little benefit. Warmer temperatures will be detrimental to production, except to species which are being grown outside their optimal temperature range. This will be irrespective of other resources. However, the EC changes will probably broadly cancel out. Where water is limiting, elevated $CO_2$ may have a marginal benefit by slightly decreasing Et and slowing use of soil water, and this will be important in grass lands in arid regions. However, small changes in rainfall could have a much more rapid and lasting effect. As with much of this analysis it is the detail of what EC factors are altered, by how much and when, which will determine the outcome.

Suggesting ways of adjusting crops and agronomy to optimize production in the face of EC is best based on quantitative analysis and modelling rather than informed guesses. Never-the-less, improved agronomy (including nutrition, water, pest control etc) for crops across the globe would be the single most effective way to utilize the supply of $CO_2$, and to decrease the impact of rising temperature. Modifying plants by selection breeding to increase their ability to capture resources, and to use them more efficiently, is required: the advantage is that all aspects of EC and of GxE interactions are included, even those currently poorly understood or not apparent. Also, plant breeding has an impressive record in developing crops with substantially better growth and yield characteristics. Agronomy has, over a very long period of practical experiment and application based on scientific analysis, also adopted effective methods. Continuous improvement is required for developed - and particularly for poorly developed agriculture. Input of research at all stages and to all parts of the crop production system will increase the efficiency with which water and nutrients are applied and used by crops. It is the combination of these different elements which has resulted in such large yields in `industrial' agriculture. However, breeding has not greatly increased the efficiency of basic metabolic processes, e.g. potential Ps rate per unit leaf area. Rather capacity has been altered, e.g. the potential number and size of leaves and the structure of the crop canopy to maximize light interception and Ps. Thus, to benefit from elevated $CO_2$ and to minimize the effects of increased temperature, changes to various points in metabolism would be required, e.g. improved efficiency of Rubisco, slowed development and maturation, formation of more assimilate storage organs. These may be possible with genetic engineering methods, however, the potential of these has yet to be realized. Many apparently logical and easy changes to plant metabolism do not result in desired effects, and progress may be slower than advocates acknowledge. These difficulties may in part be due to the complexity of plant metabolism. To deal with current rapid growth in human population and the demand for food, fuel, fibre etc irrespective of EC, will require application of the most advanced scientific thought to tried and tested approaches. The promise of novel genetic engineering technology has altered the balance of activities towards it and away from more established approaches. This is unsatisfactory, as the benefits to be gained from proper application of established technology globally, are considerable and likely to be much faster than the rate at which plants can be genetically modified for particularly conditions, including those associated with environmental change.

## References

Amthor JS (1997) Plant respiratory responses to elevated carbon dioxide partial pressure. In: Allen Jr LH, Kirkham MB, Olszyk DM, Whitman CE (Eds) Advances in carbon dioxide effects research. American Society of Agronomy, Madison, WI, pp 35-77

Amthor JS (1998) Perspective on the relative insignificance of increasing atmospheric $CO_2$ concentration to crop yield. Field Crop Res 58:109-127

Demmers-Derks H, Mitchell RAC, Mitchell VJ, Lawlor DW (1998) Response of sugar beet (Beta vulgaris L.) yield and biochemical composition to elevated $CO_2$ and temperature at two nitrogen applications. Plant Cell Environ 21:829-836

IPCC. Climate change (2001): Working group 1. The scientific basis: summary for policy makers. http://www.ipcc.ch/pub/spm22-01.pdf

Kimball BA, Kobayashi K, Bindi M (2002) Responses of agricultural crops to free-air $CO_2$ enrichment. Advan Agron 77:293-370

Lawlor DW, Mitchell RAC (2000) Wheat. In: Reddy KR, Hodges HF (Eds) Climate change and global crop productivity. CAB International, Wallingford, pp 57-80

Lecain DR, Morgan JA, Mosier AR, Nelson JA (2003) Soil and plant water relations determine photosynthetic responses of C3 and C4 grasses in a semi-arid ecosystem under elevated $CO_2$. Ann Bot 92:41-52

Makino A, Mae T (1999) Photosynthesis and plant growth at elevated levels of $CO_2$. Plant Cell Physiol 40:999-1006

Morison JIL, Lawlor DW (1999) Interactions between increased $CO_2$ and temperature on plant growth. Plant Cell Environ 22:659-682

Theobald JC, Mitchell RAC, Parry MAJ. and Lawlor DW (1998) Estimating the excess investment in ribulose-1,5-bisphosphate carboxylase/oxygenase in leaves of spring wheat grown under elevated $CO_2$. Plant Physiol 118:945-955

Wand SJE, Midgley GF, Jones MH, Curtis PS (1999) Responses of wild C4 and C3 grass (Poaceae) species to elevated atmospheric $CO_2$ concentration: a meta-analytic test of current theories and perceptions. Glob Change Biol 5:723-741

Ziska LH, Namuco O, Moya T, Quilang J (1997) Growth and yield response of field-grown tropical rice to increasing carbon dioxide and air temperature. Agron J 89: 45-53

# Effects of elevated carbon dioxide concentration on wood structure and formation in trees

Ken'ichi Yazaki[1], Yutaka Maruyama [1], Shigeta Mori[2], Takayoshi Koike[3], and Ryo Funada[4]

[1]Forestry and Forest Products Research Institute, Matsunosato 1, Tsukuba, Ibaraki 305-8687, Japan
[2]Tohoku Research Center, Forestry and Forest Products Research Institute, Nabeyashiki 92-25, Shimo-Kuriyagawa, Morioka, Iwate 020-0123, Japan
[3]Field Science Center for Northern Biosphere, Hokkaido University, Kita-9, Nishi-9, Kita-ku, Sapporo, Hokkaido 060-0809, Japan
[4]Faculty of Agriculture, Tokyo University of Agriculture and Technology, Saiwai-Cho 3-5-8, Fuchu, Tokyo 183-8509, Japan

**Summary**. The effects of elevated carbon dioxide concentration ($[CO_2]$) on the structure of xylem cells in trees have not yet been clarified, in spite of the importance of woody plants as large, long-term carbon sinks. We review recent studies that have investigated how elevated $[CO_2]$ affects growth ring features. In general, elevated $[CO_2]$ enhances radial growth, especially when sufficient nutrients are supplied. The mean density of growth rings also increased but sometimes it decreased or remained unchanged, depending on the extent of cell division, cell expansion and cell wall thickening under elevated $[CO_2]$.

**Key words**: Elevated $CO_2$, Structure of growth ring, Wood formation, Woody plants

## 1. Introduction

Traits of a tree stem that are relevant to its role as a carbon sink are stability and longevity as compared with herbaceous species. Stems can hold carbon for a long time: not only when they are part of a living tree, but also after harvesting, as the wood components of buildings or as papers. Thus, carbon fixation by forest trees can be effective for centuries by virtue of the photosynthates accumulated in the xylem of the stem in trees.

Changes in the morphology of xylem cells under elevated $[CO_2]$ influence not only the physical properties of xylem cells, but also the extent to which the tree is a carbon sink. However, we have little knowledge of how rising atmospheric $[CO_2]$ affects cell division, cell expansion and the thickening of xylem cell walls in woody plants, in spite of the importance of woody plants. Here, we review previous researches and our studies on the structure and formation of wood in trees under elevated $[CO_2]$.

---

*Plant Responses to Air Pollution and Global Change*
Edited by K. Omasa, I. Nouchi, and L. J. De Kok ( Springer-Verlag Tokyo 2005 )

## 2. Changes in the growth ring features under elevated [$CO_2$]

### 2.1 Growth ring width

In experiments involving artificial exposure to $CO_2$, Curtis and Wang (1998) analyzed the effects of elevated [$CO_2$] on the growth and photosynthesis of woody species in more than 500 cases with a meta-analysis method, and they concluded that elevated [$CO_2$] stimulates tree growth in almost all cases, but that the response depends on the clone, species, growth stage and duration of elevated $CO_2$ exposure. These results raise the question: how does elevated [$CO_2$] alter the structure of a tree stem when they stimulate the growth of a tree? There have been several studies on the effects of elevated [$CO_2$] on tree ring width, density and cell dimensions (Table 1). Many studies reported that elevated [$CO_2$] resulted in the formation of wider tree rings in conifers (Atwell et al. 2003; Ceulemans et al. 2002; Conroy et al. 1990; Oren et al. 2001; Telewski et al. 1999; Yazaki et al. 2001) and hardwoods (Hättenschwiler et al. 1997; Norby et al. 2001). In contrast, some studies found no difference in growth ring width (Hättenschwiler et al. 1996; Kilpeläinen et al. 2003; Tognetti et al. 2000; Yazaki et al. 2004). In particular, sufficient nutrients were often required for the [$CO_2$]-induced enhancement of wood formation (e.g. Yazaki et al. 2001), but other times the growth-inducing effect of elevated [$CO_2$] depended less on nutrient supply (Hättenschwiler et al. 1996). These variations might depend on the differences in the relationship between growth stimulation and nutrient requirements among species or developmental stages.

As described above, growth rings tend, more or less, to become wider at elevated [$CO_2$] than at ambient [$CO_2$] when there are no limitations on other environmental factors. However, long-term exposure of $CO_2$ (> 10 years) results in a decline in the enhancement of radial growth induced by elevated [$CO_2$] due to some age-related factors and/or extent of the tree's acclimation to high [$CO_2$] (Adam et al. 2004; Tognetti et al. 2000). Thus, the response of seedlings to elevated [$CO_2$] should be applied cautiously to mature trees in the higher [$CO_2$] environment expected in the future (Norby et al. 2001).

### 2.2 Density of growth rings

Increases in wood density have been induced by elevated [$CO_2$] in some conifers (Table 1). However, in similar species, other studies indicate that there is little evidence of an increase in wood density under these conditions. For hardwoods, almost no change in wood density under elevated [$CO_2$] has been observed (Table 1), although hardwoods have been studied less than conifers. Furthermore, wood density in *Pinus taeda* grown at elevated [$CO_2$] is less than controls grown in ambient conditions, in Free-air-$CO_2$-enrichment (FACE) system (Oren et al. 2001). High nutrient levels can diminish wood density while high [$CO_2$] can increase it (Hättenschwiler et al. 1996). There may be negative interactions between the effects of [$CO_2$] and nutrient availability on wood formation because growth rings in conifers often have less wood density with higher rates of radial growth.

**Table 1.** Summary of changes in xylem anatomy in woody species grown at elevated [$CO_2$]

| Species | RW | WD | CD | LD | CWT | FL | Reference |
|---|---|---|---|---|---|---|---|
| **Conifers** | | | | | | | |
| Larix kaempferi | 0 | 0 | + | 0 | 0 | n.a. | Yazaki, K. et al. (2004) |
| Larix sibirica | +* | 0 | 0 | 0 | -* | n.a. | Yazaki, K. et al. (2001) |
| Picea abies | 0 (total) +(LW) | + | n.a. | n.a. | n.a. | n.a. | Hättenschwiler et al. (1996) |
|  | n.a. | 0 | n.a. | n.a. | n.a. | n.a. | Beismann et al. (2002) |
| Pinus ponderosa | n.a. | 0 | 0 | n.a. | n.a. | n.a. | Maherali and DeLucia (2000) |
| Pinus radiata | n.a. | n.a. | n.a. | 0 | 0 | 0 | Donaldson et al. (1987) |
|  | +* | + | n.a. | 0 | +* | 0 | Conroy et al. (1990) |
|  | + | + | n.a. | 0 (area) | n.a. | n.a. | Atwell et al. (2003) |
| Pinus sylvestris | + (EW) | 0 | + | n.a. | n.a. | n.a. | Ceulemans et al. (2002) |
|  | 0 | + (LW) | n.a. | n.a. | n.a. | 0 | Kilpeläinen et al. (2003) |
| Pinus taeda | + (EW) | 0 (soft x-ray) + | n.a. | n.a. | n.a. | n.a. | Telewski et al. (1999) |
|  | +* | - | n.a. | n.a. | n.a. | n.a. | Oren et al. (2001) |
| **Hardwoods** | | | | | | | |
| Arbutus unedo<br>Fraxinus ornus<br>Quercus cerris<br>Quercus ilex<br>Quercus pubescens | 0 | n.a. | n.a. | n.a. | n.a. | n.a. | Tognetti et al. (2000) |
| Fagus sylvatica | n.a. | 0 | n.a. | n.a. | n.a. | n.a. | Beismann et al. (2002) |
| Liquidambar styraciflua | + (1st year)<br>0 (2nd year) | 0 | n.a. | n.a. | n.a. | n.a. | Norby et al. (2001) |
| Populus alba<br>Populus nigra<br>Populus deltoides x P. nigra | n.a. | 0 | n.a. | n.a. | n.a. | n.a. | Calfapietra et al. (2003) |
| Populus tremuloides | + | 0 | 0 (vessel)<br>0 (fiber) | 0 (vessel)<br>0 (fiber) | 0 (vessel)<br>0 (fiber) | n.a. | Kaakinen et al. (2004) |
| Prunus avium x P. pseudocerasus | n.a. | n.a. | n.a. | 0 | n.a. | n.a. | Atkinson and Taylor (1996) |
| Quercus ilex | + | n.a. | n.a. | n.a. | n.a. | n.a. | Hättenschwiler et al. (1997) |
|  | n.a. | 0 | n.a. | + (area) | n.a. | n.a. | Gartner et al. (2003) |
| Quercus robur | n.a. | n.a. | n.a. | + | n.a. | n.a. | Atkinson and Taylor (1996) |

0, not significant difference; +, increase; -, decrease; n.a., not available.
RW, Radial growth or growth ring width; WD, Wood density or relative area of cell wall; CD, Cell diameter of tracheid (conifer) or vessel (hardwood); LD, Lumen diameter of tracheid (conifer) or vessel (hardwood); CWT, Cell wall thickness; FL, Fiber length.
*, effective only in combination with high levels of nutritents.
EW, effective mainly in earlywood; LW, effective mainly in latewood.

The effect of [$CO_2$] on wood density is more complicated than on growth ring width. For further clarification of this inconsistency, we need to turn to microscopic analysis because wood density is determined by cell dimensions (Yasue et al. 2000) (see below).

## 3. Variation in intra-ring cell dimensions under elevated [$CO_2$]

Growth ring structure depends on cambial activity and the development of xylem cells. Many environmental factors (temperature, water availability, nutrient supply, etc.) affect the development of xylem cells (e.g. Denne and Dodd 1981). It is possible that elevated [$CO_2$] also affects wood formation directly and/or indirectly (Pritchard et al. 1999).

Wood density is determined by growth ring structure, which is, itself, a consequence of changes in the dimensions of the cells in the different parts of the wood (Fig. 1). For example, the ratio of earlywood width to latewood width and/or the ratio of the amount of cell wall to inter- and intracellular spaces affect growth ring structure and wood density. Therefore, we need to divide the wood-forming effects of [$CO_2$] into those relevant to "cell division", to "cell expansion" and to the "deposition of cell wall" (Fig. 1).

### 3.1 Cell division

For conifers, the increase in the width of growth rings under elevated [$CO_2$] (compared with ambient [$CO_2$]) mainly results from a higher number of tracheids along the radius (Yazaki et al. 2001). This increase in the number of cells is brought about by two mechanisms: (1) an increase in the rate of cell division and (2) a prolongation of the cell division period. Shoot development affects the duration of cambial activity via changes in levels of endogenous plant growth regulators in cambial regions (Funada et al. 2001;

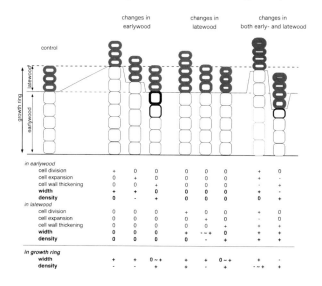

**Fig. 1.** A schematic representation of the changes in cell development and wood properties compared to control wood. 0, no change; +, increase; -, decrease.

Pritchard et al. 1999). The period of shoot elongation or radial growth could be prolonged for some species by elevated [$CO_2$] (Koike et al. 1996; Peltola et al. 2002). However, shorter periods of shoot elongation in response to elevated [$CO_2$] was also reported, in connection with accelerated leaf senescence and bud formation (Centritto et al. 1999; Sigurdsson 2001), and some studies found no differences in the timing of shoot growth cessation among various [$CO_2$] treatments (Calfapietra et al. 2003; Koike et al. 1995; Roberntz 1999). Thus, the increase in the number of cells at elevated [$CO_2$] is not always associated with specific changes in the duration of shoot elongation.

Elevated [$CO_2$] can shorten the time required for each cell division in the apical meristem of a shoot of *Dactylis glomerata*, a herbaceous species (Kinsman et al. 1996). For woody species, a $CO_2$-induced increase in relative growth rate was observed especially early in the growing season (Jach and Ceulemants 1999, Yazaki et al. 2001, 2004), implying an increase in the rate of cell division under elevated [$CO_2$]. Therefore, it is possible that elevated [$CO_2$] enhances the rate of cambial activity, although there have been very few estimates of both the response of vascular cambium to elevated [$CO_2$] and the influence of rising [$CO_2$] on growth regulators in woody plants.

## 3.2 Cell development - cell expansion and cell wall thickening

In conifers, wood properties are determined mainly by the dimensions and arrangement of tracheids in growth rings (Fig. 1). With sufficient nutrients, tracheids of *Pinus radiata* seedlings under elevated [$CO_2$] had cell walls that were 43 % thicker compared to those of control specimens, while tracheid lumen diameter was unaffected (Conroy et al. 1990). On the other hand, tracheid diameter (lumen diameter + two times the cell wall thickness) was 16 - 24 % higher at elevated [$CO_2$] than at ambient [$CO_2$] in *Pinus sylvestris* (Ceulemans et al. 2002). In contrast, no obvious change was found in the cell dimensions in *Pinus radiata* (Atwell et al. 2003; Donaldson et al. 1987) and *Larix kaempferi* (Yazaki et al. 2004) treated with elevated [$CO_2$]. Although, in *Larix sibirica*, tracheids tended to have a 10 % larger lumen diameter and a 16 % smaller cell wall thickness compared with specimens exposed to ambient [$CO_2$] where all trees received sufficient nutrients (Yazaki et al. 2001).

In some cases, the increase in growth ring width at elevated [$CO_2$] was due to a significant increase in earlywood (Ceulemans et al. 2002; Telewski et al. 1999; Table 1), implying an increase in cell expansion rate and/or a delay of thick cell wall deposition under elevated [$CO_2$]. In *Larix sibirica*, Yazaki et al. (2001) suggested that elevated [$CO_2$] had major effects on cell division and cell expansion rather than cell wall thickening, especially early in the growing season. According to this, wood density was lower in elevated [$CO_2$] than in ambient [$CO_2$] conditions.

In contrast, an increase in the growth ring width or density of latewood was also observed at elevated [$CO_2$], without significant changes in the structure of earlywood (Hättenschwiler et al. 1996; Kilpeläinen et al. 2003). Changes in latewood structure depend on the ratio of cell wall thickening to cell expansion, the first component of which can be altered by (1) prolongation of cambial activity and (2) increase in the duration or amount of secondary cell wall deposition (Denne and Dodd 1981; Funada et al. 1990; Fig. 1). However, as described previously, duration of activity of the cambium under elevated [$CO_2$] is variable and inconsistent among studies. On the other hand, the increase in wood density might be a consequence of the larger accumulation of total non-structural carbohydrates (TNC), produced in current or previous growing seasons (Hättenschwiler et al. 1996; Telewski et al. 1999). When we are able to estimate the relationship between the quantity of photosynthesis products and TNC and the process of cell wall deposition, the effects of elevated [$CO_2$] on latewood formation will become clear.

## 4. The relationship between anatomical features and physiological responses under elevated [$CO_2$]

### 4.1 Photosynthetic down-regulation and xylem structure under elevated [$CO_2$]

[$CO_2$]-induced increased rates of photosynthesis often decrease within several months, even if trees grow without any limitation of root growth or nutrient supply (e.g. Adam et al. 2004, Eguchi et al. 2004). With this reduction of photosynthetic capacity, the growth rate of the stem, which have been stimulated by elevated [$CO_2$] early in the growing season, decreases less than that of trees exposed to ambient [$CO_2$], and little difference is seen in growth ring width between [$CO_2$] treatments (Yazaki et al. 2004). Although a direct relationship between photosynthetic properties and the development of xylem has not been established, structure of growth rings might change when a $CO_2$-induced increase in photosynthesis rate is depressed by a plant's acclimation to higher ambient [$CO_2$] over time (e.g., over 3 weeks, over 1 year, etc.).

### 4.2 Water balance and vessel dimensions

The reduction of stomatal conductance and the increased water-use efficiency that are induced by elevated [$CO_2$] are well known phenomena (e.g. Wullschleger et al. 2002). Leaf water potential is also maintained at high levels by elevated [$CO_2$] owing to stomatal closure, increases in whole-plant hydraulic conductance and perhaps osmotic adjustment (Wullschleger et al. 2002).

In general, diameter of vessel elements tends to increase with xylem water potential if there are no limiting environmental factors (e.g. Doley and Leyton 1968). Thus, it is possible that the structure and dimensions of conductive cells may alter if $CO_2$-induced changes in the efficiency of water use by trees. It has been reported that mean vessel area increases by about 1.6-fold under conditions of elevated [$CO_2$] in seedlings of *Quercus robur* (ring-porous hardwood) but dose not change in seedlings of *Pruns avium* x *pseudocerasus* (diffuse-porous hardwood) (Atkinson and Taylor 1996). In contrast, our observations show no obvious difference in the mean vessel area of earlywood in seedlings of *Fraxinus mandshurica* var. *japonica* (ring-porous hardwood) and *Betula platyphylla* (diffuse-porous hardwood; Fig. 2) under various [$CO_2$] regimes. At high [$CO_2$], larger sized vessels is found in *Betula platyphylla* planted only in large pots. A close relationship between the efficiency of water use and the dimensions of the conducting cells might be expected when elevated [$CO_2$] affects the water balance of hardwoods, owing to resulting enhancement of leaf and root production.

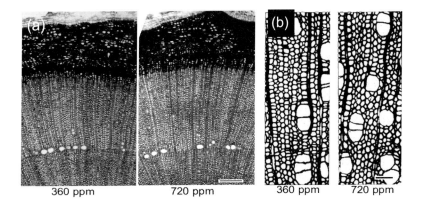

**Fig. 2.** Transverse sections of (a) *Fraxinus mandshurica* var. *japonica* and (b) *Betula platyphylla* at an ambient (360 ppm) or elevated (720 ppm) $CO_2$. Bars indicate 250 μm (a) and (b) 50 μm, respectively (Yazaki and Ishida unpublished data).

## 5. Conclusion

To sum up, according to recent studies, rising $[CO_2]$ has a less drastic effect on wood anatomical features than physiological properties of trees. In addition, the effect that elevated $[CO_2]$ does have on growth ring structure depends on species, clone, tree age and environmental factors. For the assessment of wood properties in the future, we need to investigate the response of mature trees to rising $[CO_2]$, focusing on the phenological, physiological and chemical pathways by which $[CO_2]$ affects wood formation with respect to a whole tree (Pritchard et al. 1999). The structure of a growth ring reflects the growth pattern of a tree. To better understand wood formation at elevated $[CO_2]$, we need to investigate the detailed process of cell development in growth rings and the variations in cell dimensions in the intra-growth ring at elevated $[CO_2]$ in many species and developmental stages.

Acknowledgements. The authors thank Drs. J. Ohtani, S. Fujikawa, Y. Sano, S Ishida, T. Kawagishi, E. Fukatsu and N. Eguchi, Hokkaido University, and Drs. M. Kitao and H. Tobita, Forestry and Forest Products Research Institute, for their valuable comments. This work was supported in part by Grants-in-Aid for Scientific Research (Nos. 11460061, 11460076 and 11691162) and for Research Revolution 2002 (RR-2002) Project for Sustainable Coexistence of Human, Nature and the Earth Parameterization of Terrestrial Ecosystem for Integrated Global Modeling form the Ministry of Education, Science and Culture, Japan.

# References

Adam NR, Wall GW, Kimball BA, Idso BS, Webber AN (2004) Photosynthetic down-regulation over long-term $CO_2$ enrichment in leaves of sour orange (*Citrus aurantium*) trees. New Phytol 163:341-347

Atkinson CJ, Taylor JM (1996) Effects of elevated $CO_2$ on stem growth, vessel area and hydraulic conductivity of oak and cherry seedlings. New Phytol 133:617-626

Atwell BJ, Henery ML, Whitehead D (2003) Sapwood development in *Pinus radiata* trees grown for three years at ambient and elevated carbon dioxide partial pressures. Tree Physiol 23:13-21

Beismann H, Schweingruber F, Speck T, Körner C (2002) Mechanical properties of spruce and beech wood grown in elevated $CO_2$. Trees 16:511-518

Calfapietra C, Gielen B, Galema ANJ, Lukac M, De Angelis P, Moscatelli MC, Ceulemans R, Scarascia-Mugnozza G (2003) Free-air $CO_2$ enrichment (FACE) enhances biomass production in a shoot-rotation poplar plantation. Tree Physiol 23:805-814

Centritto M, Lee HSJ, Jarvis PG (1999) Long-term effects of elevated carbon dioxide concentration and provenance on four clones of Sitka spruce (*Picea sitchensis*). I. Plant growth, allocation and ontogeny. Tree Physiol 19:799-806

Ceulemans R, Jach ME, Van De Velde R, Lin JX, Stevens M (2002) Elevated atmospheric $CO_2$ alters wood production, wood quality and wood strength of Scots pine (*Pinus sylvestris* L) after three years of enrichment. Global Change Biol 8:153-162

Conroy JP, Milham PJ, Mazur M, Barlow EWR (1990) Growth, dry weight partitioning and wood properties of *Pinus radiata* D.Don after 2 years of $CO_2$ enrichment. Plant Cell Environ 13:329-337

Curtis PS, Wang X (1998) A meta-analysis of elevated $CO_2$ effects on woody plant mass, form, and physiology. Oecologia 113:299-313

Denne MP, Dodd RS (1981) The environmental control of xylem differentiation. In: Barnett JR (ed.) Xylem cell development, Castle House Publications, pp 236-255

Doley D, Leyton L (1968) Effects of growth regulating substances and water potential on the development of secondary xylem in *Fraxinus*. New Phytol 67:579-594

Donaldson LA, Hollinger D, Middleton TM, Souter ED (1987) Effect of $CO_2$ enrichment on wood structure in *Pinus radiata* D Don. IAWA Bulletin new series 8:285-289

Eguchi N, Fukatsu E, Funada R, Tobita H, Kitao M, Maruyama Y, Koike T (2004) Changes in morphology, anatomy, and photosynthetic capacity of needles of Japanese larch (*Larix kaempferi*) seedlings grown in high $CO_2$ concentration. Photosynthetica 42:173-178

Funada R, Kubo T, Fushitani M (1990) Early- and latewood formation in *Pinus densiflora* trees with different amounts of crown. IAWA Bulletin n.s. 11:281-288

Funada R, Kubo T, Tabuchi M, Sugiyama T, Fushitani M (2001) Seasonal variations in endogenous indole-3-acetic acid and abscisic acid in the cambial region of *Pinus densiflora* Sieb. et Zucc. stems in relation to earlywood-latewood transition and cessation of tracheid production. Holzforschung 55:128-134

Gartner BL, Roy J, Huc R (2003) Effects of tension wood on specific conductivity and vulnerability to embolism of *Quercus ilex* seedling grown at two atmospheric $CO_2$ concentration. Tree Physiol 23:387-395

Hättenschwiler S, Miglietta F, Raschi A, Körner C (1997) Thirty years of in situ tree growth under elevated $CO_2$: a model for future forest responses? Global Change Biol 3:463-471

Hättenschwiler S, Schweingruber FH, Körner C (1996) Tree ring responses to elevated $CO_2$ and increased N deposition in *Picea abies*. Plant Cell Environ 19:1369-1378

Jach ME, Ceulemans R (1999) Effects of elevated atmospheric $CO_2$ on phenology, growth and crown structure of Scots pine (*Pinus sylvestris*) seedlings after two years of exposure in the field. Tree Physiol 19:289-300

Kaakinen S, Kostiainen K, Fredrik E, Saranpää P, Kubiske ME, Sober J, Karnosky DF, Vapaavuori E (2004) Stem wood properties of *Populus tremuloides*, *Betula papyrifera* and *Acer saccharum* saplings after 3 years of treatments to elevated carbon dioxide and ozone. Global Change Biol. 10:1513-1525

Kilpeläinen A, Peltola H, Ryyppö A, Sauvala K, Laitinen K, Kellomäki S (2003) Wood properties of Scot pines (*Pinus sylvestris*) grown at elevated temperature and carbon dioxide concentration. Tree Physiol 23:889-897

Kinsman EA, Lewis C, Davies MS, Young JE, Francis D, Thomas ID, Chorlton KH, Ougham HJ (1996) Effects of temperature and elevated $CO_2$ on cell division in shoot meristems: differential responses of two natural populations of *Dactylis glomerata* L. Plant Cell Environ 19:775-780

Koike T, Kohda H, Mori S, Takahashi K, Inoue MT, Lei TT (1995) Growth responses of the cuttings of two willow species to elevated $CO_2$ and temperature. Plant Species Biol. 10:95-101

Koike T, Mori S, Takahashi K, Lei TT (1996) Effects of high $CO_2$ on the shoot growth and photosynthetic capacity of seedlings of Sakhalin fir and Monarch birch native to northern Japan. Environ Sci 4:93-102

Maherali H, DeLucia EH (2000) Interactive effects of elevated $CO_2$ and temperature on water transport in ponderosa pine. Amer J Bot 87:243-249

Norby RJ, Todd DE, Fults J, Johnson DW (2001) Allometric determination of tree growth in a $CO_2$-enriched sweetgum stand. New Phytol 150:477-487

Oren R, Ellsworth DS, Johnsen KH, Phillips N, Ewers BE, Maier C, Schafer KVR, McCarthy H, Hendrey G, McNulty SG, Katul GG (2001) Soil fertility limits carbon sequestration by forest ecosystems in a $CO_2$-enriched atmosphere. Nature 410:469-471

Peltola H, Kilpeläinen A, Kellomäki S (2002) Diameter growth of Scots pine (*Pinus sylvestrils*) trees grown at elevated temperature and carbon dioxide concentration under boreal conditions. Tree Physiol 22:963-972

Pritchard SG, Rogers HH, Prior SA, Peterson CM (1999) Elevated $CO_2$ and plant structure: a review. Global Change Biol 5:807-837

Roberntz P (1999) Effects of long-term $CO_2$ enrichment and nutrient availability in Norway spruce. I. Phenology and morphology of branches. Trees 13:188-198

Sigurdsson BD (2001) Elevated [$CO_2$] and nutrient status modified leaf phenology and growth rhythm of young *Populus trichocarpa* trees in a 3-year field study. Trees 15:403-413

Telewski FW, Swanson RT, Strain BR, Burns J M (1999) Wood properties and ring width responses to long-term atmospheric $CO_2$ enrichment in field-grown loblolly pine (*Pinus taeda* L.). Plant Cell Environ 22:213-219

Tognetti R, Cherubini P, Innes JL (2000) Comparative stem-growth rates of Mediterranean trees under background and naturally enhanced ambient $CO_2$ concentration. New Phytol 146:59-74

Wullschleger SD, Tschaplinski TJ, Norby RJ (2002) Plant water relations at elevated $CO_2$ - implications for water-limited environment. Plant Cell Environ 25:319-331

Yasue K, Funada R, Kobayashi O, Ohtani J (2000) The effects of tracheid dimensions on variations in maximum density of *Picea glehnii* and relationships to climatic factors. Trees 14:223-229

Yazaki K, Funada R, Mori S, Maruyama Y, Abaimov AP, Kayama M, Koike T (2001) Growth and annual ring structure of *Larix sibirica* grown at different carbon dioxide concentrations and nutrient supply rates. Tree Physiol 21:1223-1229

Yazaki K, Ishida S, Kawagishi T, Fukatsu E, Maruyama Y, Kitao M, Tobita H, Koike T, Funada R (2004) Effects of elevated $CO_2$ concentration on growth, annual ring structure and photosynthesis in *Larix kaempferi* seedlings. Tree Physiol 24:941-949

# III. Plant Responses to Combination of Air Pollution and Climate Change

# Carbon dioxide and ozone affect needle nitrogen and abscission in *Pinus ponderosa*

David M. Olszyk, David T. Tingey, William E. Hogsett, and E. Henry Lee

US Environmental Protection Agency, National Health and Environmental Effects Laboratory, Western Ecology Division, 200 SW 35th Street, Corvallis, Oregon 97333, USA

**Summary.** To determine whether $CO_2$ (a key contributor to climate change), and $O_3$ (an important air pollutant), interact to affect needle N dynamics; we grew *Pinus ponderosa* Dougl. ex Laws. seedlings for three seasons (1998-2000) in outdoor chambers with ambient or elevated $CO_2$ combined with low or elevated $O_3$. Nitrogen concentration (area basis) was measured for attached and abscised needles. Nitrogen retranslocation was determined by comparing N for attached needles in July *vs.* October, and N resorption by comparing N for needles which abscised from July to October *vs.* attached needles in October. Ozone, and to a lesser extent $CO_2$, reduced needle N depending on age class and sampling date, but there were no $CO_2$ x $O_3$ interactions. There was a significant $CO_2$ x $O_3$ interaction for N retranslocation for 1998 needles in 1999 ($O_3$ reduced higher retranslocation with elevated $CO_2$), but $CO_2$ and $O_3$ had only marginal effects on N retranslocation or resorption for other needles. Abscission of year-old needles increased with elevated $CO_2$, but not $O_3$. Long-term studies are needed to determine if the responses to elevated $CO_2$ and $O_3$ in our seedlings persist as trees mature and have multiple age-classes of needles.

**Key words.** Ponderosa pine, Retranslocation, Resorption, $CO_2$, $O_3$, N

## 1. Introduction

Two important atmospheric gases whose concentrations are affected by human activities, $CO_2$ (a key contributor to climate change, IPCC 2001) and $O_3$ (a regional air pollutant, Chameides et al. 1994), can affect the normal pattern of needle N and abscission, which can impact both carbon uptake and subsequent tree growth as well as nutrient cycling through forest ecosystems (Norby et al. 1999). Decreased needle N on a mass basis ($N_{weight}$), both in attached and abscised leaves, is a common response of plants to elevated $CO_2$ (Norby et al. 1999, 2001). Elevated $CO_2$ also has been reported to reduce N resorption from leaves prior to abscission (Arp et al. 1997), and to increase needle abscission in *Pinus ponderosa* (Surano et al. 1986). In combination, these changes could affect the amount and quality of needle litter. However, recent meta-analyses indicated that needle N does not decrease with elevated $CO_2$ on an area basis ($N_{area}$) (Curtis 1996), which is a better indicator of N status of leaves since an increase in needle weight due to $CO_2$ exposure would not be a confounding factor (Norby et al. 1999). Furthermore, $CO_2$ may not affect N resorption from leaves as originally hypothesized (Norby et al. 2001), and

*Plant Responses to Air Pollution and Global Change*
Edited by K. Omasa, I. Nouchi, and L. J. De Kok ( Springer-Verlag Tokyo 2005 )

responses to $CO_2$ may differ for conifers vs. broadleaf trees, e.g., $CO_2$ had no effect on leaf abscission in broadleaf species (Herrick and Thomas 2003; Norby et al. 2000).

Effects of $O_3$ on needle N vary widely with tree species, N availability and growing conditions (Scherzer et al. 1998). In *Pinus ponderosa*, both needle and litter N increased at field sites with both high ambient $O_3$ and high atmospheric N deposition (Fenn 1991). Needle N also increased with $O_3$ in *P. ponderosa* in controlled studies (Andersen et al. 2001; Temple and Reichers 1995). Temple and Reichers (1995) also showed that $O_3$ caused premature needle abscission and increased resorption of N from older needles. However, tree response to $O_3$ may also differ with leaf type, as for the broadleaf species *Populus trichocarpa* x *maximowizii*, where Bielenberg et al. (2001) showed that $O_3$ could increase, decrease, or have no effect on leaf N depending on N supply and leaf age.

Because of continued uncertainty regarding the effects of elevated $CO_2$ or elevated $O_3$ on needle N dynamics and needle abscission on trees, and the potential for trees to be exposed concurrently to both gases; we determined the individual and combined effects of elevated $CO_2$ and $O_3$ on 1) needle N, 2) needle abscission, and 3) on retranslocation of N from attached needles during the growing season and resorption of N during abscission. This investigation was part of a larger study on the effects of elevated $CO_2$ and $O_3$ on carbon and nitrogen cycling through a *Pinus ponderosa* mesocosm (Olszyk et al. 2001).

## 2. Methods

*Pinus ponderosa* Dougl. ex Laws. seedlings were grown for three years in outdoor, sunlit chambers (Olszyk et al. 2001; Tingey et al. 1996). Each chamber was 2 m wide, 1 m front-to-back; 1.7 to 1.5 m tall on each side attached to a 1 m deep x 1 m wide x 2 m long soil lysimeter. The soil was an ashy, glassy Xeric Vitricryands soil obtained from a P. ponderosa forest on the east side of the Cascade Mountains in Oregon, USA (Olszyk et al. 2001). Ambient conditions were monitored to provide target air temperature, dewpoint and $CO_2$ values for the chambers (Olszyk et al. 2001, Tingey et al. 1996). Over nearly three years, the difference between ambient and elevated $CO_2$ treatments was 269 ppm. The $O_3$ treatments were expressed as a cumulative seasonal (mid-May to early October) exposure (SUM 06), based on all growing season hourly averages $\geq 0.060$ ppm (Lee et al. 1988). The SUM06 values with elevated $O_3$ were 9.8 ppm hr for 1998, 11.0 ppm hr for 1999 and 26.2 ppm hr for 2000 (Olszyk et al. 2002, 2001). The SUM06 values with low $O_3$ were near zero, i.e. $\leq 0.1$ ppm hr for each year.

*Pinus ponderosa* seedlings (half-sib) were obtained from seeds planted in the spring of 1997 at a U.S. Forest Service nursery near Bend, Oregon (Olszyk et al. 2001). Eleven seedlings were transplanted into each chamber on 8 April, 1998. Soil moisture was controlled weekly to reflect a typical seasonal cycle (wet winter and dry summer) found in Pacific Northwest *Pinus ponderosa* forests. Time Domain Reflectometry data were used to determine the amount of reverse-osmosis water added to a chamber each week to maintain target field level in ambient $CO_2$ and low $O_3$ chambers. All other treatment chambers also received this amount of water. No fertilizer was added to the soil.

Needles attached to trees were sampled each October, April and July between Oct. 1998 and April 2001. One needle fascicle per tree was pooled for each age class in each chamber. Needle area by age class was determined for the fresh material (planar area using a LI-COR 3100 area meter), except that average needle areas in Oct. 1999 also were used for needles sampled in July 1999. Needle mass was determined after drying at 60°C. Needle abscission was defined and measured as the amount of needles that fell from the trees in each chamber approximately every two weeks from the fall of 1998 onward. Needles still in the canopy but no longer connected to the tree were included. Prior to 1 October, 2000, abscised needles were separated by age class. From 1 October, 2000 onward, needles were separated by color (green or brown), with brown needles considered to be from the 1999 age class, and green needles from the current (2000) age class. Abscised needles were dried and weighed. Needle biomass at the end of the experiment was determined by removing, drying and weighing all needles by age class from the woody tissue and drying at 60°C, needle area was determined from dried needles. There was shrinkage of needles either while lying the chambers or during drying, but it appeared to be uniform among treatments, and only meant that N resorption may have been underestimated. Needle biomass per chamber was divided by the number of trees to provide weight per tree.

Needle % N for attached and abscised needles was measured using flash-combustion analysis (Carlo-Erba), and expressed on a needle area basis ($N_{area}$). Nitrogen retranslocation, i.e., the net N moving between attached green needles and the rest of the plant, was determined in late summer of each year by comparing the N concentration in needles in October vs. July. Nitrogen resorption, the movement of N from needles during abscission, was determined for one-year old needles (the main age class abscising each year) in each of 1999 and 2000 by comparing the concentration of N in attached needles in October vs. abscised needles falling between July and October of each year. Both retranslocation and resorption were based on the process for comparing nutrient concentrations in presenescent and senescent leaves described by Killingbeck et al. (1990).

The experiment followed a factorial design with two levels of $CO_2$ and two levels of $O_3$, resulting in four treatments: ambient $CO_2$, low $O_3$ (ACLO); elevated $CO_2$, low $O_3$ (ECLO); ambient $CO_2$, elevated $O_3$ (ACEO), and elevated $CO_2$, elevated $O_3$ (ECEO). Individual chambers were the experimental units. There were three replicate chambers per treatment; but data from only two of the ACLO treatment chambers were used for statistical analysis due to problems with $O_3$ control and insect infestations in one ACLO chamber. Measurements were made separately for different needle age classes, with successive sampled dates considered as repeated measures. Since age class had a large effect on $N_{area}$, data for each needle age class was considered separately using a MANOVA to evaluate trends in needle N over dates and with treatments. The MANOVA also was used to determine treatment effects on N retranslocation by comparing data for July vs. October in 1999 and 2000 by age class, and was used to determine effects on N resorption by comparing data for attached vs. abscised year-old needles in October 1999 and 2000. Treatment differences were evaluated on specific dates and age classes using univariate Analysis of Variance (ANOVA). The MANOVA and ANOVA used SAS® software.

## 3. Results

Attached needle $N_{area}$ was highest in the first age class (1998) emerging under the experimental treatments (Fig. 1a,d), and decreased in the succeeding 1999 (Fig. 1b,e) and 2000 (Fig. 1c,f) age classes. For 1998, and to a lesser extent 1999 needles, $N_{area}$ increased from the fall in the year of emergence to the highest level the following April, followed by a decrease in $N_{area}$ over time. For 2000 needles, $N_{area}$ remained low from emergence onward. Treatment effects were most apparent for the 1999 age class, where $O_3$ produced a significant (p<0.05) decrease in $N_{area}$ based across all dates (p=0.02), and

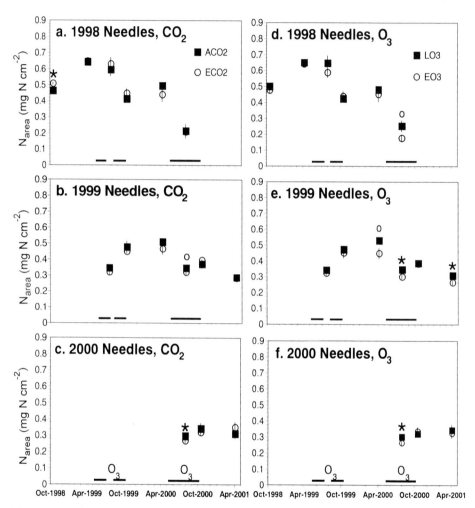

**Fig. 1a-f.** Effects of $CO_2$ (a,b,c) and $O_3$ (d,e,f) on needle nitrogen for 1998, 1999 and 2000 age classes of attached *Pinus ponderosa* needles. Abbreviations: ACO2 = ambient $CO_2$, ECO2 = elevated $CO_2$, LO3 = low $O_3$, EO3 = elevated $O_3$. An "*" or "o" above symbols indicates significant treatment effects at p<0.05 or 0.05<p<0.10, respectively, based on univariate ANOVA. There were no significant $CO_2$ x $O_3$ interactions, thus data are presented only for main effects ($CO_2$ across $O_3$ levels, $O_3$ across $CO_2$ levels). Each symbol is the mean ± standard error (vertical bars) for six replicate chambers for $CO_2$ and EO3 and five replicate chambers for ACO2 and LO3. Symbols may be behind each other if overlapping. Black bars at base of each panel are when $O_3$ exposures occurred in 1999 and 2000. There also was an $O_3$ exposure from July 27 through October 7, 1998 (not shown).

specifically for July 2000 and April 2001 (Fig. 1e). Ozone also produced a significant reduction in $N_{area}$ for 2000 needles in July 2000 (Fig. 1f). There were marginally non-significant ($0.05<p<0.10$) reductions in $N_{area}$ with $O_3$ for 1998 needles in Oct. 1999 (Fig. 1d) and 1999 needles in April 2000 (Fig. 1e). The only significant $CO_2$ effects were an increased $N_{area}$ for 1999 needles in Oct. 1998 (Fig. 1b), and a reduced $N_{area}$ for 2000 needles in July 2000 (Fig. 1c). Carbon dioxide was associated with a marginally non-significant decrease in $N_{area}$ for 1999 needles across dates (p=0.08) and in July 2000 (Fig. 1b). There were no $CO_2$ x $O_3$ interactions for attached $N_{area}$ on any date.

There was significant (p<0.02) N retranslocation between July and Oct. in both 1999 and 2000, for attached *Pinus ponderosa* needles across treatments. In 1999, retranslocation was outward from 1998 needles ($N_{area}$ decreased from 0.61 to 0.43 mg N $cm^{-2}$, Fig. 1a,d), which complimented an inward retranslocation to 1999 needles ($N_{area}$ increased from 0.33 to 0.46 mg N $cm^{-2}$, Fig. 1b,e). Overall, N retranslocation was much less in 2000 than 1999, and was inward for both the 1999 and 2000 classes. The $N_{area}$ increased from 0.32 to 0.38 mg N $cm^{-2}$ for 1999 needles (Fig. 1b,e), and from 0.28 to 0.33 mg N $cm^{-2}$ for 2000 needles across treatments (Fig. 1c,f). This increase in $N_{area}$ for needles in 2000 may have at least in part, due to N coming from abscission of nearly all the remaining 1998 needles prior to the fall needle collection. The only treatment effect on N retranslocation was for 1998 needles in 1999, where elevated $O_3$ apparently inhibited the higher retranslocation with elevated $CO_2$, resulting in a significant $CO_2$ x $O_3$ interaction (p=0.03).

Nearly all needle abscission occurred from mid-July through Oct. in both 1999 (Fig. 2a) and 2000 (Fig. 2b). In both years, abscission tended to be greatest with elevated $CO_2$, with a nearly significant effect for year-old (1999) needles (p=0.07) and total needles (p=0.06) in 2000 (Fig. 2b). There were no significant $O_3$ or $CO_2$ x $O_3$ effects on needle abscission.

There was significant (p<0.01) N resorption for abscised compared with attached needles across treatments in the fall for one-year-old *Pinus ponderosa* needles in both 1999 (Fig. 3a) and 2000 (Fig. 3b). This indicated that N moved back into the trees as needles abscised in both years. For 1998 needles in 1999, $N_{area}$ was 22% lower in abscised than in attached needles; and for 1999 needles in 2000, $N_{area}$ was 48% lower in abscised than in attached needles. There were only trends for treatment effects on N resorption: for 1998 needles in fall 1999 (Fig. 3a) when elevated $O_3$ reduced N resorption (p=0.10); and for 1999 needles in fall 2000 (Fig. 3b) when elevated $CO_2$ increased N resorption (p=0.06).

## 4. Discussion

The decrease in needle N with elevated $CO_2$ for *Pinus ponderosa* in our study was not as large or statistically significant as was the decrease in needle N observed in other studies (Grulke et al. 1993; Johnson et al. 1997). This was likely due to calculation of needle N as $N_{weight}$ in the other studies. We also saw a significant decrease in needle N on $N_{weight}$ for 1999 needles in July 1999 (Olszyk et al. 2001), which was not apparent on an area basis. Decreased needle N over time for the 1998 and 1999 age classes was consistent with other *P. ponderosa* studies (Johnson et al. 1997). Overall high needle N early in our study likely reflected N available to the trees in the nursery and/or N mineralization following soil placement in the mesocosms, as the initial soil N in the "A" horizon which we placed in the chambers was only 0.13% (data not shown). Our results in part support the conclusion that elevated $CO_2$ may not affect dynamics of N transport in leaves, as shown by most previous studies of N retranslocation (Norby et al. 2001).

Greater needle fall with elevated $CO_2$ likely was due to enhanced abscission and not needle production as there was only slightly more needles produced with elevated $CO_2$ in 1998 and 1999 (sum of abscised and on trees at final harvest, data not shown). Enhanced abscission of *Pinus ponderosa* needles with elevated $CO_2$ exposure in our study was similar to the results of Surano et al. (1986). They hypothesized that elevated needle temperatures for *P. ponderosa* needles due to stomatal closure with elevated $CO_2$, may have been injurious, thus accentuating abscission. However, in their study, the elevated air temperatures and altered needle boundary-layer conditions due to their open-top

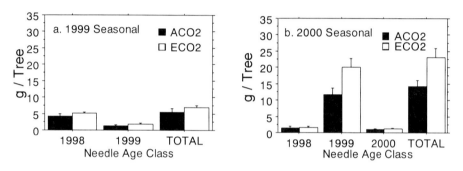

**Fig. 2.** Effect of $CO_2$ (across $O_3$ levels) on seasonal (mid-July through October) needle abscission in (a) 1999 and (b) 2000 for *Pinus ponderosa* trees. Bars are means for 5 and 6 replicate chambers ± S.E. for ambient $CO_2$ ($ACO_2$) and elevated $CO_2$ ($ECO_2$), respectively. There were small amounts (a few mg) of 1997 needles in both seasons which were too small to be analyzed as an individual age class, but which were included in the totals.

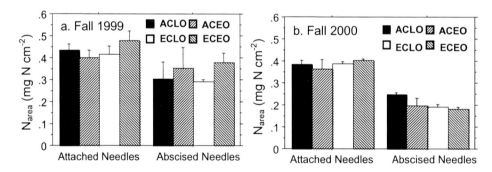

**Fig. 3.** Effects of $CO_2$ and $O_3$ on N in attached and abscised one-year-old needles, i.e., 1998 needles in Fall 1999 (a) and 1999 needles in Fall 2000 (b). Abbreviations: Abbreviations: ACLO = ambient $CO_2$, low $O_3$; ACEO = ambient $CO_2$, elevated $O_3$; ECLO = elevated $CO_2$, low $O_3$; ECEO = elevated $CO_2$, elevated $O_3$. Each bar is the mean ± standard error (vertical bars) for three replicate chambers except for 2 chambers for ACLO.

chambers may have further enhanced the adverse effects of $CO_2$ on needle temperature. Since air temperature was controlled at ambient (Olszyk et al. 2001), and there was considerable air flow circulating through our chambers, direct heat stress likely was not a factor in the abscission in our study. Furthermore, we saw only slight stomatal closure with elevated $CO_2$ (Olszyk et al. 2002). Since we only measured stomatal conductance for current year needles; it is possible that the year-old needles which abscised may have been undergoing undetected heat stress. However, other, unknown, mechanisms likely played a role in needle abscission in our study.

Decreased needle N and lack of needle abscission with $O_3$ for attached needles of *Pinus ponderosa* in our study, was in contrast to the increase in needle N and increased abscission with $O_3$ reported by others (Fenn et al. 1991; Temple and Reichers 1995; Anderson et al. 2001). Our results may be a consequence of the overall low level of N in our needles, and the moderate $O_3$ levels which occurred in our "elevated" $O_3$ treatment. Our study did not clarify the effect of $O_3$ on N retranslocation due to the interaction with elevated $CO_2$, at least for year-old needles during the 1999 season. Lindroth et al. (2001) saw a significant interaction of $CO_2$ and $O_3$ on N retranslocation for *Betula papyrifera* leaves, but in that case elevated $CO_2$ countered a decrease in N retranslocation with $O_3$.

In our study we did not observe premature needle abscission as reported previously for *P. ponderosa*, especially for older as compared with current needles (Temple et al. 1993). Our lack of an effect on abscission may be due to the fairly low $O_3$ exposures in the elevated $O_3$ treatment, especially during the first two years of the study (Olszyk et al. 2001). Additional studies are vitally needed to determine how needle N and abscission responses vary with N availability and different $O_3$ exposures.

The few studies evaluating combined effects of $CO_2$ and $O_3$ on leaf abscission and/or N concentration show no consistent interactions between the two gases. For example, in *Populus tremuloides* leaf % N was reduced significantly by both elevated $CO_2$ and elevated $O_3$, with no significant $CO_2$ x $O_3$ interaction on leaf % N or any treatment effects on N resorption efficiency (Lindroth et al. 2001). In contrast, in *Betula papyrifera*, only elevated $CO_2$ reduced leaf % N, while $O_3$ caused a reduction N resorption efficiency which was largely negated by elevated $CO_2$ (Lindroth et al. 2001). Broadmeadow & Jackson (2000) reported decreased leaf % N with elevated $CO_2$ for irrigated *Quercus petraea* and *Fraxinus excelsior*, but not *Pinus sylvestris*. Elevated $O_3$ also reduced leaf % N in *Fraxinus excelsior*, with the response alleviated to some extent with elevated $CO_2$ (Broadmeadow and Jackson 2000).

In conclusion, out study indicated responses of *Pinus ponderosa* seedlings to elevated $CO_2$ and $O_3$, which underscore the complicated relationship between needle N, needle abscission and the response of trees to $CO_2$ and $O_3$. Long-term studies are necessary to evaluate the complex responses of trees to joint elevated $CO_2$ and $O_3$ exposures, as those interactions may be dependent on cumulative responses of different needle age-classes.

Acknowledgments. The information in this document has been funded in part by the U.S. Environmental Protection Agency under contract number 68-DO-1005 to Dynamac Inc. It has been subject to the agency's peer and administrative review. It has been approved for publication as an EPA document. Mention of trade names or commercial products does not constitute endorsement or recommendation for use.

# References

Andersen CP, Hogsett WE, Plocher M, Rodecap K, Lee EH (2001) Blue wild-rye grass competition increases the effect of ozone on ponderosa pine seedlings. Tree Physiol 21:319-327

Arp WJ, Kuikman PJ, Gorissen A (1997) Climate change: the potential to affect ecosystem functions through changes in amount and quality of litter. In: Cadisch G, Giller K (Eds) Driven by nature. Plant litter quality and decomposition. CAB, Wallingford, pp 187-200

Beilenberg DG, Lynch JP, Pell EJ (2001) A decline in nitrogen availability affects plant responses to ozone. New Phytol 141:413-425

Broadmeadow MSJ, Jackson SB (2000) Growth responses of *Quercus petraea, Fraxinus excelsior* and *Pinus sylvestris* to elevated carbon dioxide, ozone and water supply. New Phytol 146:437-451

Chameides WL, Kasibhatla PS, Yienger J, Levy II H (1994) Growth of continental-scale metro-agro-plexes, regional ozone pollution and world food production. Science 264:74-77

Curtis PS (1996) A meta-analysis of leaf gas exchange and nitrogen in trees grown under elevated carbon dioxide. Plant Cell Environ 19:127-137

Fenn M (1991) Increased site fertility and litter decomposition rate in high-pollution sites in the San Bernardino Mountains. For Sci 37:1163-1181

Grulke NE, Homm JL, Roberts SW (1993) Physiological adjustment of two full-sib families of ponderosa pine to elevated $CO_2$. Tree Physiol 12:391-491

Herrick JD, Thomas RB (2003) Leaf senescence and late-season net photosynthesis of sun and shade leaves of overstory sweetgum (*Liquidambar styraciflua*) grown in elevated and ambient carbon dioxide concentrations. Tree Physiol 23:109-118

Intergovernmental Panel on Climate Change (2001) Climate change 2001: the scientific basis. World Meteorological Organization

Johnson DW, Ball JT, Walker FT (1997) Effects of $CO_2$ and nitrogen fertilization on vegetation and soil nutrient content in juvenile pine. Plant Soil 190:29-40

Killingbeck KT, May J.D, Nyman S (1990) Foliar senescence in an aspen (*Populus tremuloides*) clone: The response of element resorption to inter-ramet variation and timing of abscission. Can J For Res 20:1156-116

Lee EH, Tingey DT, Hogsett WE (1988) Evaluation of ozone exposure indices in exposure-response modeling. Environ Pollut 53:43-6

Lindroth R., Kopper BJ, Parsons WFJ, Bockheim JG, Karnosky DF, Hendry GR, Pregitzer KS, Isebrands JG, Sober J (2001) Consequences of elevated carbon dioxide and ozone for foliar chemical composition and dynamics in trembling aspen (*Populus tremuloides*) and paper birch (*Betula papyrifera*). Environ Pollut 115:395-404

Norby RJ, Wullschleger SD, Gunderson CA, Johnson DW, Ceulemans R (1999) Tree responses to rising $CO_2$ in field environments: implications for the future forest. Plant Cell Environ 22:683-714

Norby RJ, Long TM, Hartz-Rubin JS, O'Neill EG (2000) Nitrogen resorption in senescing tree leaves in a warmer, $CO_2$-enriched atmosphere. Plant Soil 224:15-29

Norby RJ, Cotrufo MF, Ineson P, O'Neill EG, Canadel JG (2001) Elevated $CO_2$, litter chemistry, and decomposition: a synthesis. Oecologia 127:153-165

Olszyk DM, Johnson MG, Phillips DL, Seidler RJ, Tingey DT, Watrud LS (2001) Interactive effects of $CO_2$ and $O_3$ on a ponderosa pine plant/litter/soil mesocosm. Environ Pollut 115:447-462

Olszyk DM, Tingey DT, Wise C, Davis E (2002) $CO_2$ and $O_3$ alter photosynthesis and water vapor exchange for *Pinus ponderosa* leaves. Phyton 42:121-134

Scherzer AJ, Rebbeck J, Boerner REJ (1998) Foliar nitrogen dynamics and decomposition of yellow-popular and eastern white pine during four seasons of exposure to elevated ozone and carbon dioxide. For Ecol Manage 109:355-366

Surano KA, Daley PF, Houpis JLJ, Shinn JH, Helms JA, Palassou RJ, Costella MP (1986) Growth and physiological responses of *Pinus ponderosa* Dougl ex. P. Laws. to long-term elevated $CO_2$ concentrations. Tree Physiol 2:243-259

Temple PJ, Riechers GH (1995) Nitrogen allocation in ponderosa pine seedlings exposed to interacting ozone and drought stresses. New Phytol 130:97-104

Temple PJ, Riechers GH, Miller PR, Lennox RW (1993) Growth responses of ponderosa pine to long-term exposure to ozone, wet and dry acidic deposition, and drought. Can J For Res 23:59-66

Tingey DT, McVeety BD, Waschmann R, Johnson MG, Phillips DL, Rygiewicz PT, Olszyk DM (1996) A versatile sun-lit controlled-environment facility for studying plant and soil processes. J Environ Qual 25:614-625

# Effects of air pollution and climate change on forests of the Tatra Mountains, Central Europe

Peter Fleischer[1], Barbara Godzik[2], Svetlana Bicarova[3], and Andrzej Bytnerowicz[4]

[1]Research Station of Tatra National Park, State Forest of TANAP, 059 60 Tatranska Lomnica, Slovakia
[2]Institute of Botany, Polish Academy of Sciences, Lubicz 46, 31 512 Krakow, Poland
[3]Institute of Geophysics, Slovak Academy of Sciences, Meteorological Observatory, Stara Lesna, 059 60 Tatranska Lomnica, Slovakia
[4]USDA Forest Service, Pacific Southwest Research Station, 4955 Canyon Crest Drive, Riverside, CA92507-6090, USA

**Summary.** Synergistic effects of air pollution, extreme weather conditions and biotic agents related to global climate change have caused serious deterioration of forest condition in the Tatra National Park since early 1990s. Atmospheric deposition of sulfate ($SO_4^{2-}$), nitrate ($NO_3^-$) and acidity ($H^+$) are above the established critical load limits for forests. In addition, ambient ozone ($O_3$) concentrations are also elevated mainly due to a long-range transport of polluted air masses. Ambient $O_3$ concentrations have been monitored since 1992 with active UV monitors and passive samplers showing significant differences along the elevational gradient. High $O_3$ values in spring indicate a potential for $O_3$ stratospheric intrusion into troposphere. Ozone AOT40 index for protection of natural forest vegetation (10,000 ppbh) is commonly exceeded in the middle of vegetation period.
Acidic impact may explain about 10-30% of a dieback observed in naturally oligotrophic Norway spruce *(Picea abies)* ecosystems probably due to acidification of soils leading to $Al^{3+}$ mobilization limiting growth of fine roots. This may cause weakened resistance of trees to wind throws frequently observed in the foothills of the Tatra Mountains. Seventy year-long observations of meteorological conditions provide a unique opportunity for comparing current status of the Tatra forests within a context of long-term trends. Extremely high spring and summer temperatures, low precipitation and relative humidity observed during last decade are new factors disturbing ecological stability of natural and man-made forests. Warm and dry vegetation periods favor frequent bark beetle *(Ips typographus)* outbreaks affecting large areas of the Tatra forests.

**Key words.** Critical loads, Ozone, Vertical gradient, Ecosystem stability

## 1. Introduction

The Tatra Mountains (Tatras) are the highest part of the Carpathian range. From a biogeographical point of view the Tatra Mts are an island with well preserved mountain and

*Plant Responses to Air Pollution and Global Change*
Edited by K. Omasa, I. Nouchi, and L. J. De Kok ( Springer-Verlag Tokyo 2005 )

alpine ecosystems containing unique flora and fauna. The Tatra Mountains are located in the geographical center of Europe, creating a natural border between Slovakia and Poland. Exceptional beauty of mountains and presence of numerous relict, endemic and endangered species lead to a declaration of Tatra national parks on the Slovak side in 1948 and in 1954 on the Polish side. As a bilateral biosphere reserve the Tatras were placed on UNESCO's list in 1992.

The Tatra Mountains as a huge mountain barrier intercept a wide spectrum of air pollutants coming from various directions. Till the late 1980s, air pollution problems in central Europe were mostly caused by industrial emissions of sulfur dioxide ($SO_2$), nitrogen oxides (NOx), fluorides and heavy metals (Guderian 1977). In that time the Tatra Mts belonged to the area with the most acid precipitation in Europe, with pH average below 4.2 (EMEP, 1990). Acidification of mountain lakes, contamination of plants and animals with heavy metals and forest decline indicate serious ecological problems. However, subsequent to the political and economical changes in that part of Europe, industrial air emissions have been significantly reduced (especially for sulfur compounds), while photochemical smog, especially ozone ($O_3$), gained in importance because of the rapidly increasing number of vehicles producing the $O_3$ precursors: nitrogen oxides, and volatile organic compounds (Bytnerowicz et al. 2004).

The area of the Tatra Mountains is mostly forested. Forest covers an area of 50,000 ha and creates almost continuous vegetation belt spreading from 700 m. a.s.l. up to timberline in 1,550 m a.s.l. Almost half of the forest remains in seminatural or even natural conditions. Harsh mountain climate significantly influences forests, e.g., tree species composition, as well as the vertical, spatial, and age structures. Norway spruce (*Picea abies*) is a dominant tree species, covering almost 70% of forest. Other coniferous species cover 20% (Scots pine - *Pinus sylvestris*, larch - *Larix europea*, fir - *Abies alba* and Swiss stone pine - *Pinus cembra*). Broadleaved tree species cover 10 % (birch - *Betula carpatica*, alder - *Alnus incana, A. viridis*, mountain ash - *Sorbus aucuparia*, willow - *Salix sp.*).

Climate is rather continental, with cool winter and relatively warm summer, with maximum precipitation during summer. Continentality is decreasing from the central part of the Tatras to the east and west. Forest in the central part of the mountains has naturally very narrow tree species diversity - besides Norway spruce, only larch, Swiss stone pine and mountain ash are present. Less continental parts provide more suitable conditions for fir and even for beech, especially on Paleogene sediments (flysh). Prevailing wind direction is SW – NW. Specific orientation of the massif to frequent strong wind is a reason for large disturbances caused by bora winds. Hundreds of hectars of mature forest are fallen every year, mostly on the SE – S slopes. Roughly once every hundred year a large scale catastrophy happens with several hundred thousand $m^3$ of uprooted and broken trees. Even under normal climate conditions both fallen and surrounding standing trees are affected by bark-beetle insect (*Ips typographus*). Recent warm and dry weather during growing period was a reason for unusually strong bark-beetle outbreaks.

Continuous degradation of forest indicates synergistic effect of various disturbance factors. This paper presents some results gained by monitoring of forest status, climate parameters and air quality, and discusses their negative effect on the Tatras' forest and possible interferences.

## 2. Methodology

For this study, delineation of the Tatra mountains study area was based on geomorphological classification and forest management jurisdiction. The study area is located in Slovakia and corresponds to the FMU Tatry (forest management unit) covering an area of nearly 50,000 ha.

Forest sampling design reflected the existence of numerous forest research plots established in past years and also adopted to the national and international forest monitoring network. Air quality and climate sampling locations represent typical morphological features – slope orientation (south, north) and forms (valley, regular slope). In order to spatially extrapolate point data, sites were located along vertical transects.

### 2.1 Forest status

Forest condition and degradation in Europe has been coordinated in framework of the International Co-operative Programme on the Assessment and Monitoring of Air Pollution Effects on Forest (ICP Forests) since the mid 1980s. Defoliation of trees is evaluated using standardized methodologies in all participating countries. Trees with defoliation >25% are considered unhealthy with an understanding that air pollution is one of the main causes of foliage loss. Assessment of crown condition was done on regular network 2x2 km derived from pan-European and national forest monitoring grids. Permanent plots for crown assessment were established in 1989-1991, with a total number of 180, each with a minimum of fifteen trees in category 1 or 2 according to the Kraft system. Trees were assessed on defoliation and discolouration in 5% step (ICP 1998). Fixed plots and individual trees served mostly for recognition of temporal changes of forest status. Aerial infrared photographs and satelitte imagery (Landsat, Spot, Iconos) were used for delineation and better spatial understanding of forest decline and health categories, using the ArcView and Erdas softwares.

### 2.2 Air quality

Air pollutants enter forest ecosystems by rain and snow precipitation, interception of fog and low clouds, and dry despotion of gases and particles. Precipitation samples (wet deposition, throughfall) were collected by polypropylene bulk collectors all year around in 2-week long periods. Stemflow and soil water samples were collected by stemflow collar and lysimeters during a growing season (May-September). Atmospheric deposition was monitored in four representative localities. Number of collectors ranged from 6 up to 16 per plot. Sampling design reflected a stand structure and tree species composition. Samples were analyzed for conductivity, pH, cations ($Al^{3+}$, $Ca^{2+}$, $Mg^{2+}$, $K^+$, $NH_4^+$) and anions ($SO_4^{2-}$, $NO_3^-$, $Cl^-$) by isotachophoretic analyzer (model EA 101, Villa Group, Slovakia). Critical loads were calculated by a method of Svedrup et al. (1990).

Gases such as $O_3$, $SO_2$ and $NO_2$ were monitored by passive samplers at 18 localities during 1998-2000 and on 24 localities since 2000 during growing seasons. Ozone was monitored by Ogawa passive samplers (Ogawa & Co. Pompano Beach, FL, USA). Cellulose filters coated with nitrite were exposed for two weaks and analyzed by ion chromatography (Dionex 2000) according to Koutrakis et al. (1993). Ozone concentration was calculated from a calibration curve derived from real time $O_3$ monitors collocated with passive samplers. Sulfur dioxide and $NO_2$ samplers were exposed for four weeks and analyzed according to Krochmal-Kalina (1997). Real-time $O_3$ concentrations were measured with the UV absorbtion Thermo Environmental C49 monitors at Start (1,200 m a.s.l.) and Skalnate Pleso (1,700 m a.s.l.). Monitors were placed along existing instruments of Slovak Hydrometeorological Institute (SHMI) and created the "Lomnica" vertical transect from 700 m.as.l. up to 2,633 m a.s.l. consisting of six active UV monitors.

## 2.3 Climate

Simple weather stations (OnSet Hobo) with air temerature and RH sensors, and precipitation bucket) were installed parallely to the $O_3$ passive samplers. Three mobile automatic weather stations (Campbell Scientific, MAWS Vaisala) were placed along the "Lomnica" vertical transect. Meteorological stations measured air temeparture, relative humidity, wind speed and direction, precipitation, global and UVB radiation. Data were measured in 10 minute up to one hour intervals. Based on the one hour values, diurnal, monthly, seasonal and annual averages and extreme values were calculated. The obtained values were compared with standards derived from the long term averages for the Tatra region (Koncek et al. 1974).

## 3. Results

### 3.1 Forest status

First census of crown condition in 1990s on 180 permanent plots confirmed that according to defoliation and discoloration, forest health status in the Tatras had remarkably declined. Average defoliation of trees was 28.7% (±2%) while during that time the average for Slovakia was 27% (Bucha et al. 1993). Fleischer and Koren (1995) besides air pollution identified other driving factors for defoliation: stand density (open stands were more defoliated), social rank (plus trees more defoliated), orientation (higher defoliation on NW oriented slopes), altitude (defoliation increased with altitude), and soil properties (higher defoliation on stands with low water retention capacity). The repeated census in 2002-2003 showed that average defoliation increased up to 32%. Suppisingly, the biggest difference in defoliation occured in categories described in 1990's as healthy (Fig. 1a).

Dead trees were excluded from the second assessment. Almost 20% of all trees in first census died during the ten year monitoring interval. The highest relative dieback occured in categories with former defoliation 40-60 %. Relatively stable were the trees with very high defoliation (>70%), and only a few of them died (Fig.1b). Based on the forestry records, windstorms and insect outbreaks were the main causes for the observed dieback.

Satelite imagery (Koren and Fleischer 1993) and IR aerial photopgraphs were used

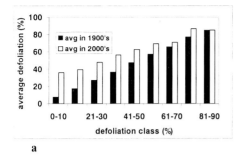 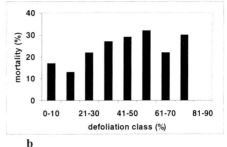

**Fig. 1a,b.** Health status in the Tatras. **a** Average defoliation in defoliation classes in 1990s and 2000s campaign. **b** Relative dieback in defoliation classes.

for spatial forest health detection. Remote sensed data confirmed highly heterogenous and mosaic health conditions, with the worst situation on the northern slopes of the Tatra mountains. Large-scale aerial photographs helped to detect vast fragmentation of forest due to incidental felling caused by windstorms and/or insect outbreaks. An analysis of incidental felling events confirmed that in 80% of events the new gap size was below 0.3 ha. Gaps were sparcely distributed and hardly detected. Under favorable conditions, disturbing factors such as insect spread influence the surrounding stands.

## 3.2 Air quality

### 3.2.1 Wet deposition

Results of pH measurement showed that acidity on some localities is still a serious problem. Table 1 shows that since 1995 annual average acidity (measured as bulk deposition) at Popradske Pleso (1,500 m a.s.l) was permanently below the value of pH 4 and was still decreasing. Due to the very high variability, differences among individual years were not statisticaly significant. The lowest pH values were measured during springtime and maximum during summer. The UN ECE (1993) critical level pH 3 for individual precipitation events was exceeded in 2-week samples measured as througfalls and stemflows. The most acidic samples were collected from stemflow on Swiss stone pine. On the other hand, less acidic samples were found beneath birch and mountain ash crowns. Similar effects were also shown for larch, but in springtime elevated concentrations of cations

**Table 1.** Mean annual pH value of precipitation (open field sampler) at Popradske Pleso, 1500 m a.s.l.

| Year | 1999 | 2000 | 2001 | 2002 | 2003 |
|---|---|---|---|---|---|
| pH | 3.96 | 3.82 | 3.69 | 3.66 | 3.62 |

indicated nutrient leaching from needles. Annual deposition of $H^+$ about two-fold exceeded the critical level of 0.6 $H^+$ kg ha$^{-1}$ in open field samples and and about 2 - 4 fold in throughfall samples.

Figure 2 shows annual and seasonal $SO_4^{2-}$ averages since 1998 measured at ICP Forests Level II plot at the Lomnica-Start site (1,200 m a.s.l). Annual averages indicated decline or at least stabilisation and higher $SO_4^{2-}$ content in throughfall water when compared to the open field samples. Seasonal course showed maximum $SO_4^{2-}$ concentration in spring and minimum during summer. The ecological limit of 14 mg l$^{-1}$ $SO_4^{2-}$ in precipitation event (UN EEC, 1993) was often exceeded in two-week stemflow and throughfall samples. Annual deposition of sulfur (S) was up to 10 kg ha$^{-1}$ in open field samples and two times more in throughfall samples. According to the UN ECE (1993), critical load of S is 10 kg ha$^{-1}$.

$NO_3^-$ annual course showed oposite tendency. Concentration in precipitation water increased, see Fig. 3. Ecological limit of 5.4 mg l$^{-1}$ (UN ECE 1993) was often exceeded even in open field samples.

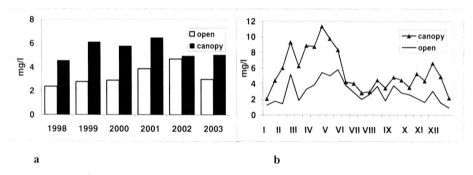

Fig. 2. Mean $SO_4^{2-}$ concentration in open field (open) and thougfall (canopy) bulk collectors for period 1998-2003 measured at ICP Forest Level II plot, Lomnica-Start in 1,200 m a.s.l. a Annual average. b Seasonal average.

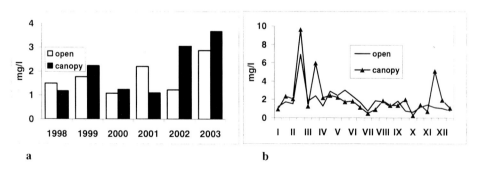

Fig. 3. Mean $NO_3^-$ concentration in open field (open) and thoughfall (canopy) bulk collectors for period 1998-2003 at ICP II plot, 1200 m a.s.l. a Annual average. b Seasonal average.

High concentrations of $Al^{3+}$ were recorded in soil solution water. Average 2-week concentration in 40 cm depth ranged from 200 up to 560 µg $l^{-1}$ with maximum up to 600 µg $l^{-1}$. These values indicated $Al^{3+}$ mobilisation from naturally weathered granite rocks deposited in old morrain material (mindel Age) by a long-term acid deposition. Notably limited root growth in neighbouring area with frequent uprootings confirmed the hypothesis on soil environment degradation by toxic $Al^{3+}$.

Based on cation exchange capacity, $Al^{3+}$ concentrations and other indicators, a critical load for selected larch-spruce and Swiss stone pine - spruce forest ecosystems was calculated. Results showed that monitored localities in higher elevation had exceeded critical loads for acidity. Up to 30% of tree dieback might be explained by overload of acidic compound (Kunca et al. 2003).

### 3.2.2 Gaseous $SO_2$, $NO_2$, and $O_3$

Elevated ambient concentrations of nitrogen oxides in the Tatras were confirmed by the results derived from the air quality monitoring network since 1998. Figure 4 shows that average $NO_2$ concentrations from May 1 untill September 30 ranged between 6 and 19 µg.$m^{-3}$ with an increasing trend. Although differences among individual months were not statistically significant, there was a trend for maximum values occurring in May and minimum values in July. The highest $NO_2$ concentration were consistently seen at 1,500 m a.s.l. of the "Lomnica" vertical transect. Sulfur dioxide concentrations measured at the same points showed rather stagnant course, with the averages arround 5 µg $m^{-3}$. Similarly to $NO_2$, the highest values occurred in May, and the lowest in July. Constantly the highest $SO_2$ concentrations were determined at the same sites as as for $NO_2$, at Start (1,200 m a.s.l.) and at timber line (1,500 m a.s.l.) of the "Lomnica" transect.

### 3.2.3 Ozone

Ozone concentrations in the Tatras showed high spatial and temporal heterogenity. Average annual $O_3$ concentrations in 1998-2003 during growing seasons was 89 µg $m^{-3}$. Table 2 shows annual averages, standard deviation, number of monitored localities and maximum value in each year of observations.

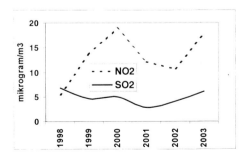

**Fig. 4.** Annual concentration of $SO_2$ and $NO_2$ in µg $m^{-3}$ by passive samplers, 1998-2003.

**Table 2.** Seasonal ozone concentrations (μg m$^{-3}$), reported as average (x) standard deviation (s$_x$), maximum 2-week average, elevation of the highest value, and number of monitoring plots (n) for the Ogawa passive sampler network.

| year | 1998 | 1999 | 2000 | 2001 | 2002 | 2003 |
|---|---|---|---|---|---|---|
| x | 93.8 | 86.1 | 93.6 | 84.9 | 81.6 | 92.0 |
| s$_x$ | 14.2 | 12.5 | 13.4 | 14.1 | 14.2 | 17.7 |
| max | 189.8 | 156.8 | 192 | 144.7 | 154.0 | 151.4 |
| altitude | 1200 | 1200 | 1200 | 1700 | 1500 | 2600 |
| n | 15 | 18 | 24 | 22 | 23 | 23 |

Seasonal course showed two peaks, first in early spring and the second in late summer. Very high O$_3$ concentrations exceeding 100 μg m$^{-3}$ were registered on the south oriented slopes. Differences in O$_3$ concentrations between the north and south oriented slopes were statisticaly significant (Godzik et al. 2003). Seasonal averages on most of the monitoring sites exceeded quality standard (UN ECE) for natural vegetation (50 μg m$^{-3}$).

Data derived from the UV absorption real-time concentration O$_3$ monitors showed all year arroud variability, diurnal changes and extreme values. Fig. 5a presents annual course of O$_3$ values from the "Lomnica" transect in 2003 smoothed by 31 day averages. It is evident that O$_3$ concentration increased with altitude and that two maxima, one in spring and the second in summer, occured. At the highest elevations spring peaks shifted to the early spring time and could be caused by stratospheric intrusion of O$_3$ to the troposphere. Diurnal curve (Fig. 5b) of O$_3$ concentrations showed also some differences among localities. At low elevation ozone reached its maximum during noon, and minimum during night. With the altitude, the diurnal curves became more flat, and difference between diurnal maximum and minimum disappeared.

Potential impact of O$_3$ on vegetation was calculated by the AOT 40 index. Critical load of 10, 000 ppb.h at low elevations was reached in mid of vegetation period. At the 1,200 m a.s.l. altitude, the critical load for O$_3$ was exceeded three-fold and in 1,700 m a.s.l. even five-fold.

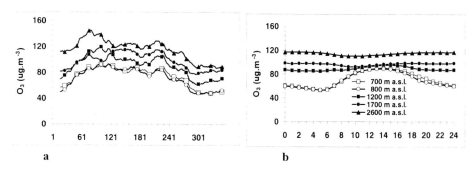

**Fig. 5.** Annual (a) and diurnal (b) distribution of O$_3$ concentrations in 2003 along the "Lomnica" vertical transect, data smoothed by 31 day average. Legend in Fig. 5b applies also to Fig. 5a.

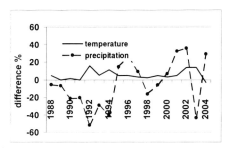

**Fig. 6.** Relative difference among recent years (1988-2004) and meteorological standard (average 1930-1960) in mean temperature and precipitation sum during May 1 and August 31.

## 3.3 Climate

Average temperature and a precipitation sum were calculated for the summer season (May 1 – August 31). Comparison with the long term averages (Koncek 1974) confirmed a presence of elevated temperature and precipitation deficit in most summers since 1988, especially on the south oriented slopes. Fig. 6 presents relative difference between individual seasons from 1988 and standard (average 1930-1960) in Tatranska Lomnica at 840 m a.s.l. It is clear, that except 2004, all monitored years had temperature above the standard. Precipitation was below the standard in 70% of the observed years. The stress causing warm and dry growing seasons occured in 1988-1994 with the most critical situation in 1992 when precipitation dropped nearly to 50% of the standard amount. During that year the average temperature for the growing season was nearly 2 °C higher than the long-term average. Another very dry and warm period was during 1998-2000.

## 4. Discussion

Increased average defoliation indicates forest health deterioration. However, a relative increase in defoliation classes did not fit a concept of a gradual rise of defoliation ending by death of trees. Alarming is a fact that almost 20% of trees assesed in first campaign in 1990s died. Aerial photographs confirmed problems caused by severe fragmentation of the formerly compact stands. Sudden exposition of spruce trunks to sunlight is a stress encouraging insect attacks and outbreaks leading to extention of gaps and further forest stands fragmentation. New forest edges are also very vulnerable to the wind damage. Also the National Forest Monitoring Office in 2003 reported the Tatra region among those with the permantly poor health condition (Bucha et al. 2003).

According to the regional air pollution and acidity of precipitation the Slovak Republic is situated within a territory with the worst environmental conditions in Europe (SHMI 2000). Air quality data confirmed a general European trend of rapid reduction of sulfur compounds concentrations and concentrations of nitrogen compounds decreasing muchless (Ågren 2003). Nitrogen oxides play an important role in $O_3$ formation. Based on the national inventory the current NOx concentrations clearly indicate that large amount of N is transported from the transboundary sources (Bicarova and Fleischer 2004). Wet deposition of total nitrogen exceeded the critical load for that element. The highest nitrate concentrations in precipitation occurred in spring when also the highest $O_3$ levels were taking place. Therefore the just developing forest vegetation was exposed to multiple stresses. Large number of plants showed injury symptoms like necrosis, stippling, bronzing, early senescence, etc. Among the forest tree species, $O_3$ injury symptoms showed *Sorbus aucuparia, Betula carpatica, Larix europea* and especially *Pinus cembra*. The most dominant tree species, *Picea abies*, showed injury symptoms only sporadically. According to Huttunen (2002), *Picea abies* belongs to the risk category trees and absence of the visible $O_3$ symptoms not necessarily means no injury. Ozone concentration data derived from the collocated passive samplers and active monitors showed rather large differences seen especially during the 1998-2000 period with limited possibilities for calibration. We expect that passive samplers might indicate besides ozone also other oxidants, e.g. OH radicals, and therefore overestimate $O_3$.

Extreme weather during warm and dry periods caused stress on trees which became attractive to bark beetles allowing enormous development of population. Situation in 2003 reminded the 1992 large insect outbreak, however, fortunately the 2004 summer 2004 was cool and wet. Despite this fact, the bark beetle population increased in central Europe and caused a huge risk for the Norway spruce population. Incidental felling caused by insect outbreaks dramatically increased during last years (Grodzki and Jachym 2004) confirming the international scope of the problem.

## 5. Conclusion

Current forest decline of the Tatra forests is caused by synergetic effect of polluted air and climate extremes on naturally very vulnerable spruce ecosystems. Dieback and fragmentation of formerly continuous forest have been reducing hydrological, erosion control, climatoterapeutical, recreational, aesthetical and nature conservation function of forests in the National park. At present more endagered are the non-native stands, with allochthonous Norway spruce and homogenous vertical and spatial structure heavily affected by bark beetle outbreaks. If dry and warm summers continue, uncontrolled outbreaks of insect may severely damage valuable natural stands.

Acknowledgements. The present study were partially financed through funds from the USDA Forest Service International Programs (grant No FG-PO-401) and Polish Scientific Research Committee (grant No 3 P04G 049 23).

## References

Ågren C (2003) The downward trend continues. Acid News, No. 4, 22
Bucha T, Mindas J, Pajtik J, Pavlenda P, Pavlendova H, Priwitzer T, Istona J, Pacutova M (2002) Forest healh status in Slovakia. Report from forest monitoring 2002. Forest Reseach Institute

Zvolen, 96 p

Bucha T, Durkovicova J, Istona J, Mankovska B, Mindas J, Pacutova M, Pajtik J, Pavlenda P, Priwitzer T, Stancikova A (2003) Forest health status in Slovakia. Report from forest monitoring 2003. Forest Reseach Institute Zvolen, 92 p

Bytnerowicz A, Badea O, Fleischer P, Godzik B, Grodzinska K (2004) Science, land management and policy in international studies on effect of air pollution on Carpathian forest ecosystems. Scand J For Res 19 (Suppl 4):129-137

Fleischer P, Koren M Jr (1993) Application of GIS and remote sensing for forest health detection in Tatra National Park. International symposium application of remote sensing in forestry. Technical University Zvolen, pp 93-100

Fleischer P, Koren M (1995) Forest health conditions in Tatra Biosphere Reserve. Ekologia (Bratislava), Vol.14, 1995/4:445-459

Godzik B, Fleischer P, Grodzinska K, Bytnerowicz A, Matsumoto Y (2003) Long term effects of air pollution on spruce forest in the Tatra Mts (Wester Carpathians) – ozone and vegetation study. Ekologia (Bratislava) 22, Supplement 1/2003:80-94

Grodzki W, Jachym M (2004) Biotic damage in forests. Proceedings from IUFRO WP 7.03.10 meeting. Matrafured, Hungary, Sept. 16, 2004, in print

Huttunen S, Manninen S, Timonen U (2002) Ozone effects on forest vegetation in Europe. In: Szaro R C, Bytnerowicz A, Oszlanzyi J (Eds) Effects of air pollution on forest health and biodiversity in forest of the Carpathian mountains. IOS Press, Amsterdam, pp 43-49

ICP (1998) Manual on methods and criteria for harmonized sampling, assessment, monitoring and analysis of the effect of air pollution on forests. UN EC for Europe, Convention on long-range transboundary air pollution. Programme Coordinating Centre, BFH, Hamburg, Germany

Koncek M, Bohus I, Briedon V, Chomicz K, Intribus R, Knazovicky L, Kolodziejek M, Kurpelova M, Murinova G, Myczkowski S, Orlicz M, Orliczowa J, Otruba J, Pacl J, Peterka V, Petrovic S, Plesnik P, Pulina M, Smolen F, Sokolowska J, Samaj F, Tomlain J, Volfova E, Wisniewski W, Wit-Jozwikowa K, Zych S, Zak B (1974) Climate of the Tatras. Veda, Bratislava (in Slovak), 855 p

Koutrakis PJ, Wolfson JM, Bunyaviroch A, Froelich SE, Hirano K, Mulik JD (1993) Measurement of ambient ozone using a nitrite coated filter. Annal Chem 65:210-214

Krochmal D, Kalina A (1997) A method of nitrogen dioxide and sulfur dioxide determination in ambient air by use of passive samplers and ion chromatography. Atmos Environ 31:3473-3480

Kunca V, Skvarenina J, Fleischer P, Celer S, Viglasky J (2003) Concept of critical loads applied in landscape ecology on an example of the geomorphological unit Tatry. Ekologia (Bratislava), Vol. 22, Supplement 2/2003:349-360

SHMI (2000) Air pollution in the Slovak republic. Slovak Hydrometeorological Institute, Bratislava, 122 p

Svedrup HU, de Vries W, Henriksen A (1990) Mapping critical loads. Nordic Council of Ministers, Copenhagen, 124 p

UN-ECE (1993) Mapping critical levels/loads. Federal environmental agency, Berlin

UNESCO (2003) Five transboundary biosphere reserve in Europe. Technical notes. UNESCO, Paris, 95 p

# IV. Genetics and Molecular Biology for Functioning Improvement

# MAPK signalling and plant cell survival in response to oxidative environmental stress

Marcus A. Samuel, Godfrey P. Miles, and Brian E. Ellis

Michael Smith Laboratories, University of British Columbia, Vancouver BC V6T 1Z4 Canada

**Summary.** Plant cells must constantly monitor and manage the levels of reactive oxygen species (ROS) accumulating within their cytosol and organelles. Extensive enzyme and metabolite-based antioxidant systems help regulate redox homeostasis, but how these systems are regulated in response to ROS challenge has remained obscure. One of the earliest responses to ROS perturbations in plants cells is the activation of specific mitogen-activated protein (MAP) kinases. Since MAPkinase cascades are involved in many aspects of eukaryotic signal transduction, these ROS-responsive kinases seem likely to form a key part of the antioxidant regulation network. Using ozone; a potent ROS generator and a phytotoxic air pollutant, we have demonstrated rapid and transient activation of MAPKs in various plant species. Genetic manipulation of the ROS-responsive MAPKs, SIPK/AtMPK6 in tobacco and Arabidopsis respectively, render the plants hypersensitive to ozone. Examination of MAPK activation profiles following ROS stress in SIPK/MPK6 transgenic plants reveals an interesting interplay between SIPK/MPK6 and another MAPK, WIPK/MPK3. Proteomic analysis of specifically suppressed transgenic lines has provided novel insights into the specific metabolic networks controlled by the ROS-responsive MAPK cascades during the cell's response to oxidative challenges. We have also been able to extend the same findings to poplar through heterologous expression of SIPK, which provides evidence for the potential of exploiting MAPK pathways for crop improvement.

**Key words.** Ozone, ROS, SIPK, WIPK, Cell death

## 1. Introduction

Plant cells function in an oxygen-rich environment where they are constantly exposed to reactive derivatives of molecular oxygen. These 'reactive oxygen species' (ROS) are generated from normal metabolic processes in plants, including the intense electron fluxes associated with both mitochondrial respiration and photosynthesis. Plant cells thus need to scavenge ROS, and their reaction products, a process that is normally achieved through constitutive accumulation of anti-oxidant metabolites (e.g. ascorbate, tocopherols, flavonoids, glutathione) and enzymic scavengers of ROS (e.g. ascorbate peroxidase, catalase, superoxide dismutase).

However, when plants are subjected to intense environmental insults such as elevated levels of ground-level ozone ($O_3$) or ultra-violet radiation (UVR) or are attacked by pathogens, abnormal amounts of ROS are produced, a process termed the 'oxidative burst'. This rapid production of ROS by challenged plant cells is a common response to many biotic and abiotic stressors (for review see Lamb and Dixon 1997), but it remains unclear whether its primary role is immediate defence or in signaling to other regions of the cell/tissue. ROS accumulation ultimately helps determine cell fate by triggering the cell's protective or cell death mechanisms. Cellular scavenging systems are usually hyper-induced to help keep the redox perturbations under control and thus protect cells against the potential injurious effects of reactions of ROS with membranes and macromolecules under stress conditions. If these systems are overwhelmed, however, and the cell cannot maintain homeostasis, a genetically programmed cell death program becomes activated. Exactly how ROS levels are monitored by plant cells, and how different ROS species mediate physiological responses, remains unclear.

To dissect this complex pattern of ROS-induced responses it is advantageous to use an elicitor that can generate ROS in plants on demand. One such elicitor is ozone. Rao and Davis (2000) demonstrated that ozone can function as an efficient cross-inducer, whereby the plant's response to ozone challenge mimics the plant responses to any stress that generates an oxidative burst. The similarity between ozone and pathogen-induced responses, including induction of an ROS burst and cell death, makes ozone a powerful tool for probing ROS-induced signalling pathways.

Apart from being a potent ROS generator, ozone is one of the most widespread phytotoxic air pollutants (Krupa and Manning 1988). It forms an important constituent of photochemical smog pollution and is the most toxic component of the oxidizing air pollutants. The detrimental effects of this pollutant on plant life include diminished photosynthesis, growth-rate retardation, altered patterns of carbon allocation, accelerated foliar senescence, and programmed cell death (PCD).

Ozone enters the leaf mesophyll via the stomata and diffuses through inner air spaces to reach the cell wall and plasmalemma (Sharma et al. 1997), where it is rapidly converted to reactive oxygen species such as $O\bullet_2^-$, $HO\bullet$, and $H_2O_2$, either by contact with water, the plasmalemma or other cellular components (Pellinen et al. 1999). Although long-term consequences of the increased ROS created by ozone have been extensively studied, little is known about the initial events that take place in plant cells in response to redox perturbations created by ozone. Identification of the components that regulate the transmission of ozone signal into the cell could not only allow us to dissect ROS signaling mechanisms during stress responses, but also reveal pathways that could be genetically manipulated to create plants that could better withstand adverse environmental conditions, including increased ozone levels.

In this paper we discuss the identification of one major component of ozone-induced signal transduction in plants, the manipulation of which was sufficient to alter the plant's sensitivity to ozone.

## 2. Ozone exposure leads to rapid activation of MAPKs in plants

It is generally accepted that the plant tissue response to ozone is initiated by the rapid accumulation of ROS (Schraudner et al. 1998). It is interesting that in mammalian models a number of early signal transduction components such as PK-C (Taher et al. 1993),

MAPK (Guyton et al. 1996), and phosphoprotein phosphatase (Caselli et al. 1998) respond rapidly to ROS stress. In plants, rapid activation of MAPKs can also be induced following treatment with various biotic and abiotic elicitors (see Zhang and Klessig, 2001 for review), all of which have been associated with induction of ROS accumulation.

The fact that a multiplicity of stresses can elicit activation of MAPK, and that many of these also involve triggering of an oxidative burst, suggests that the redox stress created by ozone in plant cells is likely to activate protein kinase cascades involving MAPKs. These, in turn, could converge on a number of targets, including enzymes, transporters and transcription factors, modifying their activities through phosphorylation (Morris, 2001).

When tobacco (Xanthi nc.) plants were exposed to ozone for a brief period and proteins were tested for MAPK activation through an in-gel kinase assay, strong MAPK activation was observed within minutes (Samuel et al. 2000) following ozone exposure. Through use of commercial anti-MAPK antibodies we showed that the responsive MAPK belonged to the ERK class of MAPKs. When these samples were probed with an antibody that specifically recognizes only active (phosphorylated) forms of MAPKs, we detected activation of two MAPKs with approximate molecular weights of 46kD and 44kD. Further characterization of this activation process using suspension cultured cells revealed that this activation was dependent on calcium influx and at least one upstream MAPK kinase. The activation by ozone was blocked completely by a free radical scavenging reagent, indicative of a requirement for ROS accumulation in the activation process, and we were able to show that hydrogen peroxide and superoxide anion radicals can substitute for ozone as the activation stimulus. Use of protein-specific antibodies allowed us to determine that the ozone-induced p46 MAPK was SIPK, a MAPK originally identified as a salicylate-induced protein kinase (Zhang and Klessig 1997). However, in this case, SIPK activation by ozone does not require salicylate as an intermediary.

Further characterization of the upstream components of the ozone-induced MAPK activation process revealed that this MAPK activation required receptor action since suramin, an inhibitor of mammalian receptor action, completely interdicted the ozone-induced signal from reaching the MAPKs (Miles et al. 2002). This indicates that the ozone-induced signal might normally be initiated at the cell membrane, possibly through oxidative activation of membrane receptors.

The observed MAPK activation of the ozone is not restricted to tobacco and its relatives, since ozone treatment of poplar and Arabidopsis, as well as plants as remotely related as conifers (*Picea*) and moss (*Physcomitrella*), resulted in rapid activation of an approximately 46 kDa kinase (Miles and Ellis, unpublished). The widespread and conserved ability of ozone to activate SIPK and its orthologs thus indicates that this protein kinase probably acts as an important transducer of redox stress signals in all plant cells.

## 3. Manipulation of SIPK leads to hypersensitivity to ozone

Despite the fact that ROS and ozone can induce MAPK activation, there has been no direct genetic evidence linking MAPK activation with the cell death induced by ozone. In order to examine the possible role of SIPK in controlling the ozone-induced cell death process in tobacco, transgenic tobacco lines were created in which SIPK was either over-expressed under the control of a CaMV35S promoter, or completely suppressed using an RNA interference-mediated approach (Samuel and Ellis, 2002). When these plants were exposed to an ozone concentration that is non-lethal in wild type (Xanthi nc.) plants, both suppression and over-expression of SIPK were found to create ozone hypersensitivity in these lines.

Analysis of the MAPK activation profiles in ROS-stressed transgenic and wild-type plants revealed a striking interplay between SIPK and a second MAPK (wound-induced protein kinase [WIPK]) in the different tobacco kinotypes. During continuous ozone exposure of SIPK-over-expressing lines, SIPK activation was abnormally prolonged (~ 4h) compared to the response in wild-type plants, where this kinase normally displays a transient activation pattern, declining to background levels after ~1 hr. The converse pattern was observed in SIPK-suppressed lines, where continuous ozone exposure resulted in weak or no activation of SIPK, as would be anticipated for a SIPK-suppressed line, accompanied by elevated and prolonged (~8h) activation of WIPK. One explanation for this pattern may be that SIPK activated by ROS in tobacco cells normally controls the inactivation of phosphorylated WIPK.

In order to better understand why both up- and down-regulation of SIPK expression should lead to increased sensitivity to ozone, we examined the ROS accumulation levels in the transgenic lines following ozone exposure. While WT plants showed little or no change in ROS accumulation at the ozone concentration used (500 ppb), the transgenic lines (both SIPK-suppressed and SIPK-overexpressing) displayed strong accumulation of ROS. This indicates that SIPK is normally involved in some manner in controlling ROS scavenging in the cell. The manipulation of SIPK expression apparently leads to redox imbalances that may render the tissue more susceptible to any impinging stress. In accordance with this idea, we have observed that expression of one major antioxidant gene (ascorbate peroxidase) is rapidly induced in SIPK-overexpressing lines in response to ozone challenge, whereas its induction is quite delayed in the SIPK-suppressed line. Transcriptome analysis of WT and SIPK-suppressed line following ozone exposure, using the 10K cDNA arrays available through The Institute of Genomics Research (TIGR), is underway to identify altered expression patterns of genes that could contribute to the observed phenotypes in these lines (Hall and Ellis, unpublished).

Recently, it was reported that activation of SIPK through a gain-of-function approach leads to rapid production of ethylene (Kim et al. 2003), a phytohormone known to be an important component/regulator of ozone-induced cell death. Plant tissues challenged with acutely toxic levels of ozone rapidly produce a burst of ethylene ("stress ethylene") (Mehlhorn and Wellburn 1987). This ethylene burst has been suggested to promote ozone-induced lesion formation by stimulating cell-to-cell propagation of cell death, since blocking ethylene production either by application of an ethylene synthesis inhibitor, or by genetic suppression of ACC synthase, can reduce ozone-induced tissue damage (Nakajima et al. 2002). The increased ozone sensitivity observed in the SIPK-overexpression line could therefore be due to enhanced production of an ethylene burst. Congruent with this hypothesis, we have observed rapid production of ethylene in SIPK-overexpressing tobacco plants challenged with ozone at concentrations that failed to stimulate any increase in ethylene release in WT plants (Samuel and Ellis, unpublished). This model is also supported by the recent report that AtMPK6, the Arabidopsis orthologue of SIPK, is able to phosphorylate and stabilize specific ACS isoforms, thus providing a biochemical rationale for the stimulation of ethylene production by an ROS-activated plant MAPK (Liu and Zhang 2004).

Whether it acts at one or multiple levels in the ozone-induced signal transduction pathway, we have shown that SIPK is a central integrator of ROS signals generated by ozone. While some of the upstream components of SIPK have been identified, identification of further downstream targets of SIPK would lead to better understanding of the complex nature of ozone signalling in plants.

## 4. Orthologous pathway to SIPK is functionally similar in Arabidopsis

The availability of a complete genome sequence, as well as tools to study the associated transcriptome and proteome, make Arabidopsis thaliana the obvious choice for a model plant to probe for components of ozone signalling. In Arabidopsis, ROS-induced signaling has been shown to activate AtMPK6 and AtMPK3, which are the orthologues of tobacco SIPK and WIPK (Kovtun et al. 2000). The phenotype observed in tobacco transgenic lines manipulated in SIPK expression prompted us to investigate whether the ozone response was transmitted through an orthologous pathway in Arabidopsis. Indeed, we observe strong activation of AtMPK6 in Arabidopsis following ozone fumigation, and when AtMPK6 expression was silenced using an RNAi-mediated approach, the plants displayed hypersensitivity to ozone similar to that observed in SIPK-suppressed tobacco plants (Miles and Ellis, unpublished). The increased damage in leaves of the MPK6-suppressed line was accompanied by accumulation of higher levels of hydrogen peroxide. The MPK6-RNAi genotype also displayed a more intense and prolonged activation of MPK3 (WIPK ortholog) compared to that seen in WT plants, again suggesting a possible MPK3-regulatory role for MPK6. Interestingly, when an MPK3 loss-of-function genotype was tested for ozone sensitivity, it was found to be even more sensitive to ozone than were the MPK6-suppressed lines. Ozone also induced prolonged activation of MPK6 in the MPK3-silenced genotype (Miles and Ellis, unpublished), indicating a possible reciprocal regulation between these two kinases following activation by ozone exposure.

A proteomic comparison of the WT and AtMPK6-suppressed Arabidopsis plants following 8 hr ozone exposure revealed significant changes in the abundance of a number of antioxidant enzymes associated with the absence of MPK6. A number of the differentially expressed proteins identified are predicted to be targeted either to the chloroplast or the mitochondria, the primary energy generation and ROS generation centres of the cell (Miles and Ellis, unpublished). This pattern further supports our theory that the ozone-induced MPK6/SIPK cascade converges on proteins that control the activity and/or transcription of antioxidant enzymes. Microarray analysis is being performed using MPK6- and MPK3-silenced plants to distinguish the transcriptome changes controlled by each of these kinases.

## 5. Suppression of cognate upstream MAPKK (AtMKK5) of ATMPK6 renders the plants ozone-sensitive

In Arabidopsis, ROS-induced signalling has been proposed to flow through a protein phosphorylation cascade involving the mitogen-activated protein kinase kinases (MAPKKs) MKK4 and/or MKK5, since these upstream kinases are capable of activating AtMPK6 and AtMPK3 (Ren et al. 2002). However, no genetic evidence is available that confirms that the ROS signal is actually transmitted through these upstream MAPKKs. To examine this question, we used RNA interference-based suppression to create stably transformed MKK5-RNAi plants. These were found to be highly susceptible to ozone injury (Miles et al. unpublished).

Fumigation with 500 ppb ozone produced extensive leaf damage and concomitant hydrogen peroxide accumulation in MKK5-RNAi plants relative to wild-type plants. When ozone-induced activation of MPK6/MPK3 was tested in this background, signal transmission to both of the MAPKs was strongly attenuated in the MKK5-suppressed plants.

Thus, although both MKK4 and MKK5 are reportedly capable of activating MPK6, suppression of MKK5 alone created an ozone-sensitivity phenotype, which leads us to conclude that, at least with respect to ozone, these closely related MAPKKs do not play fully redundant roles.

## 6. SIPK over-expression also alters sensitivity of poplar to ozone

Poplar is fast becoming a model tree system for studying stress and adaptation responses of trees. The fact that ozone pollution is considered to be partly responsible for the forest decline phenomenon seen in North America and Europe (Brown et al. 1995), makes it imperative to gain a thorough understanding of how ozone stress is sensed and transmitted in trees. This would allow us to identify various players in the pathway that could be genetically manipulated to increase overall forest health. The ROS-activated MAPK cascade components appear to be broadly functional across plant taxa. The Arabidopsis MAPKKs, AtMKK4 and AtMKK5, have been shown to be functional in tobacco (Ren et al. 2002), and recently Shou et al. (2004) demonstrated that expression of a tobacco MAPKKK, NPK1, in maize leads to enhanced freezing tolerance.

To explore this possibility in tree species, tobacco SIPK was ectopically over-expressed in poplar, and these transgenic plants were assessed for their altered sensitivity to ozone. A number of transgenic poplar lines that stably over-expressed SIPK to various levels were identified (Gill, Samuel and Ellis, unpublished). When representative lines were tested for activation of ectopic SIPK by ozone, strong activation was observed 30 min following ozone exposure, accompanied by activation of endogenous poplar MAPKs. When these lines were tested for their sensitivity to ozone, lesions appeared consistently 12 hours after initiation of ozone exposure in the SIPK-OX lines, along with concomitant accumulation of hydrogen peroxide around the lesions, while the WT plants lacked any foliar lesions.

The observed functional similarity of SIPK in both tobacco and poplar provides strong evidence for a MAPK matrix that serves as central oxidant-induced signal transducers in both herbaceous and tree species.

## 7. Conclusion

Based on work both from our laboratory and from other research groups, it is clear that the ozone signal is transmitted through plant cells via SIPK/MPK6 and WIPK/MPK3. It appears that sustained activity of either of these kinases (SIPK and WIPK), commits the cells to a cell death pathway. However, what lies between the activation of these kinases and the ensuing cell death outcome remains unclear. The combination of microarray profiling and phosphoprotein profiling approaches with specific biochemical validation should ultimately enable us to determine which other molecules play essential roles in sensing and responding to oxidant stress, as the cell attempts to maintain its homeostatic balance in the face of either chronic or acute environmental challenges.

Acknowledgements. We thank Y. Ohashi for providing tobacco MAPK antibodies. Funding and technical support for this research were provided by Natural Sciences and Engineering Research Council and Genome Canada. We also thank the Japan Society for the Promotion of Science for a travel fellowship to BEE.

## References

Brown M, Cox R, Bull KR, Dyke H, Sanders G, Fowler D, Smith R, Ashmore MR (1995) Quantifying the fine scale (1KM x 1KM) exposure, dose and effects of ozone, part 2 estimating yield losses for agricultural crops. Water Air Soil Pollut 85:1485-1490

Caselli A, Marzocchini R, Camici G, Manao G, Moneti G, Pieraccini G, Ramponi G (1998) The inactivation mechanism of low molecular weight phosphotyrosine-protein phosphatase by $H_2O_2$. J Biol Chem 273:32554-32560

Guyton KZ, Liu Y, Gorospe M, Xu Q, Holbrook NJ (1996) Activation of mitogen-activated protein kinase by $H_2O_2$. J Biol Chem 271:4138-4142

Kim CY, Liu Y, Thorne ET, Yang H, Fukushige H, Gassmann W, Hildebrand D, Sharp RE, Zhang S (2003) Activation of a stress-responsive mitogen-activated protein kinase cascade induces the biosynthesis of ethylene in plants. Plant Cell 15:2707-2718

Kovtun Y, Chiu WL, Tena G, Sheen J (2000) Functional analysis of oxidative stress-activated mitogen-activated protein kinase cascade in plants. Proc Natl Acad Sci USA 97:2940-2945

Krupa SV, Manning WJ (1988) Atmospheric ozone, formation and effects on vegetation. Environ Pollut 50:101-137

Lamb C, Dixon RA (1997) The oxidative burst in plant disease resistance. Annu Rev Plant Physiol 48:251-275

Liu Y, Zhang S (2004) Phosphorylation of 1-aminocyclopropane-1-carboxylic acidby MPK6, a stress-responsive mitogen-activated protein kinase, induces ethylene biosynthesis in Arabidopsis. Plant Cell 16

Mehlhorn H, Wellburn AR (1987) Stress ethylene formation determines plant sensitivity to ozone. Nature 327:417-418

Miles GP, Samuel MA, Ellis BE (2002) Suramin inhibits oxidant signaling in tobacco suspension-cultured cells. Plant Cell Environ 25:521-527

Morris PC (2001) MAP kinase signal transduction pathways in plants. New Phytologist 151:67-89

Nakajima N, Itoh T, Takikawa S, Asai N, Tamaoki M, Aono M, Kubo A, Azumi Y, Kamada H, Saji H (2002) Improvement in ozone tolerance of tobacco plants with an antisense DNA for 1-aminocyclopropane-1-carboxylate synthase. Plant Cell Environ 25:727-736

Rao MV, Koch JR, Davis KR (2000) Ozone, a tool for probing programmed cell death in plants. Plant Mol Biol 44:345-58

Ren D, Yang H, Zhang S (2002) Cell death mediated by MAPK is associated with hydrogen peroxide production in Arabidopsis. J Biol Chem 277:559-565

Samuel MA, Ellis BE (2002) Double jeopardy, both overexpression and suppression of a redox- activated plant mitogen-activated protein kinase render tobacco plants ozone sensitive. Plant Cell 14:2059-2069

Samuel MA, Miles GP, Ellis BE (2000) Ozone treatment rapidly activates MAP kinase signalling in plants. Plant J 22:367-376

Schraudner M, Moeder W, Wiese C, Camp WV (1998) Ozone-induced oxidative burst in the ozone biomonitor plant, tobacco Bel W3. Plant J 16:235-245

Sharma YK, Davis KR (1997) The effects of ozone on antioxidant responses in plants. Free Radical Biol Med 23:480-488

Shou H, Bordallo P, Fan JB, Yeakley JM, Bibikova M, Sheen J, Wang K. (2004) Expression of an active tobacco mitogen-activated protein kinase kinase kinase enhances freezing tolerance in transgenic maize. Proc Nat Acad Sci USA 101:3298-3303

Taher MM, Garcia JG, Natarajan V (1993) Hydroperoxide-induced diacylglycerol formation and protein kinase C activation in vascular endothelial cells. Arch Biochem Biophys 303:260-266

Zhang S, Klessig DF (1997) Salicylic acid activates a 48kD MAP kinase in tobacco. Plant Cell 9:809-824

Zhang S, Klessig DF (2001) MAPK cascades in plant defense signaling. Trends Plant Sci 6:520-527

# Expression of cyanobacterial *ictB* in higher plants enhanced photosynthesis and growth

Judy Lieman-Hurwitz[1], Leonid Asipov[1], Shimon Rachmilevitch[1], Yehouda Marcus[2] and Aaron Kaplan[1]

[1]Department of Plant and Environmental Sciences, The Hebrew University of Jerusalem, 91904 Jerusalem, Israel
[2]Department of Plant Sciences, Tel Aviv University, 69978 Tel Aviv, Israel

**Summary.** Under many environmental conditions plant photosynthesis and growth are limited by the availability of $CO_2$ at the site of ribulose 1,5-bisphosphate carboxylase/oxygenase (RubisCO). We expressed *ictB*, a gene involved in $HCO_3^-$ accumulation in *Synechococcus* sp. PCC7942, in higher plants. The transgenic *Arabidopsis thaliana* and *Nicotiana tabacum* plants exhibited significantly faster photosynthetic rates than the wild types under limiting, but not under saturating $CO_2$ concentrations. Similar results were obtained in *Arabidopsis* plants bearing *ictB* from *Anabaena* sp. PCC7120. Growth of transgenic *A. thaliana* plants maintained under low humidity was considerably faster than that of the wild type. There was no difference in the amount of RubisCO or the activity of the enzyme activated *in vitro* in the wild types and the transgenic plants. In contrast, the *in vivo* RubisCO activity, without prior activation of the enzyme, in plants grown under low humidity was considerably higher in *ictB*-expressing plants than in their wild types. The $CO_2$ compensation point in the transgenic plants was lower than in the wild types suggesting a higher $CO_2$ concentration in close proximity to RubisCO. This may explain the higher activation level of RubisCO and enhanced photosynthesis and growth in the transgenic plants. These data indicated a potential use of *ictB* for the stimulation of crop yield.

**Key words.** Growth, IctB, Inorganic carbon, Photosynthesis

## 1. Introduction

Plants that belong to the physiological $C_4$ or the Crassulacean acid metabolism groups possess biochemical $CO_2$ concentrating mechanisms (CCM, Cushman and Bohnert 2000; Hatch 1992) whereas in many photosynthetic microorganisms a biophysical CCM is functional (Kaplan and Reinhold 1999). CCM mechanisms enable these organisms to raise the concentration of $CO_2$ in close proximity to RubisCO and hence overcome, at least partly, the low affinity of the enzyme for $CO_2$. In contrast, plants that belong to the $C_3$ group, including most crop plants, do not possess this ability. Therefore, under many environmental conditions plant photosynthesis is rate-limited by the concentration of $CO_2$ at the carboxylation site and/or by the activity of RubisCO. Attempts are being made to raise the apparent photosynthetic affinity of $C_3$ plants for $CO_2$ by various biotechnological approaches. These include expression of genes involved in $C_4$ metabolism within $C_3$ plants (Matsuoka et al. 2001; Surridge 2002); a search for a RubisCO that exhibits an elevated specificity for $CO_2$ among natural photosynthetic populations (Tabita 1999); site directed modifications of the enzyme (Spreitzer and Salvucci 2002); and expression of a cyanobacterial gene encoding fructose-1,6/sedoheptulose-1,7-bisphospate phosphatase thereby raising the level of intermediates of the Calvin cycle (Miyagawa et al. 2001).

## 2. Results and discussion

High-$CO_2$-requiring mutants of the cyanobacterium *Synechococcus* sp. strain PCC 7942 (hereafter *Synechococcus*) were raised by transformation with an inactivation library. One of them, IL-2, was severely impaired in the ability to accumulate inorganic carbon internally, implicating IctB in this activity (Bonfil et al., 1998).

Analysis of the IctB sequence, well conserved among cyanobacteria, suggested that it is a hydrophobic protein presumably with 10 trans-membrane domains. Interestingly, a hydrophilic region (Fig. 1) showed very high conservation in homologous genes from various cyanobacteria. The exact role of IctB is not known since it was not possible to directly inactivate it or its homologue *slr1515* from *Synechocystis* sp. strain PCC 6803.

```
IctB                LVAVLGLEPLRVRVLSIFVGREDSSNNFRINVWLAVL
Anabaena 7120       LIAVLFVEPVRFRVLSIFADRQDSSNNFRRNVWDAVF
Nostoc punc.        LLAVIFVEPVRLRVFSIFADRQDSSNNFRRNVWDAVF
Trichodesmium       ILAVVLLEPLRDRVLSVFAGRQDSSNNFRMNVWMSVF
Slr1515             GGALIAVEPIRLRAMSIFAGREDSSNNFRINVWEGVK
Thermosynechococcus MGTIVSVPPLRERAASIFVARGDSSNNFRINVWMAVQ
Prochlorococcus 9313 VIAATQIEPIRTRITSLIAGRSDSSNNFRINVWLSSL
Synechococcus WH    ALAITQLDPIRTRVLSLVAGRGDSSNNFRINVWLAAI
                     :  : *:* *  *:.. * ******* *** .

IctB                QMIQDRPWLGIGPGNTAFNLVYPLYQQ-ARFTALS
Anabaena 7120       EMIRDRPIIGIGPGHNSFNKVYPLYQR-PRYSALS
Nostoc punc.        EMIRDRPIFGIGPGHNSFNKVYPLYQH-PRYTALS
Trichodesmium       DMIRDRPILGIGPGNDVFNKIYPLYQR-PRYSALS
Slr1515             AMIRARPIIGIGPGNEAFNQIYPYYMR-PRFTALS
Thermosynechococcus QMIWARPWLGIGPGNVAFNQIYPLYQVNVRFTALG
Prochlorococcus 9313 EMIQARPWLGIGPGNAAFNRIYPLFQQ-PKFNALS
Synechococcus WH    EMVQDRPWLGIGPGNAAFNSIYPLYQQ-PKFDALS
                    *:  **  :*****:  ** :**  :    :: **.
```

Fig. 1. Alignment of a hydrophilic region (positions 301-372) in IctB from Synechococcus PCC 7942 to homologous genes from Anabaena sp. strain PCC 7120, Nostoc punctiforme, Trichodesmium erythraeum, Slr1515 from Synechocystis PCC 6803, Thermosynechococcus elongates BP-1, Prochlorococcus marinus MIT9313 and Synechococcus WH 8102.

## 2.1 Photosynthetic performance

In an earlier study we showed enhanced photosynthesis under limiting $CO_2$ levels in *Arabidopsis* and tobacco plants expressing *ictB* from *Synechococcus* (Lieman-Hurwitz et al. 2003). In the present study we extended this analysis to *Arabidopsis* plants bearing *ictB* (all5073) from *Anabaena*. Transgenic plants bearing *ictB* from *Synechococcus* or *Anabaena* sp. strain PCC 7120 (hereafter *Anabaena*) were raised as described (Lieman-Hurwitz et al. 2003). We examined the rate of photosynthesis in the wild types and the transgenic plants as affected by intercellular $CO_2$ concentration. (Fig. 2, Lieman-Hurwitz et al. 2003).

Plants that expressed *ictB* exhibited higher apparent photosynthetic affinity to $CO_2$ than did the wild type. In most cases, at saturating $CO_2$ levels, the photosynthetic rates of transgenic and wild type plants were similar. This suggested that the ability to perform maximal photosynthesis was not affected by the expression of *ictB*. Results presented here, using the gene from *Anabaena*, were similar to those obtained in transgenic plants expressing *ictB* from *Synechococcus* (Lieman-Hurwitz et al. 2003) indicating that regardless of the source of *ictB*, its expression in *Arabidopsis* enhanced the photosynthetic activity at limiting $CO_2$ levels.

The mechanism whereby IctB stimulated photosynthesis in transgenic plants exposed to low $CO_2$ is not clear. However, two sets of independent experiments suggested a higher level of $CO_2$ at the site of RubisCO in these plants.

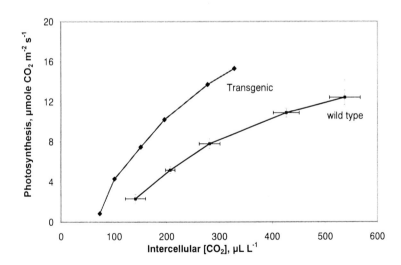

**Fig. 2.** The rate of photosynthesis as affected by intercellular $CO_2$ concentration in wild type Arabidopsis (average ± SD, n= 7) and one of the transgenic plants bearing ictB from *Anabaena*. Note, there was a large variation in the photosynthetic performance among the transgenic plants, some showed curves similar to the wild type, others showed enhanced photosynthesis, possibly related to the extent of expression of *ictB*. The experiments were performed with LicoR-6400, as in Lieman-Hurwitz et al. (2003).

**Table 1.** The $CO_2$ compensation points (in $\mu L\ L^{-1}$) in wild type (WT) and transgenic *Arabidopsis* and tobacco plants expressing *ictB* from *Synechococcus*. The data are presented as the average ± SE, n=18.

| Plant | $CO_2$ compensation point |
|---|---|
| *Arabidopsis* WT | 46.1 ± 1.1 |
| *Arabidopsis* - Transgenic | 40.1 ± 1.1 |
| Tobacco WT | 56.9 ± 1.6 |
| Tobacco –Transgenic | 47.6 ± 1.6 |

## 2.2 $CO_2$ compensation point

If *ictB* expression leads to an elevated $CO_2$ level at RubisCO sites within the transgenic plants, it would be expected to lower the $CO_2$ compensation point. We examined the $CO_2$ compensation point in wild type and transgenic *Arabidopsis* and tobacco plants by measuring the $CO_2$ exchange in plants exposed to a range of $CO_2$ concentrations between 0 and 150 $\mu l\ l^{-1}$ $CO_2$. In Table 1 we show that the average $CO_2$ compensation point was significantly ($P<0.01$) lower in transgenic *Arabidopsis* and tobacco plants than in the respective wild types. This is in agreement with the steeper initial slope of the curves relating $CO_2$ fixation to its concentration in the transgenic plants which express *ictB* than in the respective wild types (Fig. 2, Lieman-Hurwitz et al. 2003).

## 2.3 Activation state of RubisCO

The different slopes of the curves obtained in Fig. 2 could be due to a higher level of RubisCO activity in the transgenic plants (Poolman et al. 2000). We did not detect significant differences in the abundance of active sites of RubisCO per leaf surface area or per soluble proteins between the wild types and their respective *ictB*-expressing plants. To examine the possibility that RubisCO activity (per active site) was higher in the transgenic plants, we exposed neighboring leaves of wild types and of transgenic plants, of similar age, to identical ambient conditions of light intensity and orientation, temperature and relative humidity (either high or low) for several days. The leaves were excised three hours after onset of illumination in the growth chamber and placed in liquid nitrogen (see Lieman-Hurwitz et al. 2003). These experiments were performed on wild type and transgenic *Arabidopsis* and tobacco plants. As an example, we provide results obtained with the wild type and one of the transgenic tobacco plants (Fig. 3).

Following activation *in vitro* by the addition of $CO_2$ and $MgCl_2$, where RubisCO activity was close to its maximum (Spreitzer and Salvucci 2002; Marcus and Gurevitz 2000), there was no significant difference between the activities observed in the wild type and transgenic plants (Lieman-Hurwitz et al. 2003, Fig. 3) confirming that insertion of *ictB* did not alter the intrinsic properties of RubisCO. Under low humidity conditions the *in vivo* activity of RubisCO was about 40% higher in the transgenic than in the wild type plants over the entire range of $CO_2$ concentrations examined in the activity assays (Fig. 3). In Fig. 3 we show the activities of RubisCO exposed to different $CO_2$ concentrations in order to emphasize the consistency of the data, even at various $CO_2$ levels, rather than to provide a complete account of the kinetic parameters of activated and non-activated RubisCO from tobacco. Nevertheless, analysis of the kinetic parameters from experiments similar to that depicted in Fig. 3, indicated that while the Km($CO_2$) was scarcely affected by the expression of *ictB*, the Vmax of carboxylation, *in vivo*, was significantly higher in the *ictB*-expressing plants. The higher *in vivo* RubisCO activity, under the dry conditions where stomata conductance may limit $CO_2$ supply, is consistent with the steeper slope of the curve relating photosynthetic rate to intercellular $CO_2$ concentration (Fig. 2). Naturally, the *in vivo* RubisCO activities were lower than those depicted by the *in vitro* activated enzyme (Fig. 3). The reduced *in vivo* RubisCO activity in the dry vs. the high humidity grown wild type plants (Lieman-Hurwitz et al. 2003) is possibly due to lower internal $CO_2$ concentration imposed by the decreased stomatal conductance. These are also the conditions where the transgenic plants exhibited faster photosynthesis (Fig. 2) and growth (see below). In the absence of an independent method to directly

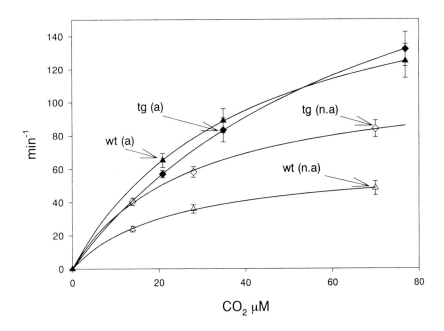

**Fig. 3.** RubisCO activity *in vivo* (non-activated, n.a, opened symbols) and following *in vitro* activation (closed symbols) in the wild type (wt) and transgenic (tg) tobacco plant as affected by the $CO_2$ concentration during the assay. Rate of carboxylation is in nmol $CO_2$ fixed nmol$^{-1}$ active sites min$^{-1}$, n=6.

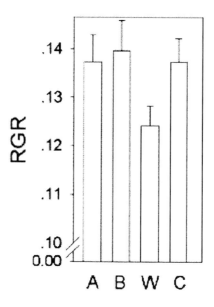

**Fig. 4.** Growth of transgenic (A, B and C) and WT (W) *Arabidopsis* plants expressing *ictB* from *Synechococcus*. Data are provided as the relative growth rate (RGR). The growth experiments were performed six times for 18 days each, n=18 (Lieman-Hurwitz et al. 2003).

determine the $CO_2$ concentration in close proximity to RubisCO, we must rely on measurements of the $CO_2$ compensation point (Table 1) and RubisCO activity *in vivo* as indicative of higher internal $CO_2$ level in the transgenic plants.

## 2.4 Growth experiments

In view of the enhanced photosynthesis in the transgenic plants under $CO_2$- limiting conditions (Lieman-Hurwitz et al. 2003), we examined their growth as affected by the relative humidity. Under low humidity (25-30%), where both wild type and transgenic plants grew slower than in humid conditions, the transgenic plants grew significantly faster ($p<0.03$) than the wild type. In Fig. 4 we provide the relative growth rates (RGR) over 18 days under dry conditions. Enhancement of growth of the transgenic plants was observed throughout the growth period and not at a particular phase of growth.

Data presented here suggested a potential use of *ictB* for the enhancement of crop yield.

Acknowledgments. Research in this laboratory is being supported by grants from the Israel Science foundation (ISF), German BMBF and the Deutch Foundation for applied sciences, the Hebrew University of Jerusalem.

# References

Bonfil DJ, Ronen-Tarazi M, Sultemeyer D, Lieman-Hurwitz J, Schatz D, Kaplan A (1998) A putative $HCO_3^-$ transporter in the cyanobacterium *Synechococcus* sp. strain PCC 7942. FEBS Lett 430:236-240

Cushman JC, Bohnert HJ (2000) Genomic approaches to plant stress tolerance. Curr Opin Plant Biol 3:117-124

Hatch MD (1992) C4 Photosynthesis: An unlikely process full of surprises. Plant Cell Physiol 33:333-342

Kaplan A, Reinhold L (1999) The $CO_2$ concentrating mechanisms in photosynthetic microorganisms. Annu Rev Plant Physiol 50:539-570

Lieman-Hurwitz J, Rachmilevitch S, Mittler R, Marcus Y, Kaplan A (2003) Enhanced photosynthesis and growth of transgenic plants that express *ictB*, a gene involved in $HCO_3^-$ accumulation in cyanobacteria. Plant Biotechnology J 1:43-50

Marcus Y, Gurevitz M (2000) Activation of cyanobacteria RuBP-carboxylase/oxygenase is facilitated by inorganic phosphate via two independent mechanisms. Eur J Biochem 267:5995-6003

Matsuoka M, Furbank RT, Fukayama H, Miyao M (2001) Molecular engineering of $C_4$ photosynthesis. Annu Rev Plant Physiol 52:297-314

Miyagawa Y, Tamoi M, Shigeoka S (2001) Overexpression of a cyanobacterial fructose-1,6/sedoheptulose-1,7-bisphosphatase in tobacco enhances photosynthesis and growth. Nature Biotech 19:965-969

Poolman MG, Fell DA, Thomas S (2000) Modelling photosynthesis and its control. J Exp Bot 51:319-328

Spreitzer RJ, ME Salvucci (2002) Rubisco: Structure, regulatory interactions, and possibilities for a better enzyme. Annu Rev Plant Biol 53:449-475

Surridge C (2002) Agricultural biotech: The rice squad. Nature 416:576-578

Tabita FR (1999) Microbial ribulose 1,5-bisphosphate carboxylase/oxygenase: A different perspective. Photosynth Res 60:1-28

# Improvement of photosynthesis in higher plants

Masahiro Tamoi and Shigeru Shigeoka

Faculty of Agriculture, Kinki University, Nakamachi 3327-204, Nara 631-8505, Japan

**Summary.** We generated and analyzed transgenic tobacco plants with enhanced activities of fructose-1,6-bisphosphatase (FBPase) and/or sedoheptulose-1,7-bisphosphatase (SBPase) in the chloroplasts. The photosynthetic $CO_2$ fixation and the final dry matter under atmospheric conditions (360 ppm $CO_2$) in the transgenic plants were increased compared with those in the wild-type plants. Levels of the intermediate compounds in the Calvin cycle and the accumulation of carbohydrates were also higher than those in the wild-type plants. The transgenic tobacco also increased in the initial activity of Rubisco compared with that in the wild-type plants. These data suggest that both FBPase and SBPase involved in the regeneration of RuBP seem to be one of the limiting factors that participate in the regulation of the carbon flow through the Calvin cycle and the determination of the partitioning of carbon to end products.

**Key words.** Calvin cycle, Transgenic plants, Fructose-1,6-bisphosphatase, Sedoheptulose-1,7-bisphosphatase, Rubisco

## 1. Introduction

In the near future, we will face with a food shortage as a result of the exploding world population and environmental deterioration. To improve production efficiency, earlier maturing and high-yield crops must be developed. Photosynthetic carbon metabolism in higher plants is thought to be one determining factor in plant growth and crop yield. The Calvin cycle (photosynthetic carbon reduction cycle) is the primary pathway for carbon fixation.

In order to determine the limiting steps of photosynthesis and factors that influence the carbon allocation, a considerable number of studies have been made on the regulation of carbohydrate metabolism in the photosynthetic $CO_2$ fixation in plant leaves (Stitt and Sonnewald 1995, Fridlyand et al. 1999). The results obtained in recent years with genetically manipulated plants are leading to a revision of ideas about the regulation of metabolism. In the antisense plants of chloroplastic FBPase or SBPase, the rate of photosynthesis was significantly diminished in proportion to the decrease in the respective enzyme activity, suggesting that the reduction in photosynthesis is due to a decrease in the capacity for RuBP regeneration in the Calvin cycle (Harrison et al. 2001, Koßmann et al. 1994, Raines et al. 2000, Ölçer et al. 2001, Woodrow and Mott 1993).

Moreover, though some of the enzymes involved in the Calvin cycle are present at levels well in excess of that required to sustain a continued rate of $CO_2$ fixation, levels of fructose-1,6-bisphosphatase (FBPase) and sedoheptulose-1,7-bisphosphatase (SBPase) are extremely low compared to those of the other enzymes in the Calvin cycle (Woodrow, I. E. & Mott 1993). These enzymes catalyze irreversible reactions, and their activities are regulated by light-dependent changes in redox-potential through the ferredoxin/thioredoxin system (Buchanan 1991). The product of the reaction catalyzed by FBPase, i.e. fructose 6-phosophate, is the branch point for metabolites leaving the Calvin cycle and moving into starch biosynthesis. From these facts, it seems likely that FBPase and SBPase in the Calvin cycle are important strategic positions to determine the partitioning of carbon to end products.

We have previously demonstrated that cyanobacterium (*Synechococcus* PCC 7942) cells contain two FBPase isozymes, designated fructose-1,6-/sedoheptulose-1,7-bisphosphatase (FBP/SBPase) and FBPase-II. FBP/SBPase can hydrolyze both FBP and SBP with almost equal specific activities (Tamoi et al. 1996, 1998). The deduced amino acid sequence of the FBPase-II gene has been found to be considerably similar to those of the cytosolic and chloroplastic forms from eukaryotic cells. The enzymatic properties of FBPase-II were more similar to those of chloroplastic FBPase than to the cytosolic form of FBPase in higher plants; AMP and fructose 2,6-bisphosphate had no effect on the FBPase-II activity. Furthermore, we have also isolated and characterized SBPase from halotolerant *Chlamydomonas* W80 (Tamoi et al. 2001). To clarify the contribution of the levels of FBPase and/or SBPase to the photosynthesis rate and the carbon flow in source and sink organs, here we generated transgenic tobacco plants expressing cyanobacterial FBP/SBPase, FBPase-II or *Chlamydomonas* SBPase in chloroplasts.

## 2. Results and discussion

### 2.1 Analysis of transgenic tobacco plants expressing FBP/SBPase

We confirmed the localization of FBP/SBPase in transgenic plants by Western blot analysis of protein extracts from isolated chloroplasts using an antibody raised against *S.* 7942 FBP/SBPase. We determined the activities of FBPase and SBPase in the fourth leaf from the top of ten transformants and wild-type plants. The total FBPase and SBPase activities derived from endogenoues plastidic enzymes and cyanobacterial FBP/SBPase in TpFS-3 and TpFS-6 was 2.3 ± 0.4 and 1.7 ± 0.1-fold higher than that in the wild-type plants, respectively.

Transgenic plants (TpFS-3 and TpFS-6) were phenotypically distinguishable from the wild-type plants grown hydroponically under growth conditions (360 ppm $CO_2$, 400 $\mu$mol photons $m^{-2} s^{-1}$). The significant differences in size and growth rate between transgenic plants and wild-type plants were evaluated by measuring the height and dry weight. Similar differences were also observed between transgenic plants and wild-type plants that were cultivated in soil under the same culture conditions. At 12 weeks after planting, transgenic plants grew significantly larger than wild-type plants. After 18 weeks, TpFS-6 and TpFS-3 increased approximately 1.4- and 1.5-fold in height compared to the wild-type plants, respectively. The total dry weight of TpFS-6 and TpFS-3 increased approximately 1.4- and 1.5-fold compared to that of the wild-type plants, respectively. It is worth noting that the leaf size, stem thickness and root size of TpFS-3 plants were larger than those of the wild-type plants. Especially, the fresh weight of roots in TpFS-3 was approximately 3-fold larger that that in wild-type plants. However, there was no difference between the TpFS-3 plant leaves and the wild-type plant leaves with regard to chlorophyll and total protein contents per square meter. Moreover, thickening of the leaf and structural changes to the chloroplast in the transgenic plants were not observed by either photomicroscopy or electron microscopy, respectively.

Total extractable activities of phosphoribulokinase (PRK), NADPH-dependent glyceraldehyde-3-phosphate dehydrogenase ($NADP^+$-GAPDH), aldolase, total Rubisco remained unaltered in the transformants. Interestingly, the initial activity of Rubisco in TpFS-3 and TpFS-6 was approximately 1.2- and 1.1-fold higher than that in wild-type plants, respectively, indicating that the in vivo activation state of Rubisco in transgenic plants was increased. There were no differences in total extractable activities of transketolase (TK), ADP-glucose pyrophosphorylase (AGPase), and sucrosephosphate synthase (SPS) between wild-type and TpFS plants.

Photosynthetic activity in the fourth leaf of the TpFS-3, TpFS-6, and wild-type plants was measured under atmospheric conditions (360 ppm $CO_2$) and various light intensities (50 to 1,500 $\mu$mol photons $m^{-2}$ $s^{-1}$) at 25°C. Under low light (50 to 200 $\mu$mol photons $m^{-2}$ $s^{-1}$), there was no significant difference in the photosynthetic activity between the TpFS-3 plants and the wild-type plants. However, the photosynthetic activity of the transgenic plants increased significantly compared with that of the wild-type plants at irradiances above 200 $\mu$mol photons $m^{-2}$ $s^{-1}$. Under 1,500 $\mu$mol photons $m^{-2}$ $s^{-1}$, the photosynthetic activities of TpFS-3 and TpFS-6 were 1.24- and 1.20-fold higher than those of wild-type plants, respectively. The response of the saturation levels of $CO_2$ assimilation to a range of intercellular $CO_2$ concentrations (Ci) was assessed by taking *in vivo* measurements of $CO_2$ uptake at irradiance (400 $\mu$mol photons $m^{-2}$ $s^{-1}$), indicating that the photosynthetic activity of TpFS-3 increased approximately 1.50-fold more than that of wild-type plants under saturated $CO_2$ conditions.

We determined the levels of phosphorylated metabolites and carbohydrates in 12-weeks-old tobacco plants. The levels of carbohydrates were measured in the forth leaf from the top (upper), third leaf from the bottom (lower), stem, and root at 12 h in the light regime and 12 h in the dark regime, while the levels of phosphorylated metabolites were measured in the fourth leaf at 6 h after illumination. The content of RuBP in TpFS-6 and TpFS-3 plants was 1.5- and 1.8-fold larger than that in wild-type plants, respectively. In TpFS-6 and TpFS-3 plants, the content of phosphoglycerate (PGA) was 1.3- and 1.4-fold higher than that in the wild-type plants, respectively. In TpFS-3 plants, the contents of dihydroxyacetone phosphate (DHAP), fructose 6-phosphate (F6P), and glucose 6-phosphate (G6P) were 1.2- to 1.5-fold higher than those in wild-type plants. In TpFS-6 plants, the contents of PGA, DHAP, F6P, and G6P were 1.0- to 1.4-fold higher than those in wild-type plants. In the upper leaves of TpFS-3 plants, hexose, sucrose, and starch contents were approximately 2.5-, 2.0-, and 1.6-fold higher than those in wild-type plant leaves, respectively. In TpFS-6 plants, these levels were approximately 1.7-, 1.5-, and 1.5-fold higher than those in wild-type plant leaves, respectively. Total starch amount in the roots of TpFS-3 was 3-fold larger than that in the root of wild-type plants because the fresh weight of roots in TpFS-3 was 3-fold larger that that in wild-type plants. In the dark regime (12 h dark), hexose, sucrose, and starch contents showed almost equal levels in all lines.

Why did the rate of photosynthesis increase under normal conditions in transgenic plants expressing cyanobacterial FBP/SBPase in chloroplasts? Furquhar et al.[14] have reported that the photosynthetic rate is limited by the Rubisco capacity at relatively low

$CO_2$ concentration. Experiments with antisense Rubisco small subunit (*rbcS*) tobacco have shown that Rubisco plays an important role in controlling the irradiance-saturated rate of photosynthesis below 400 ppm of intercellular $CO_2$ partial pressure (Hudson et al. 1992, Stitt and Schulze 1994). Interestingly, the initial activity of Rubisco in TpFS-3 plants, reflecting the degree of activation *in vivo*, was 1.2-fold higher than that in wild-type plants; however, there were no differences in the total activity of Rubisco between wild-type and TpFS plants. Rubisco is active when a specific lysil residue within its catalytic site is carbamylated (complexed with $CO_2$) and bound with $Mg^{2+}$. The changes in carbamylation state in vivo are mediated by other stromal protein, Rubisco activase (Somerville et al. 1982, Salvucci et al. 1985). Rubisco activase has no effect on the activation state of Rubisco in the absence of RuBP, suggesting that Rubisco activase may bind Rubisco only when the active site is occupied with RuBP (Portis 1995, Salvucci and Ogren 1996). The content of RuBP in TpFS-3 plants was 1.8-fold larger than that in wild-type plants. It is therefore conceivable that the increase in RuBP results in the high activity of Rubisco activase, leading to the increase in the initial activity of Rubisco in the transgenic plants.

However, these data raise a question which enzyme substantially contributes to the photosynthesis rate and the carbon flow in source and sink organs. To clarify this problem, we generated transgenic tobacco plants expressing a cyanobacterial FBPase-II or *Chlamydomonas* SBPase in chloroplasts. While further investigation is required to answer this question unequivocally, the data presently available suggest that increased level of FBPase and SBPase activities in chloroplasts may induce the increased level of RuBP followed by the activation of Rubisco and the increase in photosynthesis.

## 2.2 Analysis of transgenic tobacco plants expressing FBPase-II or SBPase

Activities of FBPase and SBPase in the fourth leaf from the top of transformants and wild-type plants were determined. The FBPase activity in chloroplasts of FBPase-II-expressed plants was approx. 1.7- (TpF-9) and 2.3-fold (TpF-11) higher, respectively, than that in wild-type plants. On the other hand, in SBPase-introduced plants, the total activities of SBPase were approx. 1.3- (TpS-2), 1.6- (TpS-11), and 4.3-fold (TpS-10) higher, respectively, than that in wild-type plants.

Total extractable activities of enzymes, except for FBPase and SBPase, involved in the Calvin cycle and carbon metabolisms were not altered in all transformants. Interestingly, the increase by 10% relative to the initial activity of Rubisco was observed in TpF-11, TpS-11, and TpS-10 compared to wild-type plants, indicating that the *in vivo* activation state of Rubisco in transgenic plants is increased. On the other hand, there were no differences in the initial activities of Rubisco between TpF-9, TpS-2, and wild-type plants.

TpF-11, TpS-11, and TpS-10 were phenotypically distinguishable from the wild-type plants in terms of growth rate, height, size, and dry weight when grown hydroponically under normal conditions (360 ppm $CO_2$, light intensity: 400 $\mu$mol photons $m^{-2}$ $s^{-1}$). The transgenic plants grew significantly larger than wild-type plants after 12 weeks after planting, and thus TpF-11, TpS-11, and TpS-10 were approx. 1.2-, 1.4-, and 1.3-fold taller, respectively, than the wild-type plants at 18 weeks after planting. The total dry weight of TpF-11, TpS-11, and TpS-10 were increased approx. 1.3-, 1.7-, and 1.4-fold, respectively, compared with that of the wild-type plants. TpF-10 plants with similar

FBPase activity than TpF-11 showed the same growth rate and photosynthesis as TpF-11. TpS-3, -1, and -4 having 2.6-, 3.1-, and 3.4-fold higher SBPase activity showed increased growth rate and photosynthesis, which are dependent on increased level of the SBPase activity. On the other hand, TpF-9 with1.7-fold higher FBPase activity and TpS-2 with 1.3-fold higher SBPase activity were phenotypically indistinguishable from the wild-type plants.

The photosynthetic activities of TpF-11, TpS-11, and TpS-10 were increased significantly compared with that of the wild-type plants when light intensities were in excess of 200 $\mu$mol photons m$^{-2}$ s$^{-1}$. At saturating irradiance (1,500 $\mu$mol photons m$^{-2}$ s$^{-1}$), the photosynthetic activities of TpF-11, TpS-11, and TpS-10 were 1.15-, 1.27-, and 1.23-fold higher, respectively, than that of wild-type plants. On the other hand, no differences were observed in the photosynthetic activities between TpF-9, TpS-2, and wild-type plants. The same data were also obtained in transgenic plants exhibiting a similar or lower level of FBPase or SBPase activities.

The levels of starch, sucrose, and hexose were determined in the fourth leaf from the top (upper) and the third leaf from the bottom (lower) at 12 h in the light period, while the levels of metabolites were determined in the fourth leaf at 6 h after illumination. There was no difference in the content of RuBP between TpF-9, TpS-2 and wild-type plants. On the other hand, in TpF-11, TpS-11, and TpS-10 plants, the contents of RuBP were 1.4-, 1.7-, and 1.7-fold larger, respectively, than those in wild-type plants. In TpF-9 plants, the content of starch was slightly increased in the upper leaves and the hexose content was increased 1.6-fold in the stem.

## 2.3 Contribution of FBPase and SBPase to photosynthetic carbon flow in the Calvin cycle

In the TpFS plants, 2.3-fold or 1.7-fold increase of FBPase and SBPase activities caused the increase of RuBP content and the activation state of Rubisco in vivo. As a result, the photosynthesis capacity and the growth rate of transgenic plants were enhanced compared with those of wild-type plants. In the TpF plants, 2.3-fold increase of FBPase activity resulted in the increase of photosynthesis rate and growth rate, like TpFS plants. These findings clearly support the possibility that the increase in the chloroplastic FBPase has a significant positive effect on the process of RuBP regeneration, leading to the enhancement of the rate of photosynthesis and the growth of tobacco plants. However, only 1.7-fold increase of FBPase activity did not accelerate photosynthesis and growth. In the TpF-9 plants, a slight increase in the FBPase activity by less than 1.7-fold seems to contribute to the starch synthesis rather than to the RuBP regeneration in chloroplasts. That is, without enhancement of RuBP level and the photosynthetic activity, an increase of the FBPase activity in chloroplasts leads to an increased amount of F6P and resultant changes in the pathway of flux into starch synthesis in chloroplasts, because F6P, the product of the reaction catalyzed by FBPase, is the branch point for metabolites leaving the Calvin cycle and moving into starch biosynthesis through the conversion into G6P. On the other hand, an increase in chloroplastic FBPase level that is correlated with an increase in the RuBP level affects the photosynthetic capacity and the growth in transgenic plants. In the TpF-11 plants, excess F6P generated by increased cyanobacterial FBPase-II may be used for the generation of SBP in the Calvin cycle and thus excess SBP may increase in vivo SBPase activity. It seems likely that in chloroplasts of higher plants, the

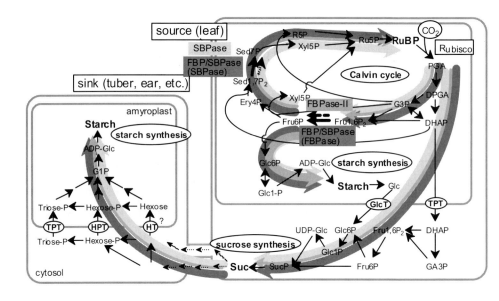

**Fig. 1.** Carbon flow in the wild-type and transgenic tobacco plants. The gray arrows indicated flow enhanced by the introductions of FBPase and/or SBPase.

level of chloroplastic FBPase controls strictly the regeneration of RuBP in the Calvin cycle and the starch synthesis. On the other hand, in the case of SBPase-transgenic plants, only 1.3-fold increase of SBPase activity did not accelerate photosynthesis and growth, but 1.6-fold increase of SBPase activity caused the increase of photosynthesis and growth rate, like TpFS plants.

These data indicate that both FBPase and SBPase seem to be one of the limiting factors that participate in the RuBP regeneration and the regulation of the carbon flow through the Calvin cycle. Moreover, it seems likely that SBPase is more important factor for RuBP regeneration than FBPase in the Calvin cycle and that FBPase contributes to partitioning of fixed carbon for RuBP regeneration or starch synthesis.

## References

Buchanan BB (1991) Regulation of $CO_2$ assimilation in oxygnic photosynthesis: the ferredoxin/thioredoxin system. Arch Biochem Biophys 288:1-9

Fridlyand LE, Backhausen JE, Scheibe R (1999) Homeostatic regulation upon changes of enzyme activities in the Calvin cycle as an example for general mechanisms of flux control. What can we expect from transgenic plants? Photosynth Res 61:227-239

Harrison EP, Olcer H, Lloyd JC, Long SP, Raines CA (2001) Small decrease in SBPase cause a linear decline in the apparent RuBP regenetration rate, but do not affect Rubisco carboxylation capacity. J Exp Bot 52:1779-1784

Hudson GS, Evans JR, von Caemmerer S, Arvidsson YBC, Andrews TJ (1992) Reduction of ribulose-1,5-bisphosphate carboxylase/oxygenase content by antisense RNA reduces photosynthesis in transgenic tobacco plnts. Plant Physiol 98:294-302

Koßmann J, Sonnewald U, Willmitzer L (1994) Reduction of the chloroplastic fructose-1,6-bisphosphatase in transgenic potato plants impares photosynthesys and plant growth. Plant J 6:637-650

Ölçer H, Lloyd JC, Raines CA (2001) Photosynthetic capacity is differentially affected by reductions in sedoheptulose-1,7-bisphosphatase activity during leaf development in transgenic tobacco plants. Plant Physiol 125:982-989

Raines CA, Harrison EP, Lloyd JC (2000) Investigating the role of the thiol-regulated enzyme sedoheptulose-1,7-bisphosphatase. Physiol Plant 100:303-308

Portis AR Jr (1995) The regulation of Rubisco by Rubisco activase. J Exp Bot 46:1285-1291

Salvucci ME, Ogren WL (1996) The mechanism of Rubisco activase: insights from studies of the properties and structure of the enzyme. Photosynth Res 47:1-11

Salvucci ME, Portis AR Jr., Ogren WL (1985) A soluble chloroplast protein catalyzes ribulosebisphosphate carbxylase oxygenase activation in vivo (technical note). Photosynth Res 7:193-201

Somerville SC, Portis AR Jr, Ogren WL (1982) A mutant of Arabidpsis thaliana which lacks activation of RuBP carbxylase in vivo. Plant Physiol 70:381-387

Stitt M, Schulze E-D (1994) Does Rubisco control the rate of photosynthesis and plant growth? An exercise in molecular ecophysiology. Plant Cell and Environ 17:465-487

Stitt M, Sonnewald U (1995) Regulation of metabolism in transgenic plants. Annu Rev Plant Physiol Plant Mol Biol 46:341-368

Tamoi M, Ishikawa T, Takeda T, Shigeoka S (1996) Molecular characterization and resistance to hydrogen peroxide of two fructose-1,6-bisphosphatses from *Synechococcus* PCC 7942. Arch Biochem Biophys 334:27-36

Tamoi M, Murakami A, Takeda T, Shigeoka S (1998) Acquisition of a new type of fructose-1,6-bisphosphatase with resistance to hydrogen peroxide in cyanobacteria: molecular characterization of the enzyme from *Synechocystis* PCC 6803. Biochim Biophys Acta 1383:232-244

Tamoi M, Kanaboshi H, Miyasaka H, Shigeoka S (2001) Molecular mechanisms of the resistance to hydrogen peroxide of enzymes involved in the Calvin cycle from halotolerant *Chlamydomonas* sp. W80. Arch Biochem Biophys 390:176-185

Woodrow IE, Mott KA (1993) Modeling C3 photosynthesis: A sensitivity analysis of the photosynthetic carbon-reduction cycle. Planta 191:421-432

# Modification of $CO_2$ fixation of photosynthetic prokaryote

Akira Wadano, Manabu Tsukamoto, Yoshihisa Nakano, and Toshio Iwaki

Department of Applied Biochemistry, Osaka Prefecture University, Gakuencho 1-1, Sakai, Osaka, Japan

**Summary.** Resemblances and differences on photosynthesis of $C_3$ plant and cyanobacterium *Synechococcus* PCC7942 were discussed with a simplified model of photosynthesis and photorespiration. Then the transformation of the cyanobacterium was shown with *Chromatium vinosum* (*Cv*) RuBisCO. Total RuBisCO activity and photosynthesis were compared with wild type *S.* 7942 and RuBisCO high expression mutant. Results were explained with RuBisCO environment of *S.* 7942 mutant. For excluding the influence of inorganic carbon concentrating mechanism (CCM) on the effect of RuBisCO, carboxysome less mutant was created with homologous recombination of *ccm operon* to be replaced with kanamycine resistant gene. And the deletion of the operon was confirmed by genomic PCR on the transformant. The confirmation was performed with three primers. Electron micrograms of *S.* 7942 show the existence of carboxysome in wild type and its absence in the mutant. Growth rate of CL mutant on different $CO_2$ concentration shows that there is a possibility to increase a photosynthetic ability of a cyanobacterium with increases the RuBisCO activity.

**Key words.** Cyanobacteria, RuBisCO, Transformation, Photosynthesis

## 1. Introduction

To decrease $CO_2$ in air, a lot of organic carbon compounds must be made from the inorganic carbon compound using the energy of the sun. The changed solar energy decreases the consumption of the other energy resulting in suppresses the production of $CO_2$ (Robinson et al. 2003). Moreover, the compounds made by the organisms can be used as the food staff in many cases, and regarded to be superior compounds.

Now, the handling of the ecosystem is indispensable to build the zero emission system for the environment of our milieu. It is important to utilize the ability of the creature, which fixes $CO_2$ by the photosynthesis to accumulate the energy of the sun. As the photosynthetic organisms, there are single cell prokaryotes like cyanobacteria, photosynthetic sulfur bacteria, and lot of the alga, the higher plant as the eukaryotes. They have been useful for the maintenance of the environment in a many fields such as the accumulation of the fossil fuels, and food, the fuel, and the building material. And the biomass reduced emissions of carbon dioxide, sulfur, and nitrogen oxides by cofiring biomass with coal in existing power plants.

Among the photosynthetic organisms, cyanobacteria, which are a single cell prokaryote as mentioned above, have been studied as the model of the chloroplast of the high plant. The complete genome sequence for this microorganism has been determined (Kotani et al. 1994) and its transformation method was established (Orkwiszewski and Kaney 1974). Moreover, the concentration mechanism of inorganic carbon was revealed and the gene for the mechanism was cloned (Shibata et al. 2002) to use for the transformation of higher plant.

In our lab, we have been tried to increase the photosynthetic ability of the organism with a foreign RuBisCO from a photosynthetic bacterium, *Chromatium vinosum* (Haranoh et al. in preparation). The carbon concentration mechanism, which is called CCM, makes difficulty for estimation of the role of RuBisCO on $CO_2$ fixation, so we produced the mutant without of CCM to check the effect of foreign RuBisCO on a cyanobacterium, Synechococcus PCC 7942 in this study.

## 2. Results and discussion

Figure 1shows the preliminary simplified photosynthetic metabolic pathway on $CO_2$ and $O_2$ fixation. $CO_2$ fixation and photorespiration pathways start on RuBisCO, and this enzyme has important role to control the both pathways. For the research of role of RuBisCO on photosynthesis, we use the cyanobacterium as the model of higher plants, because the organism has about the same function for getting the energy from sunlight with generation of $O_2$. We know also there are many different functions on the photosynthesis between the cyanobacteria and higher plants. One of significant differences is in existence of the CCM in cyanobacteria. CCM also contains carboxysome as the part of its mechanism, which contains RuBisCO. Depend on the different mechanism of photosynthesis between the cyanobacteria and higher plants; there is some different physiology of their photosynthesis. The left figure in Fig. 2 shows the dependency of higher plant photosynthesis on $CO_2$ concentration under saturated sunlight by von Caemmerer and Edmondson (1986). As increasing the $CO_2$ concentration, photosynthesis becomes a maximum level and RuBP concentration decrease to a steady state level for higher plants and cyanobacterium. Thick dotted line shows the level of RuBP and thin dotted line shows the level of RuBisCO active center concentration. As increase of $CO_2$ concentration the rate of photosynthesis became constant, because the concentration of RuBP becomes lower than the concentration of RuBisCO active center in higher plants. The shortage of energy for regeneration of RuBP is the cause of the steady state of the photosynthesis rate. Although the photosynthesis of cyanobacterium *Synechococcus* PCC 7942 depends on $CO_2$ concentration same as those of higher plant, but significant difference is in the concentrations of RuBisCO and RuBP. When the photosynthesis becomes the steady state, the level of RuBP is much higher than that of RuBisCO (Satoh 1998). It might depend on the existence of carboxysome and the ability of the RuBisCO for the fixation of $CO_2$.

If we can express high activity of RuBisCO in a cyanobacterium, then we can answer the difference between steady state of photosynthesis of higher plants and cyanobacteria. These three plasmids in Fig. 3 were used to make high RuBisCO expression mutant. pUC303 is original shuttle vector usually used for cyanobacterium PCC 7942 transformation. Selection was performed on streptomycin resistance. As high expression promoters, *Synechosistis* PCC6803 *psb*AII and Synechococcus PCC7942 psbAI promoters were

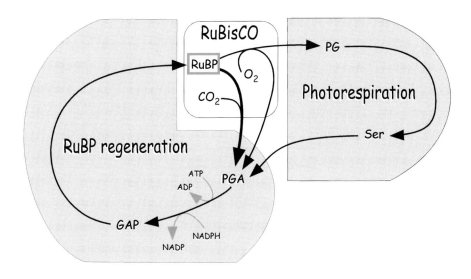

**Fig. 1.** Simplified model of photosynthesis and photorespiration
RuBP, ribulose 1,5-bisphosphate; PG, phosphoglycolate; Ser, serine; PGA, phosphoglycerate; GAP, glyceraldehyde 3-phosphate.

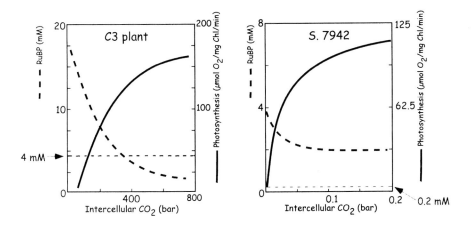

**Fig. 2.** Photosynthesis of $C_3$ plant and cyanobacterium *Synechococcus* PCC7942
Thick line, Photosynthesis; Thick dotted line, the level of RuBP; Thin dotted line, the level of RuBisCO active center concentration.

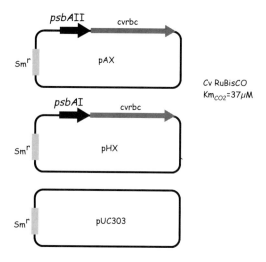

**Fig. 3.** High expression vector *Chromatium vinosum (Cv)* RuBisCO in cyanobacterium, PCC 7942.

used. The former plasmid is relatively stable but its activity is not so high, but the second was unstable although its activity was so high.

Fig. 4 shows total RuBisCO activity and photosynthesis of S7942 mutant with high RuBisCO expression. Grey and black bars show total RuBisCO activity and photosynthesis, respectively. Up to HX18 mutant the photosynthesis increase with increase of RuBisCO activity. On HX18 mutant there are two possibilities, one is down regulation by RuBisCO to photosystem as the stringent control. Another possibility is some suppression of photosystem by homologous recombination of promoter S. 7942 *psbA*I. The photosynthesis can be increased up to 1.5 times with increase of RuBisCO activity. However, any mutants could not grow faster than control mutant, which has only streptomycin resistance. It might be natural, because the foreign RuBisCO can't penetrate into carboxysome although we expected L8S8 foreign RuBisCO might penetrate into the carboxysome.

Fig. 5 shows RuBisCO environment of S.7942 mutant. Original RuBisCO exits in the carboxysome, and the foreign its RuBisCO of *Chromatium vinosum*, which is a purple photosynthetic bacterium, is outside of the carboxysome. Under the $CO_2$ saturated condition, both RuBisCO work about the same condition showing higher photosynthesis with higher enzyme content in short time measurement.

Why does the maximum photosynthesis not result in increased growth under high $CO_2$ concentration and saturated light for long time growth rate measurement? Long time experiment like growth rate check, there must be some effect of phosphoglycolate from photorespiration. So we thought that we should make the same condition against the original and the foreign RuBisCO. This means that we have to push the foreign enzyme into the carboxysome or take out the enzyme from the cellular organ. It was impossible to push the foreign enzyme into the carboxysome, carboxysome less (CL) mutant was

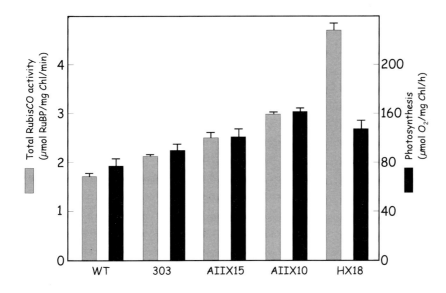

**Fig. 4.** Total RuBisCO activity and photosynthesis of *S.* 7942 RuBisCO high expression mutant Gray bar, RuBisCO activity; Black bar, Photosynthesis.

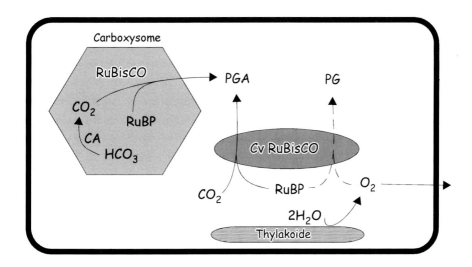

**Fig. 5.** RuBisCO environment of *S.* 7942 mutant.

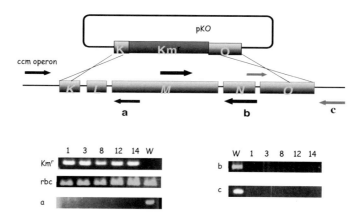

**Fig. 6.** Carboxysome less mutant was created with homologous recombination of *ccm operon* (A) and confirmed by genomic PCR (B) on the transformant. CCM operon was replaced with kanamycine resistant gene. And the confirmation was performed with three primer, a, b and c. A primer confirmed the existence of Km gene in the cyanobacterial mutant, but there is no gene of the Km tolerant in the wild. Moreover, there is no *K* and *L* genes, *M* and *N*, and *O* gene in the transformant with a, b, and c primers. As the control, RuBisCO gene, *rbc* is in both wild and transformant.

created with homologous recombination of *ccm operon* (Fig. 6). CCM operon was replaced with kanamycine resistant gene, and electron micrograms of *S.* 7942 and CL strain show the existence of carboxysome in wild type and its absence in the mutant (Fig. 7). Growth rate of CL mutant on different $CO_2$ concentration shows also that this mutant does not have any inorganic carbon concentrating activity depend on the lack of carboxysome (Fig. 8). This mutant can grow under high concentration of $CO_2$ as wild type mutant can. But under 0.5% $CO_2$ condition the mutant almost could not grew. Now we can expect the original RuBisCO also is in the cytosol (Fig. 9). This must simplify the explanation of effect by foreign RuBisCO activity on the physiology of mutant. We used again the same *Chromatium vinosum (Cv)* RuBisCO high expression vector. This vector will add second antibiotics resistant against carboxysome less mutant. The same level of total activity could not be gotten on carboxysome less mutant compared with Wild type. The activity of double mutant was about 10% of the control. But in the double mutant the promoter *psb*AI and *psb*AII express the foreign RuBisCO to make different level RuBisCO activity. Western blotting shows that the activity level correlated with the amount of foreign RuBisCO.

Increase of RuBisCO activity increased the photosynthesis of CL mutants, but the increase was not correlated with each other. Photosynthetic activity reach the steady state under the lower value than that expected with the RuBisCO activity. Both original and foreign RuBisCO are in cytosol, and they work on photorespiration depend on their specificities. Under the condition of high concentration of $CO_2$, CL303 grows faster than

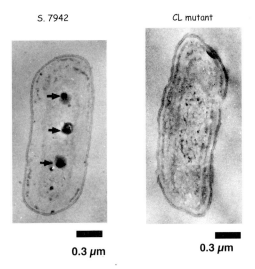

**Fig. 7.** Electron micrograms of S. 7942 and CL strain show the existence of carboxysome in wild type and its absence.

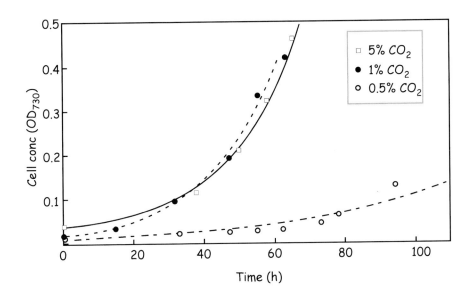

**Fig. 8.** Growth rate of CL mutant on different $CO_2$ concentration.

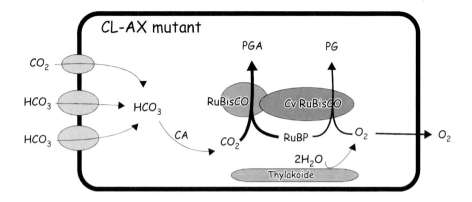

**Fig. 9.** RuBisCO environment of double mutant of S. 7942.

CL-AX7. Each mutant shows different adaptation against 5% $CO_2$+2% $O_2$ and 5% $CO_2$+air conditions. CL and CL-303 mutants don't have significant difference on their growth rate between the different $O_2$ conditions. But the CL-AX7 showed significant growth rate difference on the different conditions. The mutant could grew faster than CL mutant under the low $O_2$ condition. Thus it is reasonable to conclude there remains still the possibility of increasing the growth rate on increase of RuBisCO activity of wild type cyanobacterium.

## References

Caemmerer S von, Edmondson DL (1986) Relationship between steady-state gas exchange , in vivo ribulose bisphosphate carboxylase activity and intermediates in *Raphanus sativus*. Aust J Plant Physiol 13: 69-688

Haranoh K, Inoue N, Kojima K, Satoh R, Nishino T, Wada K, Ihara H, Tsuyama S, Kobayashi H, Iwaki T, Wadano A. Expression of foreign type ribulose 1,5-bisphosphate carboxylase/oxygenase (EC 4.1.1.39) stimulate photosynthesis in cyanobacterium *Synechococcus* PCC7942 cells, in preparation

Kotani H, Kaneko T, Matsubayashi T, Sato S, Sugiura M, Tabata S (1994) A physical map of the genome of a unicellular cyanobacterium *Synechocystis* sp. strain PCC6803. DNA Res 1:303-307

Orkwiszewski KG, Kaney AR (1974) Genetic transformation of the blue-green bacterium, *Anacystis nidulans*. Arch Mikrobiol 98:31-37

Robinson AL, Rhodes JS, Keith DW (2003) Assessment of potential carbon dioxide reductions due to biomass-coal cofiring in the United States. Environ Sci Technol 37:5081-9

Satoh R (1998) Analysis of ribulose bisphosphate carboxylase/oxygenase activity in a cyanobacterium *Synechococcus* PCC7942, in PhD thesis

Shibata M, Katoh H, Sonoda M, Ohkawa H, Shimoyama M, Fukuzawa H, Kaplan A, Ogawa T (2002) Genes essential to sodium-dependent bicarbonate transport in cyanobacteria: function and phylogenetic analysis. J Biol Chem 277:18658-18664

# Specificity of diatom Rubisco

Richard P. Haslam[1], Alfred J. Keys[1], P John Andralojc[1], Pippa J. Madgwick[1], Inger Andersson[2], Anette Grimsrud[3], Hans C. Eilertsen[3], and Martin A. J. Parry[1]

[1] Crop Performance and Improvement Division, Rothamsted Research, Harpenden, Hertfordshire AL5 2JQ, UK
[2] Department of Molecular Biology, Swedish University of Agricultural Sciences, Uppsala Biomedical Centre, Box 590, S-751 24 Uppsala, Sweden
[3] Norwegian College of Fisheries Science (NFH), University of Tromsø, N-9037 Tromsø, Norway

**Summary.** Rubisco is an inefficient enzyme; this inefficiency is greatest when $CO_2$ availability at the site of Rubisco is low. We hypothesise that the selection pressure for a more efficient Rubisco will be greatest in species growing under $CO_2$ limited conditions, particularly when low light levels reduce the cost effectiveness of a carbon concentrating mechanism. We determined the specificity factor for four marine diatoms, *Thalassiosira antarctica, Skeletonema costatum, Chaetoceros socialis,* and *Thalassiosira hyalina* adapted to the arctic environment.

**Key words.** Rubisco, Oxygenase, Diatoms, Arctic

## 1. Introduction

All nature's molecular machines are built around a central scaffolding of organic carbon and thus the assimilation of $CO_2$ underpins the existence of life. Photosynthetic rate is a major factor determining plant productivity. Understanding photosynthetic mechanisms and their regulation is essential if the efficiency of carbon metabolism is to be improved. Ribulose bisphosphate carboxylase/oxygenase (Rubisco) forms the primary bridge between life and the lifeless, creating organic carbon from the inorganic carbon dioxide in the air. Rubisco catalyses the carboxylation of ribulose-1,5-bisphosphate (RuBP) a short sugar chain with five carbon atoms. While still bound to Rubisco the lengthened chain is hydrolysed into two phosphoglycerate (PGA) molecules, each with three carbon atoms. Within the Calvin cycle the molecules of PGA are either used to generate more of the acceptor molecule RuBP or used to make sucrose that is used for production. However, compared to other enzymes within the Calvin cycle, Rubisco is such a slow catalyst that plants must invest much carbon and nitrogen in Rubisco if they are to sustain high photosynthetic rates. Rubisco alone can account for 50% of soluble leaf protein, or 25% of leaf nitrogen (Parry et al. 2003). Despite this high level of investment, under certain conditions Rubisco activity can still be limiting.

It is necessary for the $CO_2$ assimilation by Rubisco to be tightly regulated so that the leaf's capacity for RuBP regeneration (light harvesting, electron transport and photophosphorylation) is commensurate with its capacity for RuBP carboxylation and starch and sucrose synthesis. Rubisco also catalyses a competing and wasteful reaction with $O_2$. This oxygenase reaction leads to photorespiration which consumes energy and releases fixed $CO_2$ and $NH_3$ (Ogren and Bowes 1971; Laing et al. 1974). Up to 60% of carbon fixed by Rubisco in phytoplankton can be lost in this way e.g. 64% in the haptophycean *Phaeocystis pouchetii* (Guillard and Helleburst 1971). In higher plants this inefficiency is greatest at elevated temperature and under drought (i.e., when stomata are closed and, thus, $CO_2$ availability at the site of Rubisco is low). In the ocean $CO_2$ is plentiful due to the bicarbonate reservoir, but since the dissociation to $CO_2$ is slow, inefficiency is at a maximum when a species has a poor ability to concentrate $CO_2$ or when growth rate (and hence irradiance) is high and local depletions of $CO_2$ occur (Thomas et al. 2001). $CO_2$ and $O_2$ compete directly for reaction with an enzyme bound enediol intermediate of RuBP rather than having a formal binding site in the protein. Once the enediol has reacted with either gas the enzyme is committed to form products. The ratio for partitioning between the two reactions is dependent on the relative concentration of the two gases. The lower the $CO_2$ concentration the greater the losses through the oxygenase reaction and *vice versa*. Environmental stresses typically decreases carboxylase activity relative to oxygenase. For example under drought stress the $CO_2$ within a leaf becomes depleted since the stomata are closed, and at high temperatures the solubility of $CO_2$ is decreased relative to $O_2$. Given the significance of Rubisco to carbon assimilation and crop yield in agriculture, there is considerable interest in reducing the wasteful oxygenase activity. One approach is to seek a more efficient Rubisco in species living in hostile environments where growth factors are limiting (e.g. temperature and light) and to use it or knowledge of its structure to improve crop plants.

## 2. Specificity factor

In spite of the lack of any formal enzyme binding site for the gaseous substrates one major factor determining the partitioning between carboxylase and oxygenase activities is the inherent ability of the enzyme to discriminate between $CO_2$ and $O_2$ which is described by the specificity factor ($\tau = Vc.Ko/Vo.Kc$). Significantly, the specificity factor is not constant for Rubisco from different species, there being a tremendous variation (see Fig. 1). Indeed, data for 32 enzymes compiled by Badger et al. (1998) from diverse cyanobacteria, algae and higher plants show a 25-fold range in specificity factor. Increased specificity factor is dictated by evolutionary pressure towards the more efficient utilisation of $CO_2$ and indicates that subtle differences in enzyme structure can influence the ability of the gaseous substrates to access and react with the enediol intermediate. The lowest specificity factor is found for Rubisco from *R. rubrum*, intermediate values in Rubisco from cyanobacteria and green algae and higher values in crop plants (Fig. 1). However, the highest values are found in marine red algae.

The variation in specificity factor between species indicates that the structure can influence the ability of the gaseous substrates to access and react with the enediol intermediate. The enzyme has been the subject of intense analyses, including structural, mechanistic and mutagenesis studies and a vast amount of data is now available on Rubisco. However, the molecular basis for $CO_2/O_2$ discrimination is still not fully understood. In most species Rubisco is a 550-kDa hexadecamer composed of eight large and eight small subunits ($L_8S_8$). Typically, the small subunits are nuclear encoded and do not contribute

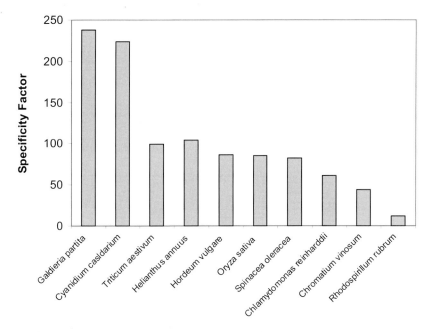

**Fig. 1.** The specificity factor (Vc.Ko/Vo.Kc) of Rubisco from different species determined at 25°C. Values are based on data from Jordan and Ogren (1981); Parry et al. (1989); Kane et al. (1994); Uemura et al. (1997).

directly to the catalytic site. The large subunits are chloroplast encoded and are arranged as four dimers. The genes encoding the small and large subunits are designated rbcS and rbcL respectively. Most of the active site residues that bind the substrates (or substrate anlaogues) are contributed by loops situated at the mouth of the α/β barrel of one large subunit with the remainder being supplied by two loop regions in the N-terminal domain of the second large subunit within the dimer. In a few species Rubisco is composed of a single homo-dimer composed of large subunits (Form 1).

## 3. Marine Rubisco

Variability of τ for Rubisco has mainly been studied in cultivated plants and has been neglected in species from other environments. For example, virtually nothing is known about Rubisco from psychrophilic organisms, i.e. organisms from arctic waters that are adapted to a life at low temperatures. Early studies indicated that diatom Rubisco is the typical plant-type $L_8S_8$ form (Fig. 2). Diatoms are unicellular, photosynthetic eukaryotic algae found throughout the world's oceans. They are at the base of important food webs and support large-scale coastal fisheries. Photosynthesis by pelagic marine diatoms generates as much as 40% of the 45 to 50 billion tons of organic carbon produced in the sea each year (Field et al. 1998). Indeed over geological time, diatoms may have influenced global climate by changing the flux of $CO_2$ into the oceans. We have chosen to look for natural variation in

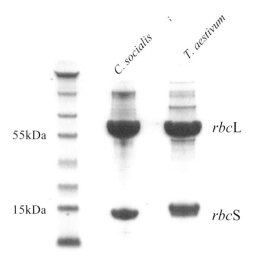

**Fig. 2.** SDS-PAGE separation of purified Rubisco.

Rubisco from diatoms found within and below the ice that thrive in a $CO_2$ and light limited environment. In diatoms both the *rbc*S and *rbc*L of Rubisco are encoded by chloroplast DNA similar to other non-green algae (Hwang and Tabita 1991). Despite the ancestral lineage of Rubisco, there has been evolution of its kinetic properties to meet some of the constraints of its chloroplast environment. Interestingly, all nongreen algal Rubiscos with high values of $\tau$, e.g. *Phaeodactylum tricornutum* see Table 1, probably come from algae that have a carbon concentrating mechanism. It would appear that the oxygenase reaction is not very relevant for the functional performance of the chloroplast, but the low affinity for external $CO_2$ is (Badger et al. 1998).

**Table 1.** Measurements of Rubisco specificity factor at 25°C determined for diatoms (bold this study where n = 5 ± se, others from Badger et al. 1998).

| Rubisco Lineage | Rubisco Specificity Factor ($\tau$) |
|---|---|
| *Phaeodactylum tricornutum* | 114 |
| *Cylindrotheca fusiformis* | 110 |
| *Cylindrotheca sp. N1* | 106 |
| *Olisthodiscus luteus* | 101 |
| ***Thalassiosira antarctica*** | **90 ± 3.2** |
| ***Skeletonema costatum*** | **72 ± 2.2** |
| ***Chaetoceros socialis*** | **92 ± 2.1** |
| ***Thalassiosira hyalina*** | **97 ± 3.4** |

## 4. Extraction and purification of Rubisco

The species were mass cultivated in a 300 litre turbidostat device in temperature controlled rooms (3 – 4 °C) at 14 h photoperiod and approximately 150 µmol quanta m$^{-2}$ s$^{-1}$. Cells were harvested, concentrated and immediately frozen in liquid nitrogen. Samples were ground with liquid nitrogen, combined with an extraction buffer (100 mM Bicine pH 8.0, 6% polyethylene glycol, 10mM dithiothretiol (DTT)), and 1mM each of benzamidine, phenylmethylsulphonylfluoride (PMSF) and ε-amino-N-caproic acid. All the subsequent extraction steps were carried out at 0 to 4 °C. The homogenate was then centrifuged at 20,000 x g for 20 min. The supernatant liquid was decanted and PEG 4000 was added as a 60% aqueous solution to produce a final concentration of 20% w/v. Also, 1 M $MgCl_2$ was added to make a final concentration of 20 mM. After 10 min the mixture was centrifuged again at 20,000 x g for 20 min. The pellet was re-suspended in Column Buffer (10 mM Tris pH 8.0 with 10 mM $MgCl_2$, 10 mM $NaHCO_3$, 1 mM EDTA, 2mM DTT and 1 mM $KH_2PO_4$) containing 1mM each of PMSF, 1mM Benzamidine and 1 mM ε-amino-N-caproic acid. The suspension was centrifuged, 10 min at 20, 000 x g, to remove insoluble material. The supernatant liquid was applied to a previously prepared sucrose step gradient (0.3 to 1.2M sucrose in column buffer) and centrifuged for three hours at 264,000 x g. The sucrose gradient was then fractionated into 1 mL aliquots and assayed for protein content and Rubisco activity (after Parry et al. 1993). Those fractions containing Rubisco were loaded on to a 1 mL HiTrapQ column (Pharmacia) equilibrated with Column Buffer and operated at about 1 ml per min. The proteins were eluted using a step gradient of 0.1 to 0.8 M in NaCl in Column Buffer and fractions were collected at 1 min intervals. Those fractions with high Rubisco activity were combined and snap frozen in liquid nitrogen prior to short term storage at –80 °C.

## 5. Specificity determinations

Measurements of specificity factor were made after Parry et al. (1989). The Wheat Rubisco was activated by incubation at 37 °C for 40 min, whilst the diatom Rubisco showed no increase in activity in response to warming and once defrosted maintained activity for 24 h when held at 4 °C (data not shown). Reaction mixtures were prepared in an oxygen electrode (Model DW1, Hansatech). RuBP oxygenation was calculated from the oxygen consumption and carboxylation from the amount of $^{14}C$ incorporated into PGA when all the RuBP was consumed (Parry et al. 1989). A number of reaction mixtures containing pure wheat Rubisco were interspersed with those containing Rubisco from diatoms. Initial concentrations of $O_2$ in solution in equilibrium with air were considered to be 356, 320, 291, and 265 µM at 10, 15, 20 and 25 °C respectively. Initial concentrations of $CO_2$ in solution were calculated from amounts of $NaHCO_3$ added using pKa values for $H_2CO_3$ of 6.25, 6.19, 6.15, and 6.11 respectively, for the four temperatures. The τ values were normalised to the average values for wheat Rubisco, of 150 at 10 °C, 135 at 15 °C 118 at 20 °C, and 100 at 25 °C.

We determined the specificity factor for Rubisco from four marine diatoms at a range of temperatures from 10 – 25 °C, (Fig. 3). In all of the species examined the specificity factor increased as temperature fell. The Rubisco from *S. costatum* and *T. hyalina* showed a linear relationship between temperature and specificity factor ($R^2$ = 0.994 and 0.979 respectively), whilst *C. socialis* with a measured specificity factor of 94 and 92, at 20 °C and 25°C respectively, tailed towards higher temperatures. This suggests that there are key structural differences between the enzymes from the different species. None of the species examined had a higher specificity factor than wheat even when values were extrapolated to 0°C. However, unlike wheat Rubisco the diatom Rubisco was not deactivated when exposed for prolonged periods (~ 24 h) to temperatures of 4°C. These observations suggest structural adaptations to the extreme environment occupied by these diatoms.

As the kinetics of Rubisco have been elucidated, the suggestion that evolutionary pressure has driven Rubisco towards more efficient utilisation of $CO_2$ over geological time has been made (Jordan and Ogren 1981). Tortell (2000) cautiously identified a relationship between the evolution of higher $CO_2$ specificity in response to a decrease in atmospheric $CO_2$. Phytoplankton have adopted carbon concentrating mechanisms (CCM) to offset the problems of $CO_2$ limitation. However, the operation of a CCM may increase photosynthetic light requirements. Thus it seems possible that the evolution of high specificity factors in diatoms may contribute to their ability to grow well in the light-limited environment typical of the early bloom or under ice. Under ice, blooms typically take place above pronounced and shallow pycnoclines (zones between waters with different densities) that form a barrier to the underlying $CO_2$ enriched water; hence $CO_2$ limitation is more likely to occur. As the catalytic efficiency of Rubisco increases one would expect that less nitrogen (as the constituent amino acids of Rubisco) should be required to maintain a given photosynthetic rate. In fact, Kristiansen et al. (1994) found very limited

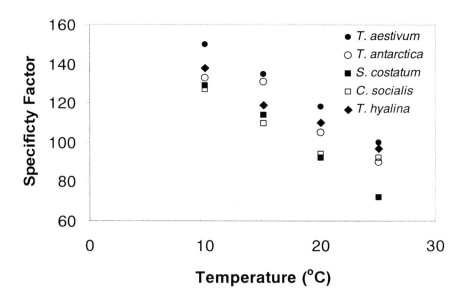

Fig. 3. Rubisco specificity factor for various diatoms and *T. aestivum* determined at temperatures from 10 to 25°C.

signs of nitrogen limitation during diatom blooms in the arctic ice zone contrary to what is common in other areas (de Baar 1994). Hobson et al. (1985) have reported high specific activities and low cellular concentrations of Rubisco in diatoms relative to other green algae, illustrating the coupling between carbon assimilation and nitrogen metabolism. Improvements in Rubisco specificity would be ecologically significant if they affected the competitive ability of a species.

In order to identify regions contributing to the higher efficiency of Rubisco, the X-ray structures of candidate enzymes will be determined and compared with the high resolution structure of relevant species (Andersson and Taylor 2003). One approach to crop improvement is the identification of amino acid residues that confer key catalytic properties (e.g. specificity factor) and are unique to these species. Such studies will be facilitated by the availability of a number of Rubisco structures at atomic resolution. We are continuing to examine more candidate species from the Arctic and are compiling a resource capable of informing rational Rubisco modification. Many authors agree that the introduction of a Rubisco with high $\tau$ into crop plants remains a realistic goal (Parry et al. 2003), and the limited sampling of available $C_3$ plants may be one of the main limitations to the achievement of this goal. Improved knowledge of the natural variation in specificity factor is also necessary for modelling purposes. Understanding the factors that influence carbon uptake in diatoms will provide vital information about factors that control oceanic primary productivity and even future climates.

Acknowledgements. Rothamsted Research receives grant aided-support from the Biotechnology and Biological Sciences Council. R.H. was funded by the fifth framework program of the European Commission; contract QLK3-CT-2002-01945.

# References

Andersson I, Taylor TC (2003) Structural framework for catalysis and regulation in ribulose-1,5-bisphosphate carboxylase/oxygenase. Arch Biochem Biophys 414:130–140

Badger MR, Andrews JT, Whitney SM, Ludwig M, Yellowlees DC, Leggat W, Price GD (1998) The diversity and co-evolution of Rubisco, plastid, pyrenoid, and chloroplast-based $CO_2$-concentrating mechanisms in algae. Can J Bot 76:1043–1071

de Baar HJW (1994) von Liebig's Law of the Minimum and Plankton Ecology (1899-1991) Prog Oceanog 33:347-386

Field CB, Behrenfeld J, Randerson JT, Falkowski PG (1998) Primary production of the biosphere: integrating terrestrial and oceanic components. Science 281:237-240

Guillard RRL, Hellebust JA (1971) Growth and production of extracellular substances by two strains of *Phaeocystis pouchetii*. J Phycol 7:330-338

Hobson LA, Morris WJ, Guest KP (1988) Varying photoperiod, ribulose-1,5-bisphosphate carboxylase and $CO_2$ uptake in *Thalassiosira fluviatilis*. Plant Physiol 79:833-837

Hwang S-R, Tabita FR (1991) Cotranscription, deduced primary structure and expression of the chloroplast-encoded rbcL and rbcS genes of the marine diatom *Cylindrotheca* sp. Strain N1. J Biol Chem 266:6271-6279

Jordan DB, Ogren WL (1981) Species variation in the specificity of ribulose bisphosphate carboxylase/oxygenase. Nature 291:513-515

Kane HJ, Viil J, Entsch B, Paul K, Morell MK, Andrews TJ (1994) An improved method for measuring the $CO_2/O_2$ specificity of Ribulose bisphosphate Carboxylase Oxygenase. Aust J Pl

Physiol 21:449-461

Kristiansen S, Farbrot T, Wheeler PA (1994) Nitrogen cycling in the Barents Sea - Seasonal dynamics of new and regenerated production in the marginal ice zone. Limnol Oceanogr 39:1630-1642

Laing WA, Ogren WL, Hageman RH (1974) Regulation of soybean net photosynthetic $CO_2$ fixation by the interaction of $CO_2$, $O_2$ and ribulose-1,5-bisphosphate carboxylase. Plant Physiol 54:678-685

Ogren WL, Bowes G (1971) Ribulose diphosphate carboxylase regulates soybean photorespiration. Nature New Biol 230:159-160

Parry MAJ, Keys AJ, Gutteridge S (1989) Variation in the specificity factor of $C_3$ higher plant Rubiscos determined by the total consumption of ribulose-$P_2$. J Exp Bot 40:317-320

Parry MAJ, Delgado E, Vadell J, Keys AJ, Lawlor DW, Medrano H (1993) water stress and the diurnal activity of ribulose-1,5-bisphosphate carboxylase in field grown Nicotiana tabacum genotypes selected for survival at low $CO_2$ concentrations. Plant Physiol Biochem 31:113-120

Parry MAJ, Andralojc PJ, Mitchell RAC, Madgwick PJ, Keys AJ (2003) Manipulation of Rubisco: the amount, activity, function and regulation. J Exp Bot 54:1321-1333

Thomas S, Pahlow M, Wolf-Gladrow DA (2001) Model of the carbon concentrating mechanism in chloroplasts of eukaryotic algae. J Theor Biol 208:295-313

Tortell PD (2000) Evolutionary and ecological perspectives on carbon acquisition in phytoplankton. Limmol Oceanogr 45:744-750

Uemura K, Anwaruzzaman K, Miyachi S, Yokota A (1997) Ribulose-1,5-bisphosphate carboxylase/oxygenase from thermophilic red algae with a strong specificity for $CO_2$ fixation. Biochem Biophys Res Comm 233:568-571

# Regulation of $CO_2$ fixation in non-sulfur purple photosynthetic bacteria

Simona Romagnoli and F. Robert Tabita

Department of Microbiology, The Ohio State University, 484 West 12th Avenue, Columbus, Ohio, 43210 USA

**Summary.** Our laboratory studies the regulation of $CO_2$ fixation in phototrophic microorganisms. The expression of genes encoding enzymes responsible for assimilating $CO_2$ is primarily regulated by a LysR-type transcriptional activator, CbbR, that functions by binding specific sites in the promoter region of both the $cbb_I$ and $cbb_{II}$ $CO_2$ fixation operons under appropriate growth conditions. This straightforward model of regulation has a further level of complexity in the non-sulfur photosynthetic bacterium *Rhodopseudomonas palustris*. The recent genome sequencing project of this organism identified a unique two-component system interposed between the structural genes of the $cbb_I$ operon (encoding form I RubisCO) and CbbR. Two-component systems are a paradigm of gene regulation in Bacteria. In general, they consist of a sensor kinase and a response regulator, that upon stimulation, transduces an external stimulus in a His-Asp phospho-relay, ultimately modulating gene expression. The system identified in *R. palustris,* in addition to its peculiar modular architecture, also has multiple PAS motifs, suggesting a complex redox sensing activity coordinating a multi-step phospho-relay which regulates $CO_2$ fixation under specific growth conditions.

**Key words.** *Rhodopseudomonas palustris*, $CO_2$ fixation, Two-component signal transduction system

## 1. Introduction

Non-sulfur photosynthetic bacteria are characterized by great metabolic versatility, which allows them to colonize and inhabit diverse environments. Their ability to grow under aerobic and anaerobic conditions, in the presence of organic carbon sources or by assimilating atmospheric $CO_2$, their ability to fix nitrogen and in some cases to oxidize reduced sulfur compounds, combined with relatively facile genetics, allows one to address several physiologically important processes in a comprehensive way.

$CO_2$ fixation occurs mainly through the CBB (Calvin-Benson-Bassham) cycle, where carbon dioxide is assimilated at the expenses of ATP and NADH. Also, in photosynthetic bacteria the CBB cycle functions as an electron sink to dissipate the excess of reducing equivalent generated by the photosynthetic metabolism, providing an efficient redox homeostasis mechanism (Tabita 1995). This important, but energetically expensive process, needs to be carefully regulated. So far, regulation of the CBB cycle has been mainly investigated in *Rhodobacter sphaeroides* and *Rhodobacter capsulatus*. These two closely related species have over the years provided reliable model systems for understanding the regulatory networks of carbon dioxide assimilation (Dubbs and Tabita 2004). The enzymes participating to the Calvin cycle reactions are organized in two distinct operons (Fig. 1). There are two sets of structural genes for ribulose-1,5-bisphosphate carboxylase/oxygenase (RuBisCO), namely *cbbLS* (form I) and *cbbM* (form II). RuBisCO catalyzes the conversion of RuBP (ribulose-1,5-bisphosphate) into glyceraldehyde-3- phosphate. The presence of two copies seems to be an evolved response to the availability of variable concentrations of atmospheric $CO_2$: Form II RuBisCO, characterized by low substrate specificity, appears to be the most prevalent form at high concentrations of $CO_2$. Some authors have postulated that this enzyme is more ancestral than form I RuBisCO. Form I, on the other hand, is enzymatically more complex than form II and is very similar in structure to the enzyme found in present day plants, algae and cyanobacteria; form I is mainly synthesized at low concentrations of $CO_2$ and for this reason is believed to be derived from form II as an adaptative response to better assimilate lower concentrations of $CO_2$ in the presence of high concentration of $O_2$ (Shively, van Keulen, Meijer, 1998 and references therein).

A master transcriptional regulator, CbbR, controls the expression of the two *cbb* operons. In analogy with other Lys-R type regulators (Shell 1993), CbbR is divergently transcribed from the *cbb* operon and essentially controls transcription by binding consensus site in the promoter region. There is one CbbR in *R. sphaeroides* responsible for the regulation of both *cbb* operons; whereas each *cbb* operon is regulated by its cognate *cbbR* gene in *R. capsulatus*.

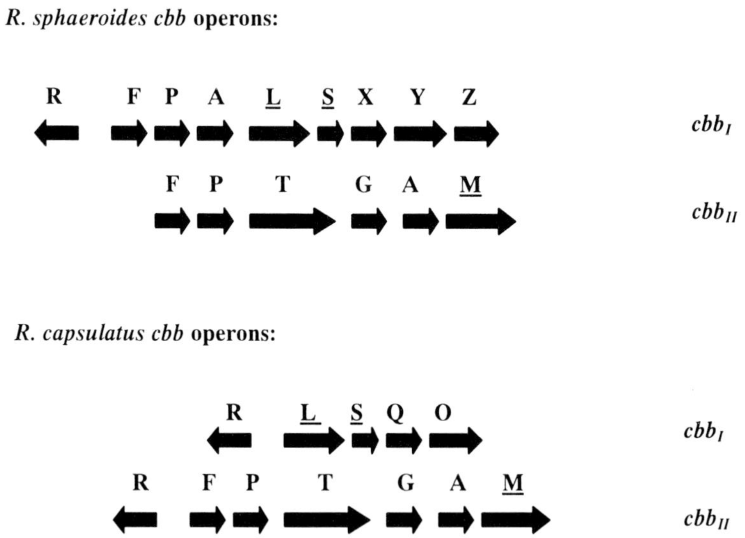

*R. sphaeroides* **cbb operons:**

R  F  P  A  L  S  X  Y  Z     $cbb_I$

F  P  T  G  A  M              $cbb_{II}$

*R. capsulatus* **cbb operons:**

R  L  S  Q  O                 $cbb_I$

R  F  P  T  G  A  M           $cbb_{II}$

**Fig. 1.** Organization of *cbb* operons in *Rhodobacter sphaeroides* and *Rhodobacter capsulatus*. The form I (*cbbLS*) and form II (*cbbM*) RuBisCO genes are underscored. Redrawn and adapted from Dubbs and Tabita 2004.

## 2. *Rhodopseudomonas palustris*: a regulatory twist on an established circuit

*R. palustris* is a photosynthetic bacterium common in soil and water. In addition to a large metabolic versatility, typical of this family of microorganisms, it is able also to carry out biodegradation of aromatic compounds, to oxidize reduced sulfur substrate, and to fix nitrogen and evolve hydrogen gas. Because of its ecological importance and potential energetic applications, *R. palustris* was the first photosynthetic microorganism to be sequenced by the Department of Energy (Larimer et al. 2004). The analysis of the $cbb_I$ RuBisCO region identified three unprecedented open reading frames juxtaposed between the divergently transcribed *cbbR* regulatory gene and the structural genes of form I RuBisCO (*cbbLS*). An extensive database search revealed no strong homologues, but identified conserved regions in the three genes likely to be involved in a His-Asp phosphorelay system. His-Asp phosphorelays are the backbone of two-component signal transduction systems (Stock, Robinson and Goudreau, 2000). Two-component systems usually consist of a sensor kinase and a response regulator. Within the sensor kinase, it is possible to identify an N-terminal sensing domain and a C-terminal transmitter domain; whereas the response regulator usually contains an N-terminal receiver domain and C-terminal effector domain that usually displays DNA-binding activity. Upon activation of the sensing domain from an internal or external system-specific signal, the sensor kinase undergoes autophosphorylation on a conserved histidine residue in the transmitter domain, and then transfers the phosphoryl group to a conserved aspartate residue in the receiver domain of the response regulator protein. This induces a conformational change of the effector domain that ultimately results in a change in expression of the target gene (s).

The signal transduction system identified in the *R. palustris* $cbb_I$ region (Fig. 2) consists of three proteins: a membrane spanning hybrid sensor kinase, with a typical kinase domain attached to a C-terminal Asp receiver domain, and two distinctive response regulator proteins. Response Regulator 1 contains an N-terminal Asp receiver domain and C-terminal HPt domain and Response Regulator 2 contains a distinct Asp receiver domain at the N- and C-termini, respectively. Interestingly, three PAS motifs were identified in the N-terminal sensory region of the sensor kinase (PAS-His KINASE). As PAS motifs are common redox and environmental sensing modules found in prokaryotes as well as eukaryotes (Taylor and Zhulin 1999), their presence suggests the potential for control of the activity of the sensor kinase by external stimuli, perhaps redox levels.

**Fig. 2.** Gene organization in the *R. palustris* $cbb_I$ region.

Because of the peculiarity of our 2- component system, we decided to address its function by first investigating the activity of the proteins involved, followed by analyzing the physiological role these proteins might play in controlling $CO_2$ metabolism. Therefore, a two-fold approach was taken: (i) $His_6$-chimeras of a truncated, soluble form of the sensor kinase and the two response regulators were produced in *E. coli* and purified. Also *E. coli* inner membranes containing the full-length $His_6$-sensor kinase were isolated, and used in phosphotransfer assays together with the purified proteins; (ii) strains were constructed whereby the *cbbR* gene and the genes encoding the two-component system were deleted, followed by noting the phenotypes of these strains after analyzing them under different growth conditions (Romagnoli and Tabita, manuscript in preparation). By analogy to other more traditional two-component systems, the full-length sensor kinase, as well as its truncated form, catalyzed rapid autophosphorylation in the presence of $[\gamma\text{-}^{32}P]ATP$. Phosphotransfer occured between the full-length kinase and both response regulators, whereas only RR1 was efficiently phosphorylated by the truncated kinase. Identificaton of the conserved His and Asp residues was carried out by sequence alignments with well characterized bacterial kinase and response regulator proteins. This allowed us to develop strategies for generating site-directed mutant proteins that could be used for biochemical assays to address the role of the multiple receiver domains present in this atypical two-component system.

The phenotype of the deletion strains so far available suggests that this two-component system finely modulates carbon metabolism under autotrophic growth conditions, possibly through a direct interaction with the main transcriptional regulator CbbR. The multi-domain organization of this two-component system underlines a complex mechanism of regulation, likely to integrate several metabolic signals. Comprehensive experiments aimed to define the details of regulation operated by this system are currently under way.

Acknowledgments. This research was supported by the Office of Science (BER) U. S. Department of Energy, Grant No. DE-FG02-01ER63241 and the U. S. National Institutes of Health, Grant GM45404.

# References

Dubbs JM, Tabita FR (2004) Regulators of nonsulfur purple phototrophic bacteria and the interactive control of $CO_2$ assimilation, nitrogen fixation, hydrogen metabolism and energy generation. FEMS Microbiol Rev 28:353-37
Larimer FW, Chain P, Hauser L, Lamerdin J, Malfatti S, Do L, Land ML, Pelletier DA, Beatty JT, Lang AS, Tabita FR, Gibson JL, Hanson TE, Bobst C, Torres JL, Peres C, Harrison FH, Gibson J, Harwood CS (2004) Nature Biotech 22:55-61
Shell MA (1993) Molecular biology of LysR family of transcriptional regulators. Ann Rev Microbiol 47:597-626
Shively JM, van Keulen G, Meijer WG (1998) Something from almost nothing: Carbon dioxide fixation in chemoautotrophs. Ann Rev Microbiol 52:191-230
Stock AM, Robinson VL, Goudreau PN (2000) Two-component signal transduction Ann Rev Biochem 69:183-215
Tabita FR (1995) The biochemistry and metabolic regulation of carbon metabolism and $CO_2$ fixa-

tion in purple bacteria. In Blankeship M, Madigan T, and Bauer CE (Eds.) Anoxygenic photosynthetic bacteria, Kluwer Academic Publisher, pp 885-914

Taylor BL, Zhulin IB (1999) PAS Domains: Internal sensors of oxygen, redox potential, and light. Microb and Mol Biol Rev 63:479-506

# V. Experimental Ecosystem and Climate Change Research

# Experimental ecosystem and climate change research in controlled environments: lessons from the Biosphere 2 Laboratory 1996-2003

Barry Osmond

Visiting Fellow, School of Biochemistry and Molecular Biology, Australian National University, PO Box 3252 Weston Creek ACT 2611, Australia

**Summary.** It is clear from the project summaries below that the Biosphere 2 Laboratory (B2L) delivered handsomely as a controlled environment facility for experimental ecosystem and global climate change research. Ironically, the short and medium term experiments with model complex systems revealed that some of the most exciting and unexpected questions involved carbon cycling in benthic and soil metabolism, the very same processes that caused the first closed mission in the facility to fail, and eventually made the apparatus available for research. The effects of elevated $[CO_2]$ on these processes in the marine and agriforest mesocosms was to stimulate flux and to reduce C-sequestration, by reduced carbonate deposition and enhanced metabolism of soil C reserves, respectively. The extent to which this and other themes that emerged from experiments were products of the initial conditions established in B2L model systems, or are general principles that prevail in natural ecosystems, remains to be seen.

**Key words.** Elevated $[CO_2]$, C-cycles, Coral reefs, Tropical forests, Agriforests

## 1. Introduction

Control of the physical and chemical environment during plant growth has been of paramount importance over the last 50 years in advancing the understanding of physiological, biochemical and genetic processes in individual plants as a foundation of crop and tree physiology, and for the emergence of plant ecophysiology as a discipline. Frits Went pioneered the concept of phytotrons in the 1950s, and these then expensive controlled environment facilities soon became indispensable for environmental plant science research and fostered the discipline of plant ecophysiology. In fact, the scale of commitment in 1960 for the Canberra Phytotron placed it in national competition for capital funding with radio telescopes (Evans 2003). As we now come down firmly to Earth and our need to understand the feedbacks between complexity in the biosphere and changing global climate systems, experimental facilities an order or more of magnitude larger than phytotrons, in size, sophistication and cost, will be needed. One such device, the $200 million Biosphere 2 facility at Oracle, Arizona (ironically, within sight of the telescopes on Kitt Peak), became available in 1996 as a prototype apparatus for large-scale experimental research with complex model ecosystems.

---

*Plant Responses to Air Pollution and Global Change*
Edited by K. Omasa, I. Nouchi, and L. J. De Kok ( Springer-Verlag Tokyo 2005 )

Harte (2002) pointed out that the polarization of research efforts into Newtonian and Darwinian camps has become a key obstruction to progress in earth systems science. The boundaries of the divide can be paraphrased as dynamic global climate models grounded in the laws of fluid dynamics on the one hand, and on the other, a fascination with biodiversity for its own sake, lists of organisms probed to the limits of molecular biology. If we are to bridge the Newtonian-Darwinian divide we must engage an expanded program of controlled experiments that will evaluate the significance of partial processes at all manageable scales (Harte 2002). Without experimental evidence it will be impossible to convince policy makers of the many dimensional responses of the biosphere to changing global climates. Indeed, can any discipline in natural science, no matter the scale or complexity, eschew the importance of controlled experiments indefinitely? In this context, the time for the Biosphere 2 Laboratory (B2L) had come (Marino and Odum 1999; Osmond et al. 2004).

## 2. How B2L became available as an apparatus for experimental ecosystem and global change research

The original focus of B2L on human life support (Walford 2002) demanded a much higher standard of engineering, with redundancy in back-up environmental control systems far exceeding that normally encountered in other systems deployed for ecological research in laboratory and field studies. Perhaps the most tightly closed building ever constructed, B2L exchanged less than 8% of its original atmosphere each year of the original closed system experiment. The original human enclosure experiments revealed that the balance and composition of soil and plants was sub-optimal for long term closed system research, principally because the $O_2$ demand of soil metabolism was greater than the $O_2$ replacement capacity of the vegetation. This itself highlighted one of the least well understood components of the Earth system. Moreover, exposed fresh structural concrete in the enclosed system became an irreversible sink for $CO_2$ (as $CaCO_3$) and rapidly lead to atmospheric $O_2$ depletion (Severinghaus et al. 1994). Subsequent reconfiguration of B2L into a series of separately controlled spaces for research gave a series of chambers that varied from 11,500 to 27,000 $m^3$, in which control and measurement of leaks, temperature and light gradients was better than in conventional growth chambers and gas exchange systems. Retrofitting for research projects in ocean, tropical forest, agriforest and wilderness mesocosms commenced immediately and within a few years distinctive programs were in place (Marino and Odum 1999; Walter and Lambrecht 2004).

Under the stewardship of the Earth Institute of Columbia University, the Biosphere 2 facility was made available in 1996 for large-scale controlled environment experiments with complex systems, as part of a research, education and public outreach effort in the context of an embryonic western campus of the University. The management contract with the owner was renewed in 2000, through 2010, with pledges of support for research leadership faculty appointments, start-up support through 2005, and a research laboratory, in the expectation that it would become self sufficient as a research facility 2006-2010. Following changes to the senior administration of the university in June 2002 it became clear that these commitments were not going to be met, and the apparatus was closed in December 2003. The site was listed for sale in January 2005. This paper

**Table 1.** Distinctive attributes of the Biosphere 2 Laboratory (B2L) for experimental ecosystem and climate-change research (Osmond et al. 2004).

---

1) CONTROLLED PHYSICAL ENVIRONMENTS in the soil-plant-atmosphere continuum with respect to $CO_2$ (and other gas) concentration, temperature, precipitation, humidity and nutrients.

2) MASS AND ISOTOPE BALANCE with all inputs, outputs and flux rates of gases, isotopes and elements measured with precision in open flow or intermittent closure mode.

3) REPLICATION OF EXPERIMENTS IN TIME using synthetic, model complex mini-ecosystems (mesocosms) was achievable within the limits of the usually reliable high light environment of Oracle, AZ.

4) ACCELERATED TESTING OF IDEAS with data available on line and experiments replicated within days to weeks vs. seasons to years to obtain comparable data sets under field conditions.

5) ACCESS TO FOREST CANOPIES via the space frame of B2L for on the leaf measurements of processes with hand held instruments.

6) CALIBRATION OF NEW TECHNOLOGIES based on isotopes and remote sensing against whole system and process measurements.

7) REPLACEMENT AND SUCCESSION because the biological composition of mesocosms can be altered at will.

8) MODEL VALIDATION AND EXTRAPOLATION from synthetic model systems of B2L to natural ecosystems using sensor arrays and defined boundary conditions.

9) INFRASTRUCTURE SUPPORT ON-SITE with engineering, accommodation, and conference facilities commensurate with the role of an inclusive, multi-user facility.

---

illustrates the attributes of B2L (Table 1) with examples of the research done 1996-2003, and in the process also illustrates the potential in future for more specific, purpose built controlled environment systems in experimental ecosystem and global change research.

## 3. Lessons from the B2L coastal marine mesocosm (CMM)

It is estimated that as much as half of the global flux of $CO_2$ into the biosphere occurs in marine ecosystems (Field et al. 1998). Unlike the terrestrial mesocosms, the B2L ocean proved to have been constructed and to have settled in a self-sustainable condition, without the occurrence of algal blooms, and able to support a small coral and fish population with a minimum of maintenance. The CCM thus required the little modification to transform it into a model experimental inshore coral reef mesocosm, and researchers moved quickly with experiments that bridged some critical aspects of the Newtonian-Darwinian divide in this mesocosms.

### 3.1 Calcification in coralline organisms is inhibited by elevated [$CO_2$]

Field scale experiments in the CCM convincingly confirmed that, at all light intensities examined, coral calcification was depressed by 40-50% in seawater with a carbonate chemistry equivalent to that in equilibrium with the atmospheric $CO_2$ concentration expected mid 21st Century (Langdon et al. 2000; Marubini et al. 2001). Moreover, there was a good correspondence between controlled environment experiments and field observations (Broecker et al. 2001). Some regard these observations as the first evidence for a direct deleterious effect of elevated $CO_2$ concentration on a key ecosystem process. When the facility was closed, Langdon was testing a new device capable of measuring coral growth rates with time resolution of hours rather than weeks.

### 3.2 Elevated [$CO_2$] accelerates biological C-influx with faster C-efflux

System level studies using an extremely low level of $^{14}CO_3^-$ (only detectable by C-dating techniques) showed that although elevated $CO_2$ accelerated photosynthesis in the model coastal ecosystem, respiratory turnover was also immediately stimulated, and there was no change in organic C-sequestration (Langdon et al. 2003). As above, the dominant effect of elevated [$CO_2$] on the carbon sink in this marine ecosystem was a 7-fold reduction of inorganic C-sedimentation. Although only a small component of total marine C-flux (Field et al. 1998), these forms of C-sequestration are of critical significance in coastal, tropical marine ecosystems.

### 3.3 Wave motion accelerates nutrient uptake and facilitates $CO_2$ transfer into the ocean during rainfall

The ability to control water flow, wave action and to simulate rainfall some 15m above the seawater surface facilitated another major line of enquiry in the B2L ocean. Demonstrations of the role of mixing in boundary layer limitation of nutrient delivery to corals were reported by Atkinson et al. (2001) and Hearn et al. (2001). Unique large-scale experimental evaluation of wave action and rain drop size on gas mixing, particularly of $CO_2$ transfer atmosphere-ocean surface, addressed one of the important unknowns of global models of $CO_2$ dynamics (Ho et al. 2004). Further reports on this project, involving scientists from 8 laboratories across the US, are in preparation, and studies on the effects of marine biota on the chemistry of ocean derived aerosols had begun before the apparatus closed (J. Allen, unpublished).

When coupled with the effects of warm water incursions on coral bleaching, it becomes clear that the sustainability of coral reefs is threatened by rising $CO_2$, ocean surface temperature and sea level change, suggesting potentially severe socio-economic consequences in coastal reef communities. As in most aspects of slowly unfolding global change, extreme events such as the December 26, 2004 tsunami in the Indian Ocean, compress time scales and almost instantly transform present ecosystems into those we may have to confront in a few decades to come.

## 4. Lessons from the B2L tropical forest mesocosms (TFM)

Difficult of access, tropical forest canopies have been estimated to comprise 90% of the functional interface between Earth's terrestrial biomass and the atmosphere (Ozanne et al. 2003). Tropical forests are thought to represent about half the global terrestrial C-sink

(Field et al. 1999; Malhi and Grace 2000), but there is concern that this sink may switch to a source of $CO_2$ in response to drought events in El Nino years. Over a decade, the model synthetic pan-tropical forest system in B2L became an excellent structural and functional approximation to a disturbed rainforest margin community (Arain et al. 2000; G. Prance 2003, unpublished).

### 4.1 Whole mesocosms responses to elevated [$CO_2$] validate the "big-leaf" model of tropical forest productivity

Lin et al. (1998, 1999), confirmed that the "big leaf" model of Lloyd et al. (1995) fitted well with the $CO_2$ response curves measured in the whole mesocosms and flux measurements from B2L provide a strong basis for improved models of the C-sequestration capacity of tropical forest canopies as atmospheric [$CO_2$] continues to rise. Depending on assumptions about temperature responses of these processes, such models suggest that this sink-capacity is already, or soon will be, saturated (Lin et al. 2001).

### 4.2 The TFM canopy responds rapidly, but reversibly, to month long droughts, but remains a net C-sink, possibly because of the functional diversity of responses among different dominant species

A set of 4 drought treatments applied over 3 years showed that mesocosm C-influx declined by 32% within days of commencement of the drought, suggesting a rapid response to water loss in the soil surface layers. The water content of deep soil was unaffected, and leaf water status of most canopy dominants was only slightly reduced. These changes were reversible within weeks (Rascher et al. 2004). Although stomatal closure reduced ecosystem conductance by 33%, individual trees responded differently and showed different levels of stress and mechanisms of stress avoidance. Taking full advantage of the space frame and climbing facilities in B2L, a team of 25 scientists from 8 institutions showed that leaf area declined by 10% due to a doubling of leaf fall and to reduction in leaf expansion growth by up to 60% in some species. Photosynthetic electron transport rate at light limitation declined by up to 70% in some species but was unaltered in others. This functional diversity introduces heterogeneity of response in the canopy with challenging implications for scaling leaf processes to ecosystem behavior.

### 4.3 Drought reduces the emission of the trace greenhouse gas $N_2O$ and isoprene

The enclosed TFM facilitated measurements of trace gases, showing that although $N_2O$ emission declined with drought, flux from the soil was stimulated 5-20-fold in 24 h after re-watering, releasing a large proportion of $N_2O$ that had not been emitted during the drought (vanHaren et al. 2005). The glass enclosure of B2L minimized the UV-catalyzed destruction of isoprene and permitted accurate estimates of leaf and mesocosm emission rate and also facilitated detection of a soil sink for isoprene of about the same magnitude as the canopy efflux (Pegoraro et al. 2005).

## 4.4 On-line monitoring of stable isotopes in the canopy atmosphere indicates changing C-flux processes during drought

With net $CO_2$-influx of the entire system readily available the TFM was an attractive setting in which to integrate and identify component processes using stable isotopes and a ratio mass spectrometer with an automated sample processing system was installed on-line to obtain routine, well replicated estimates of C, O, and N isotope fractionation. The $\delta^{13}C$ value of respiratory $CO_2$ changed markedly from -25.2 ‰ before drought, to -24.4 ‰ during drought and -28.2 ‰ after re-watering (G Lin in Osmond et al. 2004), indicating stomatal closure during the drought, and rapid respiration of more negative $\delta^{13}C$ substrates (perhaps fats and waxes in leaf litter) after re-watering.

## 4.5 Remote sensing of chlorophyll fluorescence parameters reveals functional diversity in canopy photosynthesis

Canopy access with hand held instruments remains a limiting feature for assessment of the most of the global photosynthetic system. A novel fast repetition rate laser device was developed to analyze chlorophyll fluorescence transients and thereby evaluate stress responses of photosynthetic processes in the outer canopy (Ananyev et al. 2005). Ultimately it may be possible to relate estimates of photosynthetic efficiency obtained by fluorescence (Ananyev et al. 2005; Kolber et al. 2005), or reflectance (Rascher et al. 2005), to canopy C-flux measurements (Osmond et al. 2004). Serendipitously, the arbitrary biodiversity of the TFM also facilitated evaluation of novel mechanisms of photoprotection of the photosynthetic apparatus from excess light (Matsubara et al. 2005).

The policy implications of such insights may be still more uncertain than those associated with coral calcification, but the apparatus that was the B2L TFM clearly made an important contribution to the debate on the sustainability of tropical forest ecosystems in the face of global change. As the apparatus was closed, plans were on hand to harvest and replace elements of the TFM, such as the shade belt of banana and ginger trees, to assess their edge-effect contributions to total mesocosm $CO_2$-fluxes, and to replace dead palms and their deep soil with a walk-in root observation and soil monitoring facility with early successional species. One legacy of the research may be a new generation of flux-towers fitted for isotope collection for better understanding of respiratory C-efflux processes, and with fluorescence and reflectance monitoring and imaging devices that will improve understanding of photosynthetic C-influx processes.

# 5. Lessons from the intensive forest mesocosms (IFM) in B2L

Plantation forestry using rapidly growing trees has been widely touted as a means of mitigating $CO_2$ emissions and as energy forest substitutes for fossil fuels. The decision to replace the former food growing areas of Biosphere with three replicated *Populus deltoides* stands, each exposed to different atmospheric [$CO_2$] (ambient, 2x and 3x ambient), committed this part of the apparatus to a medium term experiment of great potential. Coppiced annually and litter-free, coordination of projects in the TFM presented more problems than in the CMM and TFM, and was especially demanding of the on-site research support team.

## 5.1 Global warming increases temperature at night more than in the day, with differential effects on leaf photosynthesis and respiration

Temperature and humidity control at the stand level in the IFM opened a new range of possibilities for exploring leaf level functions in the canopy (Griffin et al. 2001) such as differential thermal effects on respiration and photosynthesis (Griffin et al. 2002 a, b; Turnbull et al. 2002, 2004; Engel et al. 2004).

## 5.2 Elevated [$CO_2$] inhibits isoprene emission but not in drought

The highly reactive chemistry of this hydrocarbon stimulates local ozone production but levels of this and other reactive oxygen species in the atmosphere is minimized under glass in B2L, facilitating precise estimates of emission and uptake. Emission from *P. deltoides* is inhibited by elevated [$CO_2$] (Rosenstiel et al. 2003) but this effect is eliminated under drought (Pegoraro et al. 2004).

## 5.3 System level measurements of stand photosynthesis and respiration differ from those projected from the leaf level

Murthy et al. (2003) first demonstrated the different effects of temperature on soil respiratory $CO_2$-efflux in the 500 $m^3$ soil blocks of the IFM compared with usual spot measurements. Subsequent experiments with a combination of atmospheric and soil water stress treatments, assessed by both leaf-level measurements and system-level $CO_2$-fluxes, revealed a transition from canopy light environment control of system C-influx to stomatal control of assimilation at the level-of the individual leaf as water stress progressed (Murthy et al. 2005).

## 5.4 Elevated [$CO_2$] promotes leaf area development and this dominates system level responses aboveground

As became obvious from greater leaf fall in response to soil and atmospheric water stress in the elevated [$CO_2$] treatment (Murthy et al. 2005), leaf area development studies by Walter et al. (2005) found larger canopy area in the elevated [$CO_2$] treatment was due to a larger population of more rapidly expanding leaves. The unusual diel pattern of expansion growth in poplar, and an afternoon depression of glucose availability under elevated [$CO_2$] may limit the extent of changes in the population distribution of growth rates.

## 5.5 Elevated [CO$_2$] promotes system respiration at all levels, increases fine root production, rhizodeposition of substrates, and accelerates decomposition of existing soil C and depletion of soil nutrients

Barron-Gafford et al. (2005) observed that although foliage was a large and variable proportion of aboveground biomass, fine root production and C-secretion belowground were key determinants of system respiration. Moreover, decoupling was evident in that stimulation of system level respiration by elevated [CO$_2$] carried over from coppicing to the next growing season initially, and was further exaggerated as the canopy developed. Soil microbial ecology was not significantly perturbed by elevated [CO$_2$], but the biomass of soil microbes increased (Lipson et al. 2005). Trueman and Gonzalez-Meler (2005) used stable isotope to show that elevated [CO$_2$] not only accelerated fine root development during the growing season and accumulation of dead roots in the soil after coppicing, but also accelerated the respiration of "old carbon" residues in the soil. Total soil-C declined over the 3-4 year experiment, and there was no evidence for greater C-sequestration in this agriforest at elevated [CO$_2$].

## 5.6 Elevated [CO$_2$] enhances expression of genes of metabolism, especially those controlling the lignin formation and the chemical composition of wood

After 3 cycles of growth at elevated [CO$_2$] with annual coppicing, *P. deltoides* leaves showed few significant changes in gene expression associated with photosynthesis or respiration. Stems showed marked enhancement of expression in genes associated with lignin biosynthesis, and repression of those associated with cell wall formation and growth (Druart et al. 2005). The implications of these results for paper production, timber quality are clear. Furthermore they present potential opportunities to engineer less biodegradable wood that might enhance C-sequestration and assist mitigation of rising [CO$_2$].

# 6. Lessons from the wilderness mesocosms in B2L

The wilderness mesocosms in B2L were in the process of reconfiguration when the facility was closed. Part of the thorn scrub was cleared of invading grasses and because it was a region of particularly steady air flow, supported studies of elevated [CO$_2$] on oviposition behavior of moths with well developed CO$_2$-sensors (Abrell et al. 2005). The response of moths to volatile chemical attractants was also explored in preparation for studies of interactions with elevated [CO$_2$] (Pophof et al. 2005). A small sonoran desert ecosystem, dominated by succulent plants with nocturnal CO$_2$-fixation was set up in the original test module of B2L. Nobel and Bobich (2002) demonstrated that root growth following rain after drought consumed all the CO$_2$ fixed as well as drawing on reserves. At the system level, this mesocosm was unable to recapture all of the CO$_2$ lost by respiration of the native soil (Rascher et al. 2005 b). This system was presented as a model of the experimental capabilities of B2L as part of the public outreach program.

## 7. Overview

It is clear from the above summaries of some high points of research at B2L that one objective for the apparatus, to operate in "telescope mode" with significant efforts from scientists at remote sites, was realized. The planned consolidation of this potential by research leadership faculty and facilities on-site was not achieved. Although dismayed by the manner and turn of events within the university that caused the above vision to fade, commitments to be withdrawn, and the facility to be closed, one hopes the experimental approaches that emerged from B2L might stimulate research on this scale elsewhere. Further insight into the interactions of temperature and elevated $[CO_2]$ on system C- and nutrient fluxes can be expected, as well as experimental definition of trophic interactions. Integration may well follow from stable isotope, remote sensing and imaging methods explored in B2L.

Acknowledgements. This manuscript is dedicated to the enthusiasm and determination of those colleagues whose work is cited below, and who contributed so much to realizing the potential of B2L. The paper is also a swansong for the author's research career in environmental plant biology that, thanks to the vision of Dr Lloyd Evans, began as graduate student attending an early international congress on the environmental control of plant growth (held to mark the opening of the Canberra Phytotron in 1962) and that ended with the challenge of the Biosphere 2 Center 2001-3. My colleagues and I are immensely grateful for support of B2L 1996-2001 from Drs Crow and Rupp (then Executive Vice-Provost and President of Columbia University, respectively), and from Mr Edward P Bass (the owner of the facility). Additional funding from the David and Lucille Packard Foundation, the Alexander von Humboldt Stiftung, NSF grant CHE-0216226, and the sustained intellectual support of a handful of remarkably visionary members of the University faculty, is gratefully acknowledged.

## References

Abrell L, Guerenstein PG, Mechaber WL, Stange G, Christensen T, Nakanishi K, Hildebrand JG (2005) Effect of elevated $[CO_2]$ on oviposition behavior of *Manduca sexta* moths. Glob Change Biol 11:1272-1282

Atkinson MJ, Falter J, Hearn J (2001) Nutrient dynamics in the Biosphere 2 coral reef mesocosm: water velocity controls $NH_4$ and $PO_4$ uptake. Coral Reef 20:12-21

Arain MA, Shuttleworth WJ, Farnsworth B (2000) Comparing micrometeorology of rain forests in Biosphere 2 and the Amazon Basin. Agr Forest Meteorol 100:273-289

Ananyev G, Kolber ZS, Klimov D, Falkowski PG, Berry JA, Rascher U, Martin R, Osmond B (2005) Remote sensing of heterogeneity in photosynthetic efficiency, electron transport and dissipation of excess light in *Populus deltoides* stands under ambient and elevated $CO_2$ concentrations, and in a tropical forest canopy, using a new laser-induced fluorescence transient (LIFT) device. Glob Change Biol 11:1195-1206

Barron-Gafford G, Martens D, Grieve K, McLain JET, Lipson D, Murthy R (2005) Growth of Eastern Cottonwoods (*Populus deltoides*) in elevated $CO_2$ stimulates stand-level respiration and rhizodeposition of carbohydrates, accelerates soil nutrient depletion, yet stimulates above and belowground biomass production. Glob Change Biol 11:1220-1233

Broecker W, Langdon C, Takahashi T, Peng T-S (2001) Factors controlling the rate of $CaCO_3$ precipitation on Grand Bahama Bank. Global Biogeochem Cycle 15:589-596

Druart N, Rodríguez-Buey M, Barron-Gafford G, Sjödin A, Bhalerao R, Hurry V (2005) Molecular targets of elevated [$CO_2$] in leaves and stems of *Populus deltoids*: implications for future tree growth and carbon sequestration. Functional Plant Biol (in press)

Evans LT (2003) Conjectures, refutations and extrapolations. Annu Rev Plant Biol 54:1-21

Engel VC, Griffin KL, Murthy R, Patterson L, Klimas CA and Potosnak MJ (2004) Growth $CO_2$ modifies the transpiration response of *Populus deltoides* to drought and vapor pressure deficit, Tree Physiol 24:1137-1145

Field CB, Behrenfield MJ, Randerson JT, Falkowski P (1998) Primary production of the biosphere: integrating terrestrial and oceanic components. Science 281:237-240

Griffin KL, Anderson OR, Gastrich MD, Lewis JD, Lin G-H, Schuster W, Seeman J, Tissue DT, Turnbull MH, Whitehead D (2001). Plant growth in elevated $CO_2$ alters mitochondrial number and chloroplast fine structure. Proceedings of the National Academy of Sciences USA 98:2473-2478

Griffin KL, Turnbull MH, Murthy R, Lin G-H, Adams J, Farnsworth B, Mahato T, Bazin G, Potosnak M, Berry JA (2002 a) Leaf respiration is differentially affected by leaf vs. stand-level night-time warming. Glob Change Biol 8:479-485

Griffin KL, Turnbull M, Murthy R (2002 b) The effect of canopy position on the temperature response of leaf respiration in *Populus deltoides*. New Phytol 154:609-619

Harte J (2002) Towards a synthesis of the Newtonian and Darwinian world views. Phys Today 55:29-37

Hearn CJ, Atkinson MJ, Falter J (2001). A physical derivation of nutrient uptake rates in coral reefs: effects of roughness and waves. Coral Reef 20:5-11

Ho DT, Zappa CJ, McGillis WR, Bliven LF, Ward, B, Dacey JWH, Schlosser P, Hendricks MB (2004) Influence of rain on air-sea gas exchange: Lessons from a model ocean. Journal of Geophysical Research 109, C08S18, doi 10.1029/2003JC001806 (15p)

Kolber Z, Klimov D, Ananyev G, Rascher U, Berry J, Osmond B (2005) Measuring photosynthetic parameters at a distance: Laser Induced Fluorescence Transient (LIFT) method for remote measurements of PSII in terrestrial vegetation. Photosynth Res (in press)

Langdon C, Takahashi T, Marubini F, Atkinson MJ, Sweeney C, Aceves H, Barnet H, Chipman D, Goddard J (2000) Effect of calcium carbonate saturation state on the calcification rate of an experimental coral reef. Global Biogeochem Cycle 14:639-654

Langdon C, Broecker W, Hammond D, Glenn E, Fitzsimmons K, Nelson SG, Peng TH, Hajdas I, Bonani G (2003) Effect of elevated $CO_2$ on the community metabolism of an experimental coral reef. Global Biogeochem Cycle 17:1-14

Lipson DA, Blair M, Barron-Gafford G, Grieve K, Murthy R (2005) Relationships between microbial diversity and soil processes under elevated atmospheric carbon dioxide and drought. (in revision for Microbial Ecology (in press))

Lin G, Marino BDV, Wei Y, Adams J, Tubiello F, Berry JA (1998) An experimental and model study of the responses in ecosystem exchanges to increasing $CO_2$ concentrations using a tropical rainforest mesocosm. Aust J Plant Physiol 25:547-556

Lin G, Adams J, Farnsworth B, Wei Y, Marino BVD, Berry JA (1999) Ecosystem carbon exchange in two terrestrial ecosystem mesocosms under changing atmospheric $CO_2$ concentrations. Oecologia 119:97-108

Lin G, Berry JA, Kaduk J, Griffin K, Southern A, Adams J, Van Haren J, Broecker W (2001) Sensitivity of photosynthesis and carbon sinks in world tropical rainforests to projected atmospheric $CO_2$ and associated climate changes. Proceedings 12th International Congress on Photosynthesis. CSIRO Publishing, Melbourne

Lloyd J, Grace J, Miranda AC, Meir P, Wong SC, Miranda H, Wright I, Gash JHC, Mc Intyre J (1995) A simple calibrated model of Amazon rainforest productivity based on leaf biochemical properties. Plant Cell Environ 18:1129-1145

Marino BDV, Odum HT, Eds (1999) Biosphere 2: Research Past and Present. Special Issue of Ecological Engineering 13, Nos.1-4, Elsevier Amsterdam, 359 p (22 chapters)

Marubini F, Barnett H, Langdon C, Atkinson MJ (2001) Interaction of light and carbonate ion on calcification of the hermatypic coral *Porites compressa*. Mar Ecol-Progr Ser 220:153-162

Matsubara S, Naumann M, Martin R, Rascher U, Nichol C, Morosinotto T, Bassi R, Osmond B (2005) Slowly reversible de-epoxidation of lutein-epoxide in deep shade leaves of a tropical tree legume may "lock-in" lutein-based photoprotection during acclimation to strong light. J Exp Bot 56:461-468

Murthy R, Griffin KL, Zarnoch SJ, Dougherty PM, Watson B, van Haren J, Patterson RL, Mahato T (2003) Response of carbon dioxide efflux from a 550$m^3$ soil bed to a range of soil temperatures. Forest Ecol Manage 178:311-327

Murthy R, Barron-Gafford G, Dougherty PM, Engel VC, Grieve K Handley L, Klimas C, Potosnak MJ, Zarnoch SJ, Zhang J (2005) Increased leaf area dominates carbon flux response to elevated $CO_2$ in stands of *Populus deltoides* (Bartr.) and underlies a switch from canopy light-limited $CO_2$ influx in well-watered treatments to individual leaf, stomatally-limited influx under water stress. Glob Change Biol 11:716-731

Nobel PS, Bobich EG (2002) Initial net $CO_2$ uptake responses and root growth for a CAM community placed in a closed environment. Ann Bot 90:593-598

Osmond B, Ananyev G, Berry JA, Langdon C, Kolber Z, Lin G, Monson R, Nichol C, Rascher U, Schurr U, Smith S, Yakir D (2004). Changing the way we think about global change research: scaling up in experimental ecosystem science. Glob Change Biol 10:393-407

Ozanne CMP, Anhuf D, Boulter SL, Keller M, Kitching RL, Korner C, Meinzer FC, Mitchell AW, Nakashizuka T, Dias PL, Stork NE, Wright SJ, Yoshimura M (2003) Biodiversity meets the atmosphere: a global view of forest canopies. Science 301:183-186

Pegoraro E, Abrell L, vanHaren J, Barron-Gafford G, Grieve K, Malhi Y, Murthy R, Lin G (2005) Effects of elevated $CO_2$ concentration and drought on plant production and soil consumption of isoprene in a temperate and tropical rainforest mesocosms. Glob Change Biol 11:1234-1246

Pegoraro E, Rey A, Murthy R, Bobich EG, Barron-Gafford G, Grieve K, Malhi YC (2004) Effect of $CO_2$ concentration and vapor pressure deficit on isoprene emission from leaves of *Populus deltoides* during drought. Functional Plant Biology 31:1137-1147

Pophof B, Stange G, Abrell L (2005) Volatile organic compounds as signals in a plant-herbivore system: electrophysiological responses in olfactory sensilla of the moth Cactoblastis cactorum. Chem Senses 30:51-68

Rascher U, Bobich EG, Lin G-H, Walter A, Morris T, Naumann M, Nichol CJ, Pierce D, Bil' K, Kudeyarov V, Berry JA (2004) Functional diversity of photosynthesis during drought in model tropical rainforest-the contributions of leaf area, photosynthetic electron transport and stomatal conductance to reduction in net ecosystem carbon exchange. Plant Cell Environ 27:1239-1256

Rascher U, Nichol CJ, Small C, Hendricks L (2005 a) Monitoring spatio-temporal dynamics of photosynthesis with a portable hyperspectral imaging system: a case study to quantify the spatio-temporal effects of drought on the photosynthetic efficiency of leaves of four tropical tree species. Photogrammetric Engineering and Remote Sensing (in revision)

Rascher U, Bobich EG, Osmond CB (2005 b) The "Kluge-Kammer": preliminary evaluation of an enclosed Crassulacean acid metabolism (CAM) mesocosm that allows separation of synchronized and desynchronized contributions of CAM plants to whole system gas exchange. Functional Plant Biology (submitted)

Rosenstiel T, Potosnak M, Griffin KL, Fall R, Monson R (2003) Elevated $CO_2$ uncouples growth and isoprene emission in a model agriforest ecosystem. Nature 421:256-259

Severinghaus JP, Broecker W, Dempster W, MacCullum T, Wahlen M (1994) Oxygen loss in Biosphere 2. Transactions of the American Geophysical Union 75:35-37

Trueman R, Gonzalez-Meler MA (2005) Accelerated belowground C losses in a managed agriforest ecosystem exposed to elevated carbon dioxide concentrations. Glob Change Biol 11:1258-1271

Turnbull MH, Murthy R, Griffin KL (2002) The relative impacts of daytime and night-time warming on photosynthetic capacity in Populus deltoides. Plant Cell Environ 25:1729-1737

Turnbull MH, Tissue DT, Murthy R, Wang X, Griffin KL (2004) Nocturnal warming increases photosynthesis at elevated $CO_2$ partial pressure in Populus deltoides. New Phytol 161:819-826

vanHaren JLM, Handley LL, Bil' K, Kudeyarov VN, McLain JET, Martens DA, Colodner DC (2005) Drought-induced $N_2O$ flux dynamics in an enclosed tropical forest. Glob Change Biol 11:1247-1257

Walford RL (2002) Biosphere 2 as a voyage of discovery: the serendipity from inside. BioScience 52:259-263

Walter A, Lambrecht SC (2004) Biosphere 2 Center as a unique tool for environmental studies. Journal of Environmental Monitoring 6:267-277

Walter A, Christ MM, Barron-Gafford G, Grieve K, Paige T, Murthy R, Rascher U (2005) The effect of elevated $CO_2$ on diel leaf growth cycle, leaf carbohydrate content and canopy growth performance of *Populus deltoides*. (Glob Change Biol, 11:1207-1219)

# Importance of air movement for promoting gas and heat exchanges between plants and atmosphere under controlled environments

Yoshiaki Kitaya

Graduate School of Life and Environmental Sciences, Osaka Prefecture University, Gakuen-cho 1-1, Sakai, Osaka 599-8531, Japan

**Summary.** The effects of the air velocity less than 1.3 m s$^{-1}$ on net photosynthetic rates, transpiration rates and water use efficiencies of plant seedlings canopies and single leaves were assessed. Control of air movement is important to enhance gas exchange between plants and the ambient air, and would consequently be important to promote plant growth. The suppression of the transpiration rate was more considerable than that of the net photosynthetic rate at low air velocities, and the water use efficiency was higher at lower air velocities. Importance of air velocity for controlling the gas exchange of plantlets in micropropagation and plants in space farming under microgravity conditions was emphasized. Control of air movement was also important to make the environmental variables uniform inside the plant canopy.

**Key words.** Air velocity, Boundary layer resistance, $CO_2$, Photosynthesis, Transpiration, Water use efficiency

## 1. Introduction

The exchange of $CO_2$ and water vapor between the plant and the atmosphere is controlled by the resistances to gas diffusion from the atmosphere to the chloroplast for $CO_2$ diffusion (photosynthesis) and from the stomatal cavity to the atmosphere for water vapor diffusion (transpiration). In general, photosynthesis and transpiration are commonly controlled by the stomatal resistance and the boundary layer resistance. A number of environmental factors including air temperature, humidity, atmospheric $CO_2$, light intensity, soil moisture, soil temperature, soil salinity have been shown to affect photosynthesis and transpiration mainly through stomatal aperture. However the report on the effects of environmental factors on photosynthesis and transpiration through the leaf boundary layer are less than that through the stomata. The leaf boundary layer resistance is dominantly controlled by the air velocity.

In agriculture, the utilization of semi-closed plant culture facilities such as environmentally controlled greenhouses becomes popular. In such facilities without any adequate air circulation systems, air movement will be extremely restricted compared with that under field conditions. Insufficient air movement around plants increases the resistance to gas diffusion in the leaf boundary layer and thus limits photosynthesis and transpiration

*Plant Responses to Air Pollution and Global Change*
*Edited by K. Omasa, I. Nouchi, and L. J. De Kok ( Springer-Verlag Tokyo 2005 )*

of plants (Yabuki and Miyagawa 1970; Monteith and Unsworth 1990, Jones 1992, Yabuki 2004), which would result in suppression of plant growth and development. Therefore the enhancement of gas exchange in leaves and growth of plant would be dependent on appropriate control of air movement.

Recently the rapid spread of desertification in semiarid areas and shortage of available water for agriculture are problems in the world. In other areas even with abundant fresh water, water pollution has become serious problems and shortage of clean water available for plant culture occurs. Therefore control of the water use efficiency (WUE) becomes important for agricultural plant production in such areas. The WUE is expected to be variable as variation in environmental factors affects the net photosynthetic rate and the transpiration rate in different ways. There have been many reports concerning the effects of environmental factors on WUE. In most of the reports, authors concluded that variable behavior of the stomata affects WUE. However there is very few reports concerning on the effect of the air velocity on WUE.

## 2. Effects of air velocity on photosynthesis, transpiration and water use efficiency of a plant canopy and plant leaves

The effects of the air velocity less than 1.3 m s$^{-1}$ on Pn, Tr and WUE of a seedlings canopy and single leaves of cucumber were assessed with assimilation chamber methods. The Pn and Tr of the plant canopy increased to 1.2 and 2.8 times, respectively, and WUE decreased to 0.4 times with increasing the air velocity from 0.02 to 1.3 m s$^{-1}$ (Fig. 1). The Pn and Tr of the single leaves increased by 1.7 and 2.1 times, respectively, and WUE decreased to 0.8 times with increasing the air velocity from 0.005 to 0.8 m s$^{-1}$ (Fig. 2). The effect of air velocity was more significant on the gas exchange in the plant canopy than that in the single leaf, because the significant reduction of the air velocity inside the plant canopy. The effect of air velocity was more significant on Tr than on Pn and thus WUE decreased with increasing the air velocity in both the plant canopy and the individual leaves.

The leaf boundary layer resistance decreased significantly as air velocities increased from 0.005 to 0.1 m s$^{-1}$ and was proportional approximately to the minus 0.37 power of the air velocity (Fig. 3). Yabuki et al. (1970) reported the same result with cucumber and cabbage leaves at wind velocities less than 2.0 m s$^{-1}$. Martin et al. (1999) estimated the boundary layer conductance from energy balance measurements in the field, which is the reciprocal of the boundary layer resistance. The value increased linearly from 0.01 to 0.15 m s$^{-1}$ as the air velocity increased from 0.1 to 2.0 m s$^{-1}$. The increases in Tr and Pn corresponded to the decrease in the leaf boundary layer resistance as the air velocity increased. The leaf resistance was almost constant during the experiment.

The $CO_2$ concentration in the sub-stomatal cavity in leaves was 43 μmol mol$^{-1}$ lower at the air velocity of 0.005 m s$^{-1}$ than at the air velocity of 0.8 m s$^{-1}$ (Fig. 2), while the water vapor pressure in the sub-stomatal cavity was considered to be constant because the leaf temperature was kept at a constant value. The decrease in the $CO_2$ concentration in the sub-stomatal cavity was a cause of lower inhibition of Pn than Tr and thus a cause of the increase in WUE under low air velocity conditions.

The net photosynthetic rate at the air velocity of 0.4 m s$^{-1}$ was 1.3 times that at 0.1 m s$^{-1}$ under 400 and 800 μmol mol$^{-1}$ $CO_2$ (Fig. 4). The net photosynthetic rate under 800 μmol mol⁻¹ $CO_2$ was 1.2 times that under 400 μmol mol⁻¹ at the air velocity ranging from 0.1 to 0.8 m s⁻¹. The results confirmed the importance of controlling air movement for enhancing the canopy photosynthesis under elevated $CO_2$ levels as well as under a normal $CO_2$ level.

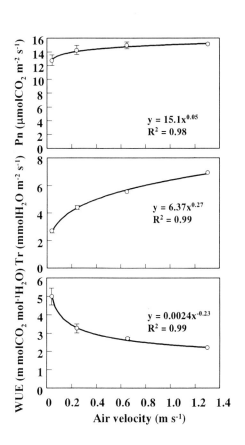

Fig.1. Effects of air velocity on Pn, Tr and WUE of the cucumber seedlings canopy (from Kitaya et al. 2005). The cucumber seedlings canopy with LAI of 1.4 and height of 0.1 m was set in a wind-tunnel installed in the assimilation chamber. The net photosynthetic rate and the transpiration rate are shown on a rooting bed area basis. The air velocity was measured 50 mm above the canopy. Vertical bars indicate standard errors.

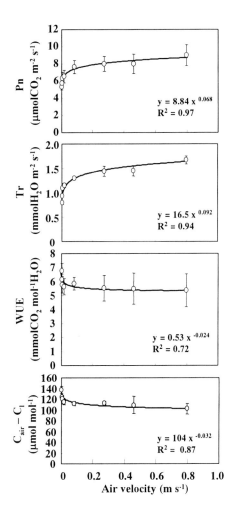

Fig. 2. Effects of air velocity on Pn, Tr, WUE and the $CO_2$ concentration difference between the ambient air ($C_{air}$) and the stomatal cavity ($C_l$) of the cucumber leaves (from Kitaya et al. 2005). The air velocity was measured 5 mm above the leaf surface. Vertical bars indicate standard errors.

188  Y. Kitaya

Fig. 4 also shows that the net photosynthetic rates increased more significantly with increasing $CO_2$ levels in the plant canopy having higher LAI and at lower air velocity. At the air velocity of 0.4 m s$^{-1}$, the net photosynthetic rates at 800 µmol mol$^{-1}$ $CO_2$ were 1.2 times and 1.3 times that at 400 µmol mol$^{-1}$ $CO_2$ when the LAI was 0.6 and 2.5, respectively. At the air velocity of 0.1 m s$^{-1}$, the net photosynthetic rates at 800 µmol mol$^{-1}$ $CO_2$ were 1.2 times and 1.4 times that at 400 µmol mol$^{-1}$ $CO_2$ when the LAI was 0.6 and 2.5, respectively.

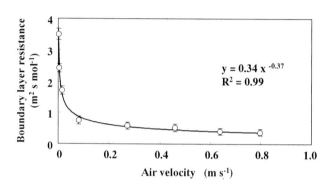

**Fig. 3.** Effects of air velocity on the boundary layer resistance of a wet paper surface. The air velocity was measured 5 mm above the paper surface. Vertical bars indicate standard errors.

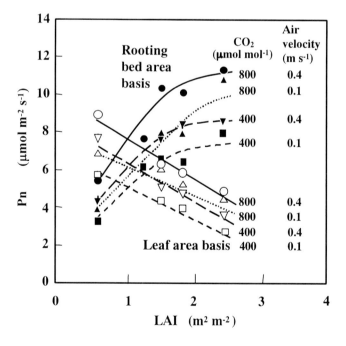

**Fig. 4.** Effects of the leaf area index (LAI) on Pn based on the rooting bed area (shown with solid symbols) and the leaf area (shown with blank symbols) of tomato seedlings canopies under different air current speeds above the canopies and atmospheric $CO_2$ concentrations (from Kitaya et al. 2004). PPFD: 250 µmol m$^{-2}$ s$^{-1}$, air temperature: 23 °C, RH: 55%, Plant heights: 0.05-0.2 m.

## 3. Importance of air velocity for controlling the gas exchange of plantlets in micropropagation

Widespread commercial use of micropropagated transplants is still restricted due to relatively high production costs. Low growth rates of in-vitro plantlets during multiplication and poor survival of the plantlets during acclimatization were the main reasons of the high costs (Kozai et al. 1992). Recently photoautotrophic micropropagation technique with sugar-free rooting medium has been developed (Kozai 1991). The photoautotrophic micropropagation technique has advantages over the conventional one since it has less necessity for keeping aseptic environment, higher growth rates, less physiological disorders and higher survival of plantlets, and thus lower transplants production costs than conventional heterotrophic or photomixotrophic micropropagation techniques with rooting medium containing sugar (Kozai and Smith 1994). The growth rates of plantlets are, however, still lower under the photoautotrophic in-vitro condition than plants under growth chamber, greenhouse and field conditions (e.g. Kozai et al. 1991). The low growth rate in-vitro could be caused by limited photosynthesis, transpiration, water uptake and nutrient uptake of the plantlets in culture vessels.

Air movement in the culture vessels and its effect on the net photosynthetic rates of the in-vitro plantlets were examined. Upward air currents were observed around the plantlets in the central part of the culture vessel and downward air currents were observed near inside walls in the culture vessel (Fig. 5). The temperature distribution inside the vessel created such a pattern. The air current speeds in culture vessels containing a single 10-mm-tall plantlet and a single 60-mm-tall plantlet were, respectively, 1/3 and 1/6 times that in the culture vessel containing no plantlet. The air movements were restricted by the presence of the plantlet in the culture vessel. The air current speed decreased more remarkably with the lager plantlet in the culture vessel mainly due to a resistance to the air movement created by the plantlet.

The net photosynthetic rates of the plantlets increased linearly from 2.0 to 2.5 µmol m$^{-2}$ s$^{-1}$ as the air current speed in the culture vessel increased from 2.0 to 8.0 mm s$^{-1}$ (Fig. 6). Enhancement of air movement inside the culture vessel promoted the photosynthesis of the plantlets. Nakayama et al. (1991) reported that net photosynthetic rates of potato plantlets in-vitro were greater in a forcedly ventilated culture vessel than in a naturally ventilated culture vessel even at the same $CO_2$ concentration inside culture vessels. The increase in the net photosynthetic rate would be due to the enhanced air movement in the culture vessel by the forced ventilation.

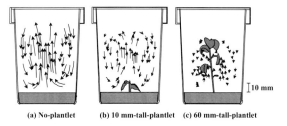

(a) No-plantlet    (b) 10 mm-tall-plantlet    (c) 60 mm-tall-plantlet

**Fig. 5.** The tracer movement in the culture vessels containing no plantlet (a), a single 10-mm-tall plantlet (b) and a single 60-mm-tall plantlet (c) (from Kitaya et al. 2005).

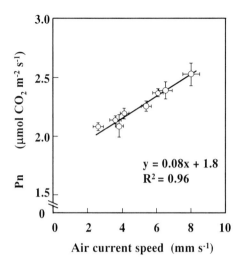

Fig. 6. Effects of the air current speed on the net photosynthetic rate of plantlets in the culture vessel (from Kitaya et al. 2005).

## 4. Importance of air velocity for controlling heat/gas exchanges of plants under microgravity conditions in space

To clarify the effects of gravity on heat/gas exchange between plant leaves and the ambient air, the leaf temperatures and net photosynthetic rates of plant leaves were evaluated at gravity levels of 0.01, 1.0, 1.5 and 2.0 g for 20 seconds each during parabolic airplane flights.

The thermal images of sweet potato leaves 20 seconds after exposure to the gravity levels of 1.0 and 0.01 g showed increase in the temperature by 1.9 °C at the central region in the leaf with decreasing gravity levels (Fig. 7). The mean leaf surface temperatures of sweet potato and barley decreased by 0.4 and 0.3 °C, respectively, for 20 seconds after gravity levels increased from 1.0 to 2.0 g, and increased by 1.4 and 1.2 °C, respectively, for 20 seconds after gravity levels decreased from 2.0 to 0.01 g at the air velocity of 0.2 m s$^{-1}$ (Fig. 8). The leaf temperature increase was approximately 1.0 °C as the gravity levels decreased from 1.0 to 0.01 g. The higher air velocity lowered the leaf temperatures (Fig. 8). The higher air velocity also reduced the surface temperature differences among the different gravity levels.

The net photosynthetic rate was 13% and 20% smaller at a 0.01 g level than at 1.0 g and 2.0 g levels, respectively (Fig. 9). Gas exchange between leaves and the ambient air was also retarded by the microgravity condition as well as heat exchange.

Plant growth chambers used in space must be highly reliable for a long-term space experiment over at least a plant life cycle with a little manual control. Proper control of air movement, therefore, would be essential to enhance the heat/gas exchange between plants and the ambient air and thus promote growth of healthy plants under microgravity conditions in space farming.

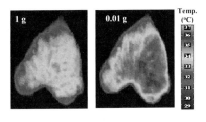

Fig. 7. Thermal images of the sweet potato leaf as affected by gravity level change (from Kitaya et al. 2003). Air temperature: 26 °C, Air velocity: 0.2 m s$^{-1}$, Irradiance: 260 W m$^{-2}$.

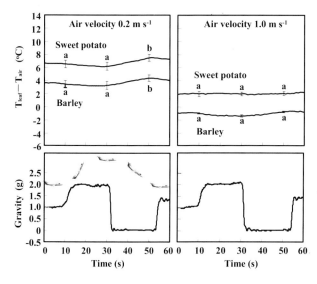

Fig. 8. Time courses of temperatures of sweet potato and barley leaves with changing gravity levels as shown by differences between leaf surface temperatures ($T_{leaf}$) and ambient air temperature ($T_{air}$) at air velocities of 0.2 and 1.0 m s$^{-1}$ (from Kitaya et al. 2003). Bars indicate standard deviations. In each curve the values with different letter indicate a significant difference at a significant level of 95%.

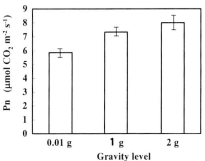

Fig. 9. Effects of gravity levels on Pn of barley leaves 20 seconds after exposure to each gravity level (Kitaya et al. 2001). Bars indicate standard deviations.

## 5. Effects of the light intensity and the air velocity on the air temperature, the water vapor pressure and the CO₂ concentration inside a plant canopy

Low air movement induced spatial variations of environmental variables, such as the air temperature, water vapor pressure and CO$_2$ concentration inside an eggplant seedlings canopy (Kitaya at al. 1998). Under a photosynthetic photon flux density (PPFD) of 500 μmol m$^{-2}$ s$^{-1}$, 2-3°C higher air temperature, 0.6 kPa higher water vapor pressure and 25-35 μmol mol⁻¹ lower $CO_2$ concentration were observed at around the canopy height than at the height 60 mm above the canopy (Fig. 10). The air temperature and the water vapor pressure increased and the $CO_2$ concentration decreased inside the canopy with increasing PPFD. Lower air temperature and higher $CO_2$ concentration inside the canopy were observed at an air velocity of 0.3 m s⁻¹ than that of 0.1 m s⁻¹ (Fig. 11).

Precise control of environmental variables around plants is important for scheduling of crop production and obtaining high yields with a rapid turnover rate in plant production and for increasing repeatable accuracy in experiments with growing plants. Appropriate air movement can make the environmental variables uniform inside the plant canopy.

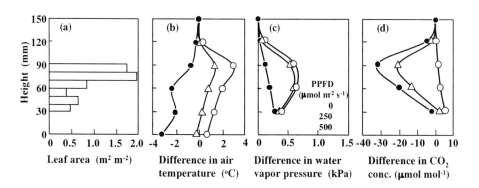

Fig. 10. Profiles of leaf area (a), air temperature (b), water vapor pressure (c) and $CO_2$ concentration (d) inside and outside the canopy of eggplant seedlings as affected by PPFD at an air velocity of 0.1 m s⁻¹ (Kitaya et al. 1998).

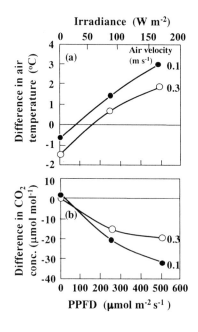

Fig. 11. Effects of PPFD on differences in the air temperature (a) and in the $CO_2$ concentration (b) between two heights at the canopy surface and 60 mm above the canopy at air velocities of 0.1 and 0.3 m s⁻¹ (Kitaya et al. 1998).

# References

Jones HG (1992) Plants and microclimate. Cambridge University Press, Cambridge, UK, 428 p

Kitaya Y, Kawai M, Tsuruyama J, Takahashi H, Tani A, Goto E, Saito T, Kiyota M (2001) The effect of gravity on surface temperatures and net photosynthetic rates of plant leaves. Adv Space Res 28:659-664

Kitaya Y, Kawai M, Tsuruyama J, Takahashi H, Tani A, Goto E, Saito T, Kiyota M (2003) The effect of gravity on surface temperatures of plant leaves. Plant Cell Environ 26:497-503

Kitaya Y, Ohmura Y, Kubota C, Kozai T (2005) Air movement and photosynthesis in micropropagation. Plant Cell Tissue Organ Culture (83:251-257)

Kitaya Y, Shibuya T, Kozai T, Kubota C (1998) Effects of light intensity and air velocity on air temperature, water vapor pressure and $CO_2$ concentration inside a plants stand under an artificial lighting condition. Life Support Biosphere Sci 5:199-203

Kitaya Y, Shibuya T, Tsuruyama J, Kiyota M (2005) Effects of air velocity on photosynthesis, transpiration and water use efficiency of a plant canopy and plant leaves. (submitted)

Kitaya Y, Shibuya T, Yoshida M, Kiyota M (2004) Effects of air velocity on photosynthesis of plant canopies under elevated $CO_2$ levels in a plant culture system. Adv Space Res 34:1466-1469

Kozai T (1991) Autotrophic micropropagation. In: Bajaj YPS (Ed) Biotechnology in agriculture and forestry, Springer-Verlag, NY, pp 313-343

Kozai T, Fujiwara K, Hayashi M, Aitken-Christie J (1992) The in vitro environment and its control in micropropagation. In: Kurata K, Kozai T (Eds) Transplant production systems, Kluwer Academic Publishers, Dordrecht, Netherland, pp 247-282

Kozai T, Smith MAL (1994) Environmental control in plant tissue culture. In: Aitken-Christie J, Kozai T, Smith MAL (Eds) Automation and environmental control in plant tissue culture, Kluwer Academic Publishers. Dordrecht Netherland, pp 301-318

Martin TA, Hinckley TM, Meinzer FC, Sprugel DG (1999) Boundary layer conductance, leaf temperature and transpiration of Abies amabilis branches. Tree Physiol 19:435-443

Monteith JL, Unsworth HM (1990) Principles of environmental physics, Edward and Arnold Publishing Co, London, 291 p

Nakayama M, Kozai T, Watanabe K (1991) Effect of the presence/absence of the sugar in the medium and natural/forced ventilation on the net photosynthetic rates of potato explants in vitro. Plant Tissue Culture Letters 8:105-109 (in Japanese with English summary)

Yabuki K (2004) Photosynthetic rate and dynamic environment. Kluwer Academic Publishers, Dordrecht Netherland, 121 p

Yabuki K, Miyagawa H (1970) Studies on the effect of wind speed on photosynthesis (2) The relation between wind speed and photosynthesis. Japan J Agric Met 26:137-141 (in Japanese with English summary)

Yabuki K, Miyagawa H, Ishibashi A (1970) Studies on the effect of wind speed on photosynthesis (1) Boundary layer near leaf surfaces. Japan J Agric Met 26:65-70 (in Japanese with English summary)

# Pros and cons of $CO_2$ springs as experimental sites

Elena Paoletti[1], Hardy Pfanz[2], and Antonio Raschi[3]

[1]IPP-CNR, Via Madonna del Piano, 50019 Sesto Fiorentino, Italy
[2]Institut für Angewandte Botanik, University Duisburg-Essen, Campus Essen, Universitätsstraße 5, 45117 Essen, Germany
[3]IBIMET-CNR, Via Madonna del Piano, 50019 Sesto Fiorentino, Italy

**Summary.** The increase of atmospheric $CO_2$ concentrations has stimulated research activity at natural $CO_2$ springs, i.e. $CO_2$-emitting vents mostly occurring at sites of former volcanic action. Besides a number of valuable benefits (long-term $CO_2$ enrichment; $CO_2$ gradients over space; natural conditions; free $CO_2$; cheap experiments; possibility to introduce selected vegetation; source of plant material for controlled-condition experiments), emission from the vents may induce variability in atmospheric (composition, temperature, vertical $CO_2$ gradients, short-term $CO_2$ fluctuations) and soil conditions (pH, temperature, $CO_2$ concentrations), and create an environment differing from the $CO_2$-enrichment scenarios. Biological investigations at $CO_2$ springs should previously record all relevant environmental factors and their co-variance. Here we review pros and cons of $CO_2$ springs with the aim to help the selection of the best $CO_2$ springs and control sites to investigate plant responses to $CO_2$ enrichment in natural conditions.

**Key words.** $CO_2$ enrichment, $CO_2$ exhalations, $CO_2$ vents, Elevated $CO_2$, Mofettes

## 1. Introduction

The most common types of facilities used in $CO_2$-enrichment studies on plants are controlled environment chambers (53.6%), open-top chambers (35.7%), free-air $CO_2$-enrichment facilities (7.1%), and natural $CO_2$ springs (3.6%) (Gielen and Ceulemans 2001). Similar to solfatares (sulphur emitting vents) and fumaroles (water vapour vents), $CO_2$ emitting vents (mofettes) mostly occur at sites of former volcanic action, even if several inorganic sources of $CO_2$ in the soil have been listed (Mörner and Etiope 2002). Carbon dioxide of deep mantle origin is released into the atmosphere by gas vents, soil cracks and microseepage. The most biological studies with $CO_2$ springs have been carried out in Italy. $CO_2$ springs have been also reported in Germany, Iceland, Slovenia, Czech Republic, Hungary, Romania, Spain, Greece, Portugal, Austria, France, USA, South Africa, New Zealand, Japan (Pfanz et al. 2004; Onoda et al. 2005), and are assumed to be a worldwide phenomenon. Proper mofette fields (without any sulphur contamination) are rare and exciting habitats for studying wildlife adaptations to a very peculiar but extreme environment. A number of plants more or less adapted to soil and air $CO_2$ extremes grow at $CO_2$ sites. Some of them are hypoxia-adapted species that are also found along rivers and creeks, and in swamps and bogs, whereas others seem to be highly adapted autochthonous and endemic species (e.g. *Agrostis canina* ssp. *monteluccii*) (Selvi 1997).

*Plant Responses to Air Pollution and Global Change*
*Edited by K. Omasa, I. Nouchi, and L. J. De Kok* ( Springer-Verlag Tokyo 2005 )

When $CO_2$ springs are regarded solely as model systems for elucidating the probable effects of future atmospheric $CO_2$ enrichment, several advantages but also disadvantages have to be enumerated. Here we resume pros and cons of $CO_2$ springs as experimental sites, with the aim to help in selecting the best $CO_2$ and control sites for future research.

## 2. Advantages

Natural $CO_2$ springs offer several advantages for investigating plants responses to elevated $CO_2$. 1) Natural vegetation around the vents is supposed to have a multiple-generation or at least a life-time exposure to $CO_2$ enrichment, resulting in long-term adaptations of plants to a naturally $CO_2$-enriched environment. 2) Atmospheric $CO_2$ concentrations are decreasing with distance from the vents and allow to compare plants along natural $CO_2$ gradients (Vodnik et al. 2002). Super-elevated concentrations close to the vent may result in growth inhibition, so that gradients in physiological and biochemical parameters - from depression to stimulation - may be investigated. If mofettes occur in ditches even linear $CO_2$ gradients may be formed and used for ecophysiological studies on plant adaptation (Turk et al. 2001). 3) All environmental conditions are natural, even if emission from the vents can modify some parameters (see below). 4) There is no extra cost: $CO_2$ is free and experiments are technically not challenging, as no sophisticated installation is required for $CO_2$ exposure. 5) Experiments may be carried out with natural or introduced vegetation. Planted and potted plants can be exposed to known $CO_2$ concentrations for short-term investigations. 6) Plant material (seeds, seedlings, herbaceous species, tree branches) may be gathered and used for transfer to experiments under controlled conditions. 7) Plant succession experiments can be performed (e.g. after ploughing). 8) The photosynthetic reaction of mofette plants may be used to determine $CO_2$ emissions from remote (Pfanz et al. 2005). 9) Gas composition may include other substances, such as sulphur-containing compounds (mainly $H_2S$ and $SO_2$) or radioactive substances (radon), and allow to study the interactive effects on plants. Nevertheless, a variable gas composition is also one of the main problem that complicates the study of plant responses to single $CO_2$ (see below).

## 3. Disadvantages

Since the first investigations at $CO_2$ springs, research may benefit of the increased knowledge about possible disadvantages.

1) $CO_2$ springs are assumed to exist for hundreds of years or even longer. However, no historical record is usually available. A few exceptions concern $CO_2$ vents formed by earthquakes in modern times (Rogie et al. 2001). Information can also be derived from local oral traditions. Despite geological processes are much slower than biological ones, trees may live even several hundred years and it is questionable that they have been exposed to $CO_2$ gradients for more than one generation.

2) Most $CO_2$ springs show variability in number and type of vents. When gas is associated with salt rich water, Ca bicarbonate can block the soil cracks. The stability over time in the amount of the gas emitted is usually unknown (Mörner and Etiope 2002).

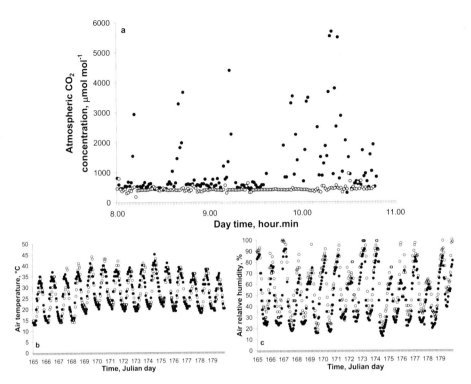

**Fig. 1.** Trends of atmospheric $CO_2$ concentrations (a), air temperature (b), and relative humidity of the air (c) at the Laiatico $CO_2$ spring in Italy, recorded at the canopy height (7 m above ground) at 5 and 130 m far from the $CO_2$ vent (closed and open symbols, respectively).

In the long term, $CO_2$ emission or the position of vents on a given area may have changed, as a result of deep modifications occurring in the feeding system. Some vents may increase emission, others may stop activity. Weak vents may be intermittent. Soil properties like compactness or water uptake capacity (pore size etc.) may interfere or temporarily block gas emission. Such a variable $CO_2$-emitting activity can alter the $CO_2$ levels as recorded in previous measurements. Nevertheless, the distribution of the vents is usually rather stabile. Short-term experiments have also shown that the individual flux from singular vents is rather constant over time (Pfanz et al. 2004).

3) Atmospheric $CO_2$ concentrations strongly fluctuate in the short-term (Fig. 1a), creating an environment that differs from the $CO_2$ enrichment scenarios. Such oscillations (in the order of seconds or minutes) depend on the atmospheric conditions (mainly wind and air temperature), site topography and vegetation cover. Variations in atmospheric $CO_2$ concentrations naturally occur also in other environments, e.g. close to the floor in humus-rich soils or under closed canopies at night (Pfanz et al. 2004). Not only the existence of $CO_2$ fluctuations, but also their frequency and amplitude can be critical in determining time-average stomatal conductance, and thus water loss and carbon gain by plants (Cardon et al. 1994). One of the main problems of the short-term variations in $CO_2$ con-

centrations is that gas exchange measurements cannot be performed by using ambient $CO_2$. Steady-state porometer, chlorophyll *a* fluorescence and $A/C_i$ curves, however, allow a direct investigation of stomatal and photosynthetic responses at $CO_2$ springs.

4) Significant nocturnal accumulation of $CO_2$ may occur, especially under calm air conditions, because of the establishment of the inversion layer and the heavier-than-air nature of $CO_2$ (Badiani et al. 2000). Elevated $CO_2$ concentrations in the dark may reduce respiration (Reuveni and Gale 1985). Leaf isotopic and gas exchange measurements, however, suggest that plants respond to the concentrations during the day rather than the much higher night-time concentrations (van Gardingen et al. 1995). Night hyper-accumulation does not occur at any site, e.g. at Laiatico (central Italy) the location of the vent allows permanent distribution of the gas. Therefore, also topographical variations should be taken into account when choosing a $CO_2$ spring as experimental site.

5) Air temperature may vary in comparison to control sites, when stable atmospheric conditions favour $CO_2$ hyper-accumulation and induce a small-scale "greenhouse"effect. This effect has been recorded at Bossoleto (central Italy), where the topography of the site - an 80m-diameter doline with $CO_2$ vents at the bottom - causes the gas to stagnate (van Gardingen et al. 1997). No significant differences in air temperature and relative humidity (Fig. 1b,c) at 5 and 130 m far from the vent has been recorded at Laiatico, where the vent is located along the slope of a hill.

6) Vertical gradients in atmospheric $CO_2$ concentrations also arise (Fig. 2) and depend both on site topography (van Gardingen et al. 1995), strength of the vents, weather conditions, and vegetation structure. Closed forest canopies reduce air circulation and create sharper vertical gradients than those of non-woody herbs or grasses. When

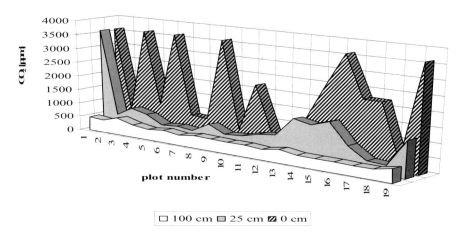

**Fig. 2.** $CO_2$ concentrations in a meadow dominated by *Echinochloa crus galli* located at a mofette field in N-Slovenia. Measurements were taken with a Licor 6400 at 0–200 cm above the ground level. Data are given for three heights (0, 25, 100 cm). Measurements were performed on a sunny and calm day (Pfanz et al. unpublished).

vegetation grows in height, $CO_2$ gradients across the crown also vary. Therefore, tree leaves are likely to have been exposed to differing $CO_2$ regimes over their life. By tracing $^{14}C$ in the wood, Saurer et al. (2003) calculated that the amount originating from the spring decreased over time, from 40% in 1950 to 15% at present, and suggested changes in stand structure - i.e. growth of the canopy into regions with lower concentrations - and decreasing emission from the spring, as possible reasons for the decrease.

7) Depending on the number and distribution of vents within a mofette field, there is a great spatial heterogeneity of $CO_2$ levels within the soil and thus the rooting zone of plants (Pfanz et al. 2004). Concentrations may be so high that root hypoxia may occur.

8) In some springs, saline ground waters occur (mineral water springs). If the water table is close to the soil surface, roots are in contact with a supersaturated salt solution at a relatively elevated temperature, e.g. 38 °C at Bossoleto (van Gardingen et al. 1997).

9) Due to the acidifying properties of $CO_2$ and $H_2S$, the soil may be often very acid (down to pH 2.4-3.7) and the content of soluble aluminium may exceed the toxicity thresholds (Selvi 1997).

10) Variations in soil pH and temperature, besides those in $CO_2$ concentrations, may alter microbial populations and nutrient availability, consequently affecting plant nutrition and soil structure. Ross et al. (2000), however, showed that neither soil pH nor moisture correlated with $CO_2$ concentrations at a $CO_2$ spring in New Zealand. Contrasting results about litter decomposition at $CO_2$ springs have been reported (Ineson and Cotrufo 1997; Gahrooee et al. 1998), possibly depending on site-to-site or even within-site variability.

11) When working with natural vegetation, variability in characteristics within and between populations can exceed any response to $CO_2$. Natural variations in plant age, size, and species distribution may limit the selection of sampling plants and species, and reduce the number of replications per $CO_2$ level.

12) The selection of a comparable control, with similar environmental and biological characteristics, is often the most critical point in working at $CO_2$ springs (Badiani et al. 2000). A control location at ambient $CO_2$ concentration may be as far from the vent as the microclimatic, soil, exposure, floristic, and canopy conditions may vary.

13) Sulphur-containing compounds, mostly $H_2S$ and $SO_2$, are often associated with $CO_2$ emissions. Usually researchers work at sites where the $CO_2$ from the vents is over 99% pure (van Gardingen et al. 1997). The fact that other gases are virtually undetectable does not mean they have no biological effects in the long-term. Acorns from $CO_2$ springs contained significantly higher sulphur and glutathione concentrations than controls from a comparable field site (Grill et al. 2004). When acorns were germinated and seedlings were grown under ambient air conditions far from the $CO_2$ springs, leaves of seedlings grown from $CO_2$ spring acorns still showed significantly higher glutathione concentrations, suggesting a permanent stress situation of the plants originating from $CO_2$ springs.

14) Working at a $CO_2$ spring may be dangerous. Atmospheric $CO_2$ concentrations may be as high as oxygen is lacking and animals die. Despite the $CO_2$ toxicity threshold is high as compared to other gases, it can cause irritations (5% concentration), loss of consciousness (10%), rapid breathing and heart rate, dizziness, visual disturbances, and sometimes death (30% upwards) (Pfanz et al. 2004).

15) Finally, $CO_2$ springs are often located at sites difficult to reach, with no water and electricity. Nevertheless, this is a common problem in open-field experiments.

## 4. Conclusions

The first research paper dealing with biological investigations at $CO_2$ springs and published in an English-speaking scientific journal, was in 1991 (Woodward et al. 1991). The number of papers peaked in 1998 and since then has been quite stabile (Fig. 3), suggesting there is still research interest for $CO_2$ springs. Investigations on ecological processes at $CO_2$ springs should be justified on scientific grounds and not solely focused on the issue of $CO_2$ enrichment in the atmosphere. Obviously, however, this is the reason for increasing research activity at these sites.

Choosing the best $CO_2$ springs and control sites are essential prerequisites to investigate adaptations to $CO_2$ enrichment itself rather than to the special environment of $CO_2$ springs. Sites where the gas is cold and the rock substrate does not contain sulphur should be preferred. Preliminary surveys of $CO_2$ vents' number and location, and of the spatial distribution of $CO_2$ in soil and air should be carried out. Sites with open topography and sparse canopy may favour air circulation and limit $CO_2$ variations over time and space. Sites with homogeneous environmental and biological characteristics allow to select comparable control vegetation at an atmospheric $CO_2$ concentration as close as possible to the ambient one.

Plant responses to vertical gradients and strong short-term fluctuations in $CO_2$ concentrations still remain open questions. $CO_2$ concentrations can vary in time also in natural environments (Blanke 1997). Nevertheless, adaptation mechanisms to fluctuating $CO_2$ might be more complex than one can assume on the basis of average values. Until unravelling this aspect, $CO_2$ concentration gradients should be considered only indicative. For ex., measurements, performed in *Quercus ilex* leaves by using the average $CO_2$ concentration of 1500 $\mu$mol mol$^{-1}$ close to the vent, showed a considerable photosynthetic stimulation (17.0 vs. 5.6 $\mu$mol m$^{-2}$ s$^{-1}$) as compared to a far site at a lower average $CO_2$ concentration (400 $\mu$mol mol$^{-1}$). This stimulation contrasts with the down-regulation of photosynthetic assimilation recorded in long-term $CO_2$ enrichments (Medlyn et al. 1999). If the plants had been really adapted to 1500 $\mu$mol mol$^{-1}$ $CO_2$, as suggested from mapping (Tognetti et al. 2000), photosynthetic rate should have been similar to that at 400 $\mu$mol mol$^{-1}$. Experiments performed in open air fumigation systems, however, have shown temporal variability in the down-regulation patterns (Tognetti et al., 1999). Some studies at $CO_2$ springs suggest no photosynthetic down-regulation, maybe reflecting the need of adequate flexibility to cope with huge and frequent short-term $CO_2$ fluctuations (Badiani et al. 2000). Research performed in mofettes in Slovenia, however, showed a down-regulation of PS or at least an inhibition of photosynthesis (Pfanz et al. 2004).

Because of the difficulties to simulate complex environmental conditions, unrealistic experimental approaches may be accepted by the scientific community. For ex., ozone pollution is characterised by fluctuations in concentration and by affecting mostly the outer crown parts, while artificial exposure is often carried out by fumigating a constant concentration in form of square wave, by injecting $O_3$ from below the crowns, and by forced ventilation to obtain similar concentrations across the crowns. The response to elevated $CO_2$ has been for long time assessed in open top chambers, whose limitations are widely known, or in single leaf chambers, and the extrapolation of data to the response in natural environments may suffer of major faints. Any experimental facility shows its own limitations and possibilities: growth chambers, OTCs, branch enclosures, even FACE facilities, in which the spatial distribution of gas is seldom

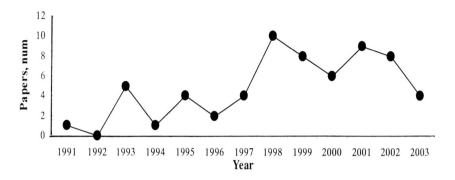

**Fig. 3.** Number of research papers dealing with biological investigations at $CO_2$ springs and published in English-speaking scientific journals.

homogeneous. All these systems are short-termed, with a maximum length of no more than one decade. Despite $CO_2$ springs are not ideal in all respects, they are the only natural areas where life under elevated $CO_2$ can be investigated, provided the selection of $CO_2$ springs and control sites follows a detailed description of all relevant environmental factors and of their co-variance. Inevitable constraints suggest caution in interpreting the results.

## References

Badiani M, Raschi A, Paolacci AR, Miglietta F (2000) Plants responses to elevated $CO_2$; a perspective from natural $CO_2$ springs. In: Agrawal SB, Agrawal M (Eds) Environmental Pollution and Plant Responses. Lewis Publishers, Boca Raton, pp 45-81
Blanke MM (1997) $CO_2$ fluctuations and $CO_2$ fluxes in a fruit tree orchard. In: Mohren GM, Kramer K, Sabaté S (Eds) Impacts of Global Change on Tree Physiology and Forest Ecosystems, Kluwer Academic Publishers, Dordrecht, pp 372
Cardon ZG, Berry JA, Woodrow IE (1994) Dependence of the extent and direction of average stomatal response in *Zea mays* L. and *Phaseolus vulgaris* L. on the frequency of fluctuations in environmental stimuli. Plant Physiol 105:1007-1013
Gahrooee FR (1998) Impacts of elevated atmospheric $CO_2$ on litter quality, litter decomposability and nitrogen turnover rate of two oak species in a Mediterranean forest ecosystem. Global Change Biol 4:667-677
Gielen B, Ceulemans R (2001) The likely impact of rising atmospheric $CO_2$ on natural and managed *Populus*: a literature review. Environ Pollut 115:335-358
Grill D, Müller M, Tausz M, Strnad B, Wonisch A, Raschi A (2004) Effects of sulphurous gases in two $CO_2$ springs on total sulphur and thiols in acorns and oak seedlings. Atmos Environ 38:3775–3780
Ineson P, Cotrufo MF (1997) Increasing concentrations of atmospheric $CO_2$ and decomposition processes in forest ecosystems. In: Raschi A, Miglietta F, Tognetti R, van Gardingen PR (Eds) Plant responses to elevated $CO_2$. Evidence from natural springs. Cambridge University Press,

Cambridge, pp 242-267

Medlyn BE, Badeck FW, de Pury DGG, Barton CVM, Broadmeadow MSJ, Ceulemans R, de Angelis P, Forstreuter M, Jach ME, Kellomaki S, Laitat E, Marek M, Philippot S, Rey A, Strassemeyer J, Laitinen K, Liozon R, Portier B, Roberntz P, Wang K, Jarvis PG (1999) Effects of elevated [$CO_2$] on photosynthesis in European forest species: a meta-analysis of model parameters. Plant Cell Environ 22:1475-1495

Mörner N-A, Etiope G (2002) Carbon degassing from the lithosphere. Global Planet Change 33:185-203

Onoda Y, Hikosaka K, Hirose T (2005) Natural $CO_2$ springs in Japan: a case study of vegetation dynamics. Phyton, in press

Pfanz H, Vodnik D, Wittmann C, Aschan G, Raschi A (2004) Plants and geothermal $CO_2$ exhalations – Survival in and adaptation to a high $CO_2$ environment. In: Progress in Botany, vol 65, Springer-Verlag Berlin Heidelberg, pp 499-538

Pfanz H, Tank V, Vodnik D (2005) Physiological reactions of plants at $CO_2$ emitting mofettes and thermal effect of emerging gas - probable use for remote sensing. Phyton, in press

Reuveni J, Gale J (1985) The effect of high levels of carbon dioxide on dark respiration and growth of plants. Plant Cell Environ 8:623-629

Rogie JD, Kerrick DM, Sorey ML, Chiodini G, Galloway DL (2001) Dynamics of carbon dioxide emission at Mammoth Mountain, California. Earth and Planetary Science Letters 188:535-541

Ross DJ, Tate KR, Newton PCD, Wilde RH, Clark H (2000) Carbon and nitrogen pools and mineralization in a grassland gley soil under elevated carbon dioxide at a natural $CO_2$ spring. Global Change Biol 6:779-790

Saurer M, Cherubini P, Bonani G, Siegwolf R (2003) Tracing carbon uptake from a natural $CO_2$ spring into tree rings: an isotope approach. Tree Physiol 23:997-1004

Selvi F (1997) Acidophilic grass communities of $CO_2$-springs in central Italy: composition, structure and ecology. In: Raschi A, Miglietta F, Tognetti R, van Gardingen PR (Eds) Plant responses to elevated $CO_2$. Evidence from natural springs. Cambridge University Press, Cambridge, pp 114-133

Tognetti R, Longobucco A, Miglietta F, Raschi A, Fumagalli I (1999) Responses of two *Populus* clones to elevated atmospheric $CO_2$ concentration in the field. Ann For Sci 56:493-500

Tognetti R, Cherubini P, Innes JL (2000) Comparative stem-growth rates of Mediterranean trees under background and naturally enhanced ambient $CO_2$ concentrations. New Phytol 146:69-74

Turk B, Pfanz H, Vodnik D, Batič F, Sinkovic T (2001) The effects of elevated $CO_2$ in natural $CO_2$ springs on bog rush (*Juncus effusus* L.) plants. I. Effects on shoot anatomy. Phyton 42:13-23

van Gardingen PR, Grace J, Harkness DD, Miglietta F, Raschi A (1995) Carbon dioxide emissions at an Italian mineral spring: measurements of average $CO_2$ concentration and air temperature. Agric For Meteorol 73:17-27

van Gardingen PR, Grace J, Jeffree CE, Byari SH, Miglietta F, Raschi A, Bettarini I (1997) Long-term effects of enhanced $CO_2$ concentrations on leaf gas exchange: research opportunities using $CO_2$ springs. In: Raschi A, Miglietta F, Tognetti R, van Gardingen PR (Eds) Plant responses to elevated $CO_2$. Evidence from natural springs. Cambridge University Press, Cambridge, pp 69-86

Vodnik D, Pfanz H, Wittmann C, Maček I, Kastelec D, Turk B, Batič F (2002) Photosynthetic acclimation in plants growing near a carbon dioxide spring. Phyton 42:239-244

Woodward FI, Thompson GB, McKee IF (1991) The effects of elevated concentrations of carbon dioxide on individual plants, populations, communities and ecosystems. Ann Bot 67(suppl 1):23-38

# VI. Global Carbon Cycles in Ecosystem and Assessment of Climate Change Impacts

# Carbon dynamics in response to climate and disturbance: Recent progress from multi-scale measurements and modeling in AmeriFlux

Beverly Law

AmeriFlux Science Chair, College of Forestry, Oregon State University, Corvallis, OR 97331-5752, USA

**Summary.** The $CO_2$ flux network, AmeriFlux, aims to quantify and understand the role of terrestrial ecosystems in the global carbon cycle. The network has grown to over 100 sites, with about 25 cluster sites in different disturbance classes or different vegetation types within a climate zone. This chapter summarizes the network objectives, recent findings of AmeriFlux research, and future directions necessary to meet global climate change research goals. The information gained from flux sites, multi-factor experiments on processes, and multi-observation networks can help to improve and parameterize models that are applied to quantify and understand carbon budgets across regions and continents.

**Key words.** Carbon budgets, Ecosystem fluxes, Remote sensing, Process modeling

## 1. Introduction

About half of the $CO_2$ emitted by human activities accumulates in the atmosphere, and half is taken up by oceans and land systems. We lack quantitative understanding of where the sinks are, how they function, and whether there is anything we can do to enhance carbon sequestration to counter emissions long enough for technological improvements to have a significant effect by reducing emissions.

The AmeriFlux network was established in 1996 to understand the role of terrestrial systems in the global carbon cycle (Law et al. 2002; AmeriFlux Strategic Plan; http://public.ornl.gov/ameriflux/about-strat_plan.shtml). The network consists of over 100 sites (Fig. 1), which make continuous meteorological and micrometeorological measurements using the eddy covariance method, core biological measurements (e.g. photosynthesis, respiration) for understanding processes that control fluxes, participate in intercalibration activities with a roving system to ensure high quality data are collected, submit their data to a central archive, and participate in synthesis activities across sites. About 25 of the research teams have cluster sites in different disturbance classes or vegetation types within climate zone to examine the effects of disturbance and climate on carbon stocks and fluxes. A Science Chair is responsible for leading scientific activities of the network, such as coordination and quality assurance of measurements across sites, leading cross-network data analysis and synthesis of results, and communicating AmeriFlux results to the scientific community and other users. A Steering Committee of

*Plant Responses to Air Pollution and Global Change*
Edited by K. Omasa, I. Nouchi, and L. J. De Kok ( Springer-Verlag Tokyo 2005 )

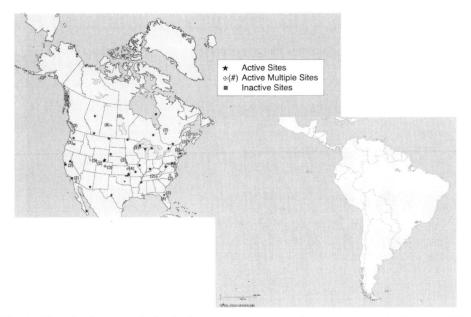

**Fig. 1.** Map of active research sites in the AmeriFlux network (as of September 2004). Many of the sites are clusters in different disturbance classes or vegetation types within a climate zone. The inactive sites have data in the AmeriFlux archive (http://public.ornl.gov/ameriflux/).

lead scientists and agency program managers works with the Science Chair by providing technical and policy advice. The science questions of AmeriFlux are:

- What is the spatial and temporal variability in $CO_2$ and $H_2O$ exchange and how does this vary with disturbance history, land use, and climate?
- What is the spatial and temporal variation in continental $CO_2$ and how does this vary with topography, vegetation, and climate?
- What is the relative effect of these factors?

AmeriFlux plays a key role in the North American Carbon Program (NACP), which was recently developed to meet goals of the US Global Change Research Program (USGCRP; http://www.usgcrp.gov). Science questions of the NACP Implementation Strategy include (1) What is the carbon balance of North America and adjacent oceans? What are the geographic patterns of fluxes of $CO_2$, $CH_4$, and CO? How is the balance changing over time? (*Diagnosis*); (2) What processes control the sources and sinks of $CO_2$, $CH_4$, and CO, and how do the controls change with time? (*Attribution/Processes*); (3) Are there potential surprises (changes in sources or sinks)? (*Prediction*); (4) How can we enhance and manage long-lived carbon sinks, and provide resources to support decision makers? (*Decision support*). AmeriFlux sites are actively participating in the development of the NACP, and in developing new approaches to estimating stocks and fluxes at relevant temporal and spatial scales.

This chapter describes some recent highlights from AmeriFlux research, and the future directions of the network in support of global climate change research.

## 2. Highlights from AmeriFlux research

### 2.1 Strong correlation between ecosystem exchange of carbon dioxide and water vapor

Tower flux data from 37 AmeriFlux and CarboEurope sites in different biomes showed a strong correlation between monthly gross photosynthesis and water vapor exchange (transpiration plus evaporation) at the ecosystem scale, consistent with leaf-level physiological control (Law et al. 2002). The water-use efficiency (WUE = slope of relation between GPP and water vapor exchange) was ranked highest to lowest: evergreen coniferous forests (4.2 g $CO_2$/kg $H_2O$) > grasslands (3.4) > deciduous broadleaf forests (3.2) > crops (3.1) > tundra vegetation (1.5). Thus, evergreen coniferous forests took up more carbon per unit of water loss than other biomes, and three of the biomes had lower but similar efficiencies (grasslands, deciduous broadleaf forests, and crops).

### 2.2 An average of 83% of gross photosynthesis is respired to the atmosphere

Among 37 sites in different biomes, an average of 83% of the total amount of carbon taken up by the terrestrial systems in photosynthesis was respired back to the atmosphere (Law et al. 2002). The ratio of annual ecosystem respiration to gross photosynthesis ranged from 0.55–1.2, with lower values for grasslands, presumably because of less investment in respiring plant tissue compared with forests. Values >1 were observed for boreal forests (i.e., net loss of $CO_2$), ecotonal temperate/boreal forests, some northern temperate forests, and a cropland. The ratio includes effects of both short- and long-term processes. Some of the variation is from differences in heterotrophic respiration, which is strongly influenced by labile carbon pools (Ryan and Law 2005).

### 2.3 Site water balance and temperature are primary abiotic controls on GPP

Across biomes, mean annual temperature and site water balance explained about 65% of the variation in gross photosynthesis (Law et al. 2002). Water availability limits leaf area index over the long term (Waring and Running 1998), and inter-annual climate variability can limit carbon uptake below the photosynthetic capacity of the ecosystems. The simple correlations, as well as biome-specific water-use efficiencies may be useful for large-scale atmospheric modeling of land surface influences on variation in atmospheric $CO_2$.

## 2.4 Ecosystem net carbon uptake is greater when the diffuse fraction of incident radiation is high

Volcanic aerosols from the 1991 Mt. Pinatubo eruption greatly increased diffuse radiation worldwide for the following two years. Long-term flux data in a deciduous forest showed that the increase in diffuse radiation enhanced noontime photosynthesis by 23% the year following eruption (1992) and 8% in the second year (Gu et al. 2003). This finding indicates that the aerosol-induced increase in diffuse radiation contributed to the enhanced terrestrial carbon sink and the temporary decline in the growth rate of atmospheric carbon dioxide ($CO_2$) following the eruption. It points to the need for investigation of the roles of variability of cloudiness and aerosol concentrations in global carbon cycle dynamics.

## 2.5 Soil respiration is strongly linked with labile pools

Soil respiration is the combination of root and rhizosphere respiration, and microbial respiration. It has become the focus of many process studies at flux sites because tower flux studies have shown that soil respiration accounts for 60 to 70% of ecosystem respiration (Law et al. 1999; Goulden et al. 1996; Janssens et al. 2001), and it is strongly linked to photosynthesis (Irvine et al. 2005; Bowling et al. 2002; Janssens et al. 2001), fine root mass (Campbell et al. 2004), and litterfall (Davidson et al. 2002). Experiments at flux sites have shown that roots and associated mycorrhizae produce roughly half of soil respiration (Law et al. 2001a), with much of the remainder derived from decomposition of recently produced litter, which confirms a synthesis of earlier work (Hanson et al. 2000).

Recent flux site studies have shown that there is a large increase in soil respiration following pulse rain events that occur when soils are relatively dry (Lee et al. 2004; Kelliher et al. 2004). A question is whether or not this has a significant effect on annual net carbon uptake by the ecosystem (balance between photosynthesis and respiration).

A synthesis across biomes showed that the lowest to highest rates of soil respiration averaged over the growing season were grassland and woodland/savanna < deciduous broadleaf forests < evergreen needleleaf, mixed deciduous/evergreen forests with growing season soil respiration significantly different between forested and non-forested biomes. This is the inverse of trends in water-use efficiency by biome (Law et al. 2002), and the results of the synthesis suggest that we need to put more effort into investigating the effects of seasonal changes in aboveground and belowground labile pools on soil respiration (Ryan and Law 2005; Hibbard et al. 2005). A synthesis of the current state of our knowledge on soil respiration, and future research needs is presented in Ryan and Law (2005).

## 3. A case study on combining measurements and modeling for regional estimates of carbon stocks and fluxes

The NACP implementation strategy identifies a priority of development of "bottom-up" designs that use multiple scales of observations and process modeling to quantify carbon

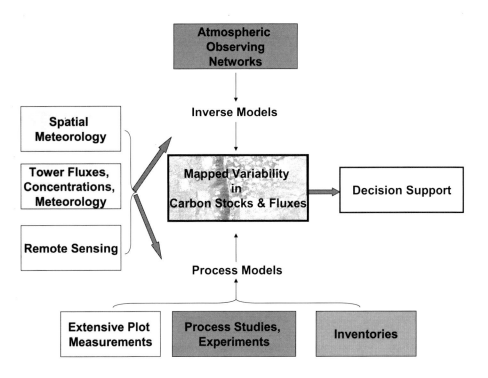

Fig. 2. A conceptual model of the major components of the North American Carbon Program, which will rely on a variety of data sets, remote sensing data and modeling to map variation in carbon stocks and fluxes across North America.

fluxes and understand the influence of climate and disturbance on stocks and fluxes(Denning et al. 2004). It also recommends top-down approaches to quantify and spatially resolve continental $CO_2$ fluxes (Fig. 2). The various approaches are to be developed and compared to reduce uncertainty in estimates of stocks and fluxes.

In Oregon, a regional scaling study was conducted using a spatially nested hierarchy of observations (flux measurements, inventories, remote sensing data) and a biogeochemistry model, Biome-BGC, to map carbon stocks and net ecosystem production (NEP) across western Oregon (Law et al. 2004,2005). Observations of forest age, LAI, and vegetation type were used to develop and test Landsat ETM+ remote sensing algorithms (Law et al. 2005). The Biome-BGC model was parameterized with some of the observations (e.g. foliar C:N, remote sensing LAI), and tested with others (e.g. NEP, GPP from flux data, NPP from extensive sites, ANPP from inventories) to iteratively improve model performance before final application of the model. Spectral regressions with half of the field data and various spectral bands reduced uncertainty in remote

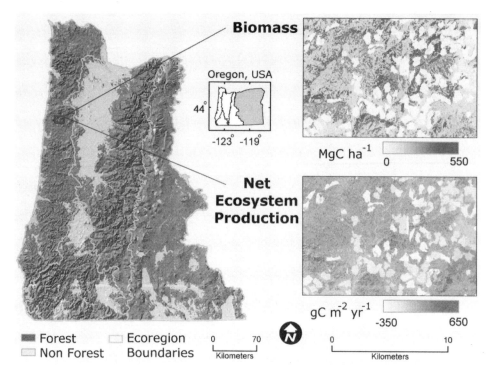

**Fig. 3.** Mapped forest biomass and net ecosystem production (NEP) in western Oregon, where a bottom-up approach was tested for applications of a spatially nested hierarchy of observations, including satellite remote sensing data, and a process model, Biome-BGC to provide realistic estimates of carbon stocks and fluxes across regions (From cover image of September 2004 issue of Global Change Biology).

sensing estimates of LAI, which were used to control the water balance (soil depth) in the model for each pixel. Field data were also used to evaluate remote sensing estimates of stand age and land cover type that are needed for model input. The inventory data were used to determine changes in carbon allocation to wood with forest age.

Western Oregon forests grow across a strong climatic gradient where annual rainfall ranges from 2100mm in the coastal forests to 300mm in the eastern Oregon juniper woodlands. For the forested part of the region (8.2 million hectares), simulated NEP was 168 g C m$^{-2}$ yr$^{-1}$, with the highest mean uptake in the Coast Range ecoregion (226 g C m$^{-2}$ yr$^{-1}$), and the lowest mean NEP in the Eastern Oregon ecoregion (88 g C m$^{-2}$ yr$^{-1}$) (Law et al. 2004; Fig. 3). Carbon stocks averaged 33.7 Kg C m$^{-2}$, with wide variability among ecoregions. This information was combined with harvest records and forest fire information to estimate net biome production (NBP) for western Oregon. The study suggested that averaged over 5 years, the forest NBP of western Oregon offset about 50% of fossil fuel $CO_2$ emissions in the state, and this was reduced by half in the year of a historically-large and severe wildfire. The estimates will be refined when new data on carbon transformations from the forest fires in the region have been analyzed (Campbell et al. in prep.). The study cautions that losses of carbon stocks and fluxes in wildfires are not well quantified in some regions, and that assumptions of carbon losses and transformations by modelers will need to be improved with new data where they are currently lacking.

In developing and testing the prognostic version of the model, Biome-BGC, we found that we need to better characterize the changes in carbon stocks and fluxes following forest stand replacing disturbance for parameterizing the model. In terms of ecosystem processes represented in the model, the dynamics of carbon allocation patterns through forest stand development should be addressed, and these change with climate zone and biome. We also need to characterize successional species mixes in forests following stand replacing disturbance, rather than assume that the disturbed areas are solely a uniform regrowth of the same tree species, and we need better representation of respiration processes (Law et al. 2001b; Thornton et al. 2002; Law et al. 2003). For other biomes, remote sensing land cover should distinguish crop type for improving model estimates; this isn't currently available in commonly used land cover products.

While there is room for improvement in the model and data assimilation approaches, this demonstrates the value of multiple data sources for model ingestion to ultimately quantify variation in the carbon balance. Likewise, the information gained from flux sites, multi-factor experiments on processes, and multi-observation networks can help to improve and parameterize models that are applied to quantify and understand carbon budgets across regions and continents.

## 4. Future directions

AmeriFlux and other flux networks provide multi-year data on half-hourly micrometeorology and meteorology, and they are the focal points of intensive process studies and thorough carbon accounting (e.g. biometric estimates of multi-annual NEP). These data should be collected at sites that represent the major biomes and ecoregions of a network (Hargrove et al. 2003). The data are critical for model development, testing, and parameterization.

Process studies are needed to improve modeling of respiration across landscapes. Although temperature and moisture data have been used to model soil respiration within a site, the relationships do not hold across sites. This requires knowledge about the labile pools, and flux sites will need to measure these pools more rigorously in the future. Long-term experimental manipulations at flux sites will also advance understanding of interactive effects of water, nitrogen and increased atmospheric $CO_2$ for improving carbon cycle models (e.g. Free Air $CO_2$ Enrichment sites that are also measuring whole ecosystem fluxes) (Oren et al. 2001).

Data assimilation methods are relatively new to the flux and biological community, and there is a need to develop and test such methods for regional and continental estimates of stocks and fluxes. The methods are challenging, and rely on both short-term (e.g. eddy flux data, seasonal changes in leaf area index) and long-term process data (e.g. soil carbon pools measured every 5 years). Early tests of these approaches show that more frequent measurements of pools and reduced uncertainty in field biological data and eddy flux data are critical for applying these methods (Williams et al. 2005). In the North American Carbon Program, it is expected that data assimilation methods will be applied in both top-down and bottom-up approaches, and that the approaches will be iteratively compared and improved to reduce uncertainty in carbon balance estimates. To be successful, this requires modelers and field researchers to work together to determine the minimum set of measurements needed with specificity on methods (e.g. soil C pools, acceptable uncertainty levels) and acceptable temporal frequency of measurements.

There is more emphasis now on quantifying and reducing uncertainty for making policy decisions that rely on robust estimates of carbon stocks and fluxes. The various networks, CarboEurope, AsiaFlux, AmeriFlux, and OzFlux, need to work together to improve data quality, data management and data sharing to advance the science that relies on flux network data (Baldocchi et al. 2001).

Acknowledgements. Thanks to the AmeriFlux and CarboEurope investigators for their contributions to the results presented in this chapter.

# References

Baldocchi D, Falge E, Gu L, Olson R, Hollinger D, Running S, Anthoni P, Bernhofer C, Davis K, Evans R, Fuentes J, Goldstein A, Katul G, Law B, Lee X, Mahli Y, Meyers T, Munger W, Oechel W, Paw U K, Pilegaard K, Schmid H, Valentini R, Verma S, Vesala T, Wilson K, Wofsy S (2001) FLUXNET: A new tool to study the temporal and spatial variability of ecosystem-scale carbon dioxide, water vapor and energy flux densities. B Am Meteorol Soc 82:2415-2434

Bowling D, McDowell N, Bond B, Law B, Ehleringer J (2002) $^{13}$C content of ecosystem respiration is linked to precipitation and vapor pressure deficit. Oecologia 121:113-124

Campbell J, Sun O, Law B (2004) Supply side controls on soil respiration among Oregon forests. Glob Change Biol 10:1857-1869

Davidson E, Savage K, Bolstad P, Clark D, Curtis P, Ellsworth D, Hanson P, Law B, Luo Y, Pregitzer K, Randolph J, Zak D (2002) Belowground carbon allocation in forest ecosystems estimated from annual litterfall and IRGA-based chamber measurements of soil respiration. Agr Forest Meteorol 113:39-51

Denning S, Oren R, McGuire D, Sabine C, Doney S, Paustian K, Torn M, Dilling L, Heath L, Tans P, Wofsy S, Cook R, Andrews A, Asner G, Baker J, Bakwin P, Birdsey R, Crisp D, Davis K, Field C, Gerbig C, Hollinger D, Jacob D, Law B, Lin J, Margolis H, Marland G, Mayeus H, McClain C, McKee B, Miller C, Pawson S, Randerson J, Reilly J, Running S, Saleska S, Stallard R, Sundquist E, Ustin S, Verma S Science Implementation Strategy of the North America Carbon Program. http://www.carboncyclescience.gov

Goulden M, Munger J, Fan S, Daube B, Wofsy S (1996) Measurements of carbon sequestration by long-term eddy covariance: methods and a critical evaluation of accuracy. Glob Change Biol 2:169-182

Gu L, Baldocchi D, Wofsy S, Munger J, Michalsky J, Urbanski S, Boden T (2003a) Response of a Deciduous Forest to the Mount Pinatubo Eruption: Enhanced Photosynthesis. Science 299:2035-2038

Hanson P, Edwards N, Garten C, Andrews J (2000) Separating root and soil microbial contributions to soil respiration: A review of methods and observations. Biogeochemistry 48: 115-146

Hargrove W, Hoffman F, B Law B (2003) New Analysis Reveals Representativeness of AmeriFlux Network. Earth Observing System Transactions, American Geophysical Union 84(48):529

Hibbard K, Law B, Reichstien M, Sulzman J (2005) An analysis of soil respiration across northern hemisphere temperate ecosystems. Biogeochemistry 73:29-70.

Irvine J, Law B, Kurpius M (2005) Coupling of canopy gas exchange with root and rhizosphere respiration in ponderosa pine: correlations or controls? Biogeochemistry 73:271-282.

Janssens I, Lankreijer H, Matteucci G, Kowalski A, Buchmann N, Epron D, Pilegaard K, Kutsch W, Longdoz B, Grunwald T, Montagnani L, Dore S, Rebmann C, Moors E, Grelle A, Rannik U, Morgenstern K, Oltchev S, Clement R, Gudmundsson J, Minerbi S, Berbigier P, Ibrom A, Moncrieff J, Aubinet M, Bernhofer C, Jensen N, Vesala T, Granier A, Schulze E, Lindroth A, Dolman A, Jarvis P, Ceulemans R, Valentini R (2001) Productivity overshadows temperature in determining soil and ecosystem respiration across European forests. Glob Change Biol 7: 269-278

Kelliher F, Ross D, Law B, Baldocchi D, Rodda N (2004) Carbon and nitrogen mineralization in litter and mineral soil of young and old ponderosa pine forests during summer drought and after wetting. Forest Ecol Manag 191:201-213

Law B, Ryan M, Anthoni P (1999) Seasonal and annual respiration of a ponderosa pine ecosystem. Glob Change Biol 5:169-182

Law B, Kelliher F, Baldocchi D, Anthoni P, Irvine J (2001a) Spatial and temporal variation in respiration in a young ponderosa pine forest during a summer drought. Agr Forest Meteorol 110:27-43

Law B, Thornton P, Irvine J, Van Tuyl S, Anthoni P (2001b) Carbon storage and fluxes in ponderosa pine forests at different developmental stages. Glob Change Biol 7:755-777

Law B, Falge E, Baldocchi D, Bakwin P, Berbigier P, Davis K, Dolman A, Falk M, Fuentes J, Goldstein A, Granier A, Grelle A, Hollinger D, Janssens I, Jarvis P, Jensen N, Katul G, Mahli Y, Matteucci G, Monson R, Munger W, Oechel W, Olson R, Pilegaard K, Paw U K, Thorgeirsson H, Valentini R, Verma S, Vesala T, Wilson K, Wofsy S (2002) Environmental controls over carbon dioxide and water vapor exchange of terrestrial vegetation. Agr Forest Meteorol 113:97-120

Law B, Sun Campbell J, Van Tuyl S, Thornton P (2003) Changes in carbon storage and fluxes in a chronosequence of ponderosa pine. Glob Change Biol 9:510-524

Law B, Turner D, Campbell J, Sun O, Van Tuyl S, Ritts W, Cohen W (2004) Disturbance and climate effects on carbon stocks and fluxes across western Oregon USA. Glob Change Biol 10:1429-1444.

Law B, Turner D, Lefsky M, Campbell J, Guzy M, Sun O, Van Tuyl S, Cohen W (2005) Carbon fluxes across regions: Observational constraints at multiple scales. In: Wu J, Jones B, Li H, Loucks O, (Eds) Scaling and Uncertainty Analysis in Ecology: Methods and Applications. Columbia University Press, New York, USA(In press).

Lee X, Wu H-J, Sigler J, Oishi C, Siccama T (2004) Rapid and transient response of soil respiration to rain. Glob Change Biol 10:1017-1026

Oren R, Ellsworth D, Johnson K, Phillips N, Ewers B, Mahr C, Schafer K, McCarthy H, Hendrey G, McNulty S, Katul G (2001) Soil fertility limits carbon sequestration by forest ecosystems in a $CO_2$-enriched atmosphere. Nature 411:469-472

Ryan M, Law B (2005) Interpreting, measuring and modeling soil respiration. Biogeochemistry 73:3-27

Thornton P, Law B, Gholz H, Clark K, Falge E, Ellsworth D, Goldstein A, Monson R, Hollinger D, Falk M, Chen J, Sparks J (2002) Modeling and measuring the effects of disturbance history and climate on carbon and water budgets in evergreen needleleaf forests. Agr Forest Meteorol 113:185-222

Waring R, Running S (1998) Forest Ecosystems – Analysis at Multiple Scales. Academic Press, San Diego, USA

Williams M, Schwarz P, Law B, Irvine J, Kurpius M (2005) An improved analysis of forest carbon dynamics using data assimilation. Glob Change Biol 11:89-105.

# Synthetic analysis of the $CO_2$ fluxes at various forests in East Asia

Susumu Yamamoto[1], Nobuko Saigusa[1], Shohei Murayama[1], Minoru Gamo[1], Yoshikazu Ohtani[2], Yoshiko Kosugi[3], and Makoto Tani[3]

[1] Institute for Environmental Management Technology, National Institute of Advanced Industrial Sciences and Technology, Onogawa 16-1, Tsukuba, Ibaraki 305-8569, Japan
[2] Forestry and Forest Products Research Institute, Matsunosato 1, Tsukuba, Ibaraki 305-8687, Japan
[3] Graduate School of Agriculture, Kyoto University, Kitashirakawa, Sakyo-ku, Kyoto, 606-8502, Japan

**Summary.** The preliminary results of long-term $CO_2$ flux measurements at forest sites in East Asia are explained and compared with each other. The features of seasonal variation of $CO_2$ fluxes are different among deciduous-broadleaf, evergreen-coniferous, deciduous-coniferous and tropical forests in East Asia, and the causes of difference are discussed. The integrated yearly NEP (net ecosystem production) estimated from the $CO_2$ flux by eddy covariance method in various forests of East Asia has a notable difference in the range of 2 to 8 tC ha$^{-1}$ year$^{-1}$. The main factors of this difference are the annual mean temperature and tree species. Furthermore, remaining issues are discussed, such as the quantitative estimation of the $CO_2$ flux by the eddy covariance method and the synthetic analysis of the carbon budget under collaborations with biological survey.

**Key words.** Net ecosystem production, Forest sites, Carbon budget, Synthetic analysis

## 1. Introduction

The elucidation of the budgets of carbon, water, and other masses at various terrestrial ecosystems has become an important subject of the global warming issue. For this type of study, the data from precise and long-term observations of carbon dioxide, water vapor, and heat flux taken at various land surfaces and ecosystems have become increasingly important. Furthermore, cooperative activities among researchers of various fields such as micrometeorology, hydrology, forestry, and plant-ecology are required; synthetic analysis according to the conjunction of various data and modeling including atmosphere, vegetation, and soil conditions is necessary as well.

The environmental conditions in East Asia under a monsoon climate differ from North America and Europe. The precipitation in East Asia is larger than that on the North American continent, in particular, the precipitation in the mid-latitude of East Asia, which is three times that of North America. Due to specific climatic feature, the carbon budgets in East Asia are different from those of North America and Europe.

*Plant Responses to Air Pollution and Global Change*
Edited by K. Omasa, I. Nouchi, and L. J. De Kok ( Springer-Verlag Tokyo 2005 )

From this point of view, in this paper, the preliminary results of long-term $CO_2$ flux measurements at forest sites in East Asia are explained and compared with each other about following points:
(1) Seasonal variations of the $CO_2$ flux and their relation with meteorological conditions,
(2) Comparison of the results of Net Ecosystem Production (NEP) with each other, and
(3) Synthetic analysis for the quantitative estimation of the $CO_2$ flux by the eddy covariance method.

## 2. Measurements sites of the fluxes in forest ecosystems

Under a monsoon climate with high precipitation and latitudinal variation, forest ecosystems in East Asia have notable diversity. In this paper, the results of long-term $CO_2$ flux measurements at forest sites in East Asia were examined. The sites were: Takayama (Japan, a cool-temperate deciduous broad-leaf forest), Fujiyoshida (Japan, an evergreen needle-leaf forest), Tomakomai Site (Japan, a deciduous needle-leaf forest), Sakaerat (Thailand, a tropical-evergreen seasonal forest), Pasoh (Malaysia, a tropical natural rain forest), and Bukit Soeharto (Indonesia, a tropical secondary rain forest). Fig. 1 is a map of the measurement sites, types of forests, and meteorological conditions, including main tree species, mean tree age, tree height, elevation above sea level, mean air temperature and annual precipitation.

## 3. Seasonal variations of $CO_2$ flux in the various forests

The features of seasonal variation of $CO_2$ fluxes (NEE= -NEP) are different among deciduous-broadleaf, evergreen-coniferous, deciduous-coniferous, and tropical forests in East Asia, as shown in Fig. 2. The seasonal variation of daytime NEE in tropical rain forests is small, but NEE at Sakaerat, a tropical-evergreen seasonal forest, has a small absolute value in the dry season. The value of NEE in a deciduous forest has a clear seasonal variation and a maximum absolute value in July. The value of ecosystem respiration, RE, was estimated from NEE at nighttime, and parameters of the relationship between RE and air temperature, T, were determined from these data. At the Takayama, Tomakomai and Fujiyoshida sites, the values of RE were approximated by the exponential functions of T, but the relation was not clear at the tropical site due to the small variation in the T range through the year. Due to the high temperature throughout the year, the values of the ecosystem respiration (RE) and gross primary production (GPP) in tropical sites should be large, therefore, the quantitative estimation of NEP (=GPP-RE) is not easy in a tropical area.

From the analysis of the relationship of the $CO_2$ flux and the meteorological conditions, the RE, GPP, and NEP of the forest ecosystems were parameterized as a function of T and PPFD (the photosynthetic photon flux density) as follows:

$$GPP = NEP + RE,$$
$$RE = R_1 \exp[R_2 T],$$
$$GPP = \varphi \times G_{max} \times PPFD / (G_{max} + \varphi \times PPFD).$$

The daily values of GPP were calculated using the net ecosystem exchange (NEE= -NEP) and RE. In these equations, parameters R1 and R2 were estimated using the data under the nearly neutral atmospheric condition in the nighttime, and parameters $\varphi$ and $G_{max}$ were estimated from the regression lines between PPFD and GPP.

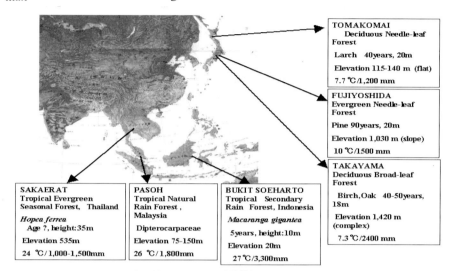

Fig. 1. A map of the measurement sites in present study with type of forest and meteorological conditions; main tree species, mean tree age, tree height, elevation above sea level, mean air temperature and annual precipitation.

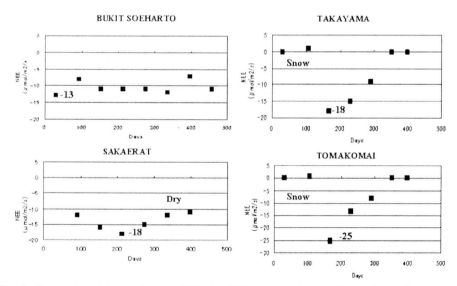

Fig. 2. Seasonal variations of mean NEE (= -NEP) at the photosynthetic photon flux density PPFD=1000 $\mu$ mol m$^{-2}$ s$^{-1}$ in every two-months at four forest sites.

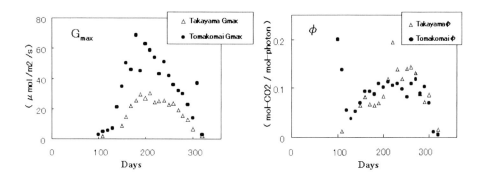

**Fig. 3.** Difference of seasonal variations of photosynthetic efficiency at Takayama and Tomakomai sites.

The seasonal variations of photosynthetic efficiency at the Takayama and Tomakomai sites are shown in Fig. 3. The value of Gmax at Tomakomai has a peak in the beginning of July, and the value of Gmax is about twice of that at Takayama. On the other hand, the value φ related to the photosynthesis in small PPFD has a peak in September, and there is no difference between the two sites.

## 4. Synthetic analysis of the integrated yearly NEP at the various types of forests

The integrated yearly NEP estimated from the $CO_2$ flux by the eddy covariance method in various forests of East Asia is shown in Table 1. The integrated yearly NEP (uptake of carbon to forests) has a notable difference in the range of 2 to 8 tC $ha^{-1}$ $year^{-1}$. The main factors of this difference are the annual mean temperature and tree species.

The relationship between the annual mean air temperature (T) and NEP, and as well as that between T and GPP in world forests, which were cited from this study (Table 1), Valentini et al. (2000); Baldocchi et al. (2001) and other papers in the references, is shown in Fig. 4a,b.

There is remarkable scattering in these figures. This large scattering may be the result of variance in the ages of the forests. However, GPP increases linearly with the temperature, and NEP has a peak in the mid-range of air temperature at around 20 °C. The decrease of NEP in the high air-temperature range may be related to the exponential increase of RE.

The relationship between the mean age of trees and NEP in various forests of the world is shown in Fig. 5. The values of NEP decrease as the age of the forest increases.

**Table 1.** Yearly Integrated NEP in this study (preliminary results) : NEP=GPP − RE.

| Site (Researchers) | Forest Type | Precipitation (mm/year) | Temperature (°C) | NEP | GPP | RE |
|---|---|---|---|---|---|---|
| | | | | (tC ha$^{-1}$ year$^{-1}$) | | |
| Takayama (Yamamoto et al.) | Cool-temperate broad-leaf, deciduous Snow(Dec-Mar) | 2400 | 7 | 2.4 | 8.4 | 6 |
| Tomakomai (Saigusa et al.) | Cool-temperate needle-leaf, deciduous Snow(Dec-Apr) | 1300 | 7 | 3.2 | - | - |
| Fujiyoshida (Ohtani et al.) | Cool-temperate needle-leaf, ever-green Less Snow | 1500 | 10 | 3.3 | - | - |
| Sakaerat (Gamo et al.) | Tropical-seasonal broad-leaf, ever-green Dry(Nov-Mar) | 1500 | 24 | 2.5 | - | - |
| Bukit Soeharto (Gamo et al.) | Toropical-rain broad-leaf, ever-green No Season | 3300 | 27 | 4 | 17 | 13 |
| Pasoh (Tani, et al) | Tropical rainforest broad-leaf, ever-green No Season | 1800 | 26 | 7.7 | - | - |

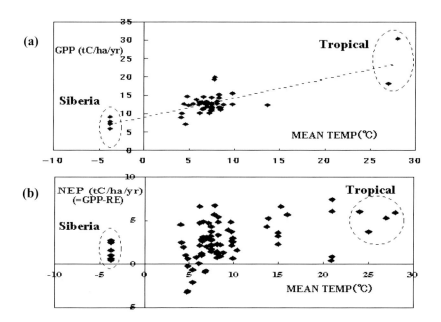

**Fig. 4.** Synthesis of the present data with FLUXNET data by Valentini et al. 2000, Baldocchi et al. 2001 and the other papers in the references. (a) Relation between annual mean air temperature and GPP in various forests. (b) Relation between annual mean air temperature and NEP in various forests.

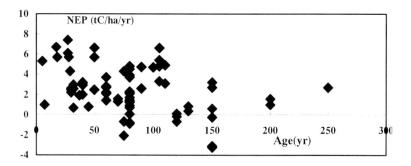

**Fig. 5.** Relation between mean age of trees and NEP in various forests.

## 5. Uncertainty of the annual net ecosystem production

There are several issues for the quantitative estimation of the $CO_2$ flux by the eddy covariance method. The most important issue may be the advection effect that occurs in stable atmospheric conditions due to topographical effects. The reason for the underestimation of NEP based on the eddy covariance method in stable nights is still unclear, but it may be related to the horizontal drainage along the topography. If the observation tower is located on a ridge, it should be possible to observe the flow divergence along the slope during stable nights. Therefore, the nighttime correction of NEP depending on the atmospheric and topographic conditions has a large impact on the annual NEP at each site. In the present study, the nighttime NEP was estimated with and without correction depending on the degree of turbulence indicated by the value of friction velocity, $u^*$.

In this correction, measurement value of RE under low-turbulent conditions was substituted by the estimated value of RE using an empirical equation of RE and the air temperature with the data under high-turbulent conditions indicated by a $u^*$ value larger than 0.2 m s$^{-1}$ or 0.5 m s$^{-1}$. The reasonable value of $u^*$ for the indicator of turbulent conditions is still not clear, but the value of $u^*$ between 0.2 and 0.3 m s$^{-1}$ is used in most analysis of NEP correction. The values of NEP at the Takayama site for 1999, 2000, and 2001 were estimated as 198, 309 and 290 gC m$^{-2}$ year$^{-1}$, respectively, with the correction ($u^*$=0.2 m/s) and 251, 376, and 342 gC m$^{-2}$ year$^{-1}$, respectively, without the correction. The nocturnal correction is responsible for the uncertainty in the annual NEP of about 52-67 gC m$^{-2}$ year$^{-1}$.

## 6. Comparison of the results of the $CO_2$ flux measurement with those from other studies

The results of the $CO_2$ flux measurement can be examined in collaborations with biological surveys, satellite estimates and terrestrial ecosystem models through comparisons of

the following points:
(1) Direct measurements of plant photosynthesis and respiration,
(2) Biometric based estimates of the annual carbon budget from stand surveys,
(3) Leaf Area Index (LAI) and Net Primary Production (NPP) estimated with the use of satellite data and terrestrial ecosystem models, and
(4) Carbon budgets from compartment models of carbon pools and flows.

From these issues, point (2) is discussed briefly using the results from the Takayama site. The annual integrated value of NEP estimated with the eddy covariance method can be compared with biometric-based estimates of the annual carbon budget from stands survey. In this study, many researchers contributed to the biometric-based NEP (Ohtsuka 2003; Nishimura et al. 2004; Mo et al. 2003; Jia et al. 2003; Lee et al. 2003). The value of NPP can be estimated from the increase (or decrease) of the biomass of trees and sasa (bamboo bush), and CWD (coarse wood debris) and litter fall measurement (leaf and small branch). Carbon loss (HR=SR-RR) was estimated through the measurement of soil respiration, SR, and root respiration, RR (trenching method). Table 2 shows the estimated NEP, 0.4 tC ha$^{-1}$ year$^{-1}$, from NPP and HR by the biometric method at Takayama. However, there are large uncertainties with regard to the litter measurement of large branches or CWD as well as the soil uptake of carbon through root turnover. These uncertain factors can contribute to the increase of NEP. On the contrary, NEP estimated from the eddy covariance method was 2.4 tC ha$^{-1}$ year$^{-1}$, but the significant lack of precision is attributed to the u* correction. At present stage, the NEP estimations obtained with the use of the biometric and eddy covariance methods have a significant degree of ambiguity.

Comparison studies about the $NEP_E$ (eddy covariance method) and $NEP_B$ (biometric method) have been conducted at the sites in North America. In Table 3, several results of $NEP_E$ and $NEP_B$ obtained at the present sites and in the North American sites are shown. From this table, the values of $NEP_E$ are larger than those of $NEP_B$ in general, but there are notable differences in the results at each site. As mentioned above, both estimations of NEP have large uncertainties such as the u* correction in $NEP_E$ or the CWD estimation in $NEP_B$.

## 7. Remaining issues

Several issues for the quantitative estimation of the $CO_2$ flux by the eddy covariance method remain. Important issues include achieving reasonable estimations of the advection effect in the stable atmospheric condition and solving the problem of lack of closure in energy balance, in particular, the large imbalance that occurs in complex topography. Most of the Asian sites are in complex topographical areas; therefore, a method for the analysis and estimation of the NEP under complex topographical conditions should be established. Furthermore, physiological and stand structure studies will be necessary to interpret and validate the results of the tower flux measurement. The results obtained from the current studies should be synthetically analyzed in precise comparisons with direct ecological surveys of photosynthesis, respiration, growth rates, and litter fall in forest ecosystems.

For extrapolating results-based tower-flux data on a spatial and temporal scale, more studies will be required to develop satellite remote sensing and numerical ecosystem models that are truly useful for ecosystem monitoring and carbon budget.

**Table 2.** The estimation of NPP and NEP at the Takayama forest by biometric method (W. Mo et al. 2003, Ohtuka 2003 and Saigusa et al. 2002).

| Carbon uptake (NPP) | NPPt = - 0.3 (Biomass increase: tree) +1.8 (CWD)+1.7 (leaf litter) = 3.2 t C ha$^{-1}$ y$^{-1}$ <br> NPPs = $\pm 0$ (Biomass increase: bamboo) + 1.2 (litter) = 1.2 t C ha$^{-1}$ y$^{-1}$ |
|---|---|
| Carbon loss (HR) | SR (Soil R.)=7.3, RR (Root R.) =3.3 t C ha$^{-1}$ y$^{-1}$ <br> HR = (SR – RR) = 4.0 t C ha$^{-1}$ y$^{-1}$ |
| Biometric based NEP | NEP$_B$ = NPP – HR = (3.2 +1.2) – 4.0 = 0.4 t C ha$^{-1}$ y$^{-1}$ |
| Eddy-covariance based NEP (= -NEE) | NEP$_E$ = 2.4 t C ha$^{-1}$ y$^{-1}$ <br> NEP$_E$ - NEP$_B$ = 2.0 t C ha$^{-1}$ y$^{-1}$ |

**Table 3.** NEP$_E$ (Eddy covariance method) and NEP$_B$(Biometric method) at the sites in present study and North American sites.

| Site | WS$^a$ | UMBS$^a$ | Hf$^a$ | OR$^b$ | MMSF$^a$ | Wb$^a$ | Takayama$^c$ (present study, preliminary results) | Pasoh$^d$ | Sakaerat$^e$ |
|---|---|---|---|---|---|---|---|---|---|
| Country | USA | USA | USA | USA | USA | USA | Japan | Malaysia | Tailand |
| Latitude ($^\circ$N) | 45 | 45 | 42 | 44 | 39 | 35 | 36 | 3 | 14 |
| Annual mean Ta ($^\circ$C) | 5 | 6 | 7 | 8 | 11 | 14 | 7 | 26 | 24 |
| NEP$_E$ (tC ha$^{-1}$ y$^{-1}$) | 2.2 | 1.7 | 2.0 | 3.0 | 2.4 | 5.8 | 2.4 | 7.7 | 2.5 |
| NEP$_B$ (tC ha$^{-1}$ y$^{-1}$) | 1.1 | 0.7 | 1.7 | 0.3 | 3.5 | 2.5 | 0.4 | -0.5 | -1.0 |
| (NEP$_E$- NEP$_B$) | 1.1 | 1.0 | 0.3 | 2.7 | -1.1 | 3.3 | 2.0 | 8.2 | 3.5 |

$^a$Curtis et al. 2002; $^b$Law et al. 2001; $^c$Ohtsuka et al. 2003; $^d$Tani et al. 2004; $^e$Gamo et al. 2003

For advances in the synthetic analysis of the carbon budget, it is crucial that the flux research network in Asia be expanded through cooperation among AsiaFlux (Japan), KoFlux (Korea), and ChinaFLUX (China). An effective system should be established for the exchange of data obtained from these individual networks and FLUXNET (global) according as the progress made by each network.

Acknowledgements. The authors would like to express their appreciation to related researchers for the use of the results at Takayama and the other sites in East Asia. This study has been supported and financed by the Global Environment Research Fund of the Ministry of the Environment, Japan.

# References

Anthoni PA, Law BE, Unsworth MH (1999) Carbon and water vapor exchange of an open-canopied ponderosa pine ecosystem. Agr Forest Meteorol 95:151-168

Anthoni PM, Unsworth MH, Law BE, Irvine J, Baldocchi DD, Tuyl SV, Moore D (2002) Seasonal differences in carbon and water vapor exchange in young and old-growth ponderosa pine eco-

systems. Agr Forest Meteorol 111:203-222
Arneth A, Kurbatova J, Kolle O, Shibistova OB, Lloyd J, Vygodskaya NN, Schulze E-D (2002) Comparative ecosystem-atmosphere exchange of energy and mass in a European Russian and a central Siberian bog 11. Inter-seasonal and inter-annual variability of $CO_2$ fluxes. Tellus, 54B:514-530
Baldocchi D, Falge E, Gu L, Olson R, Hollinger D, Running S, Anthoni P, Bernhofer C, Davis K, Evans R, Fuentes J, Goldstein A, Katul G, Law B, Lee X, Malhi Y, Meyers T, Munger W, Oechel W, Paw UKT, Pilegaard K, Schmid HP, Valentini R, Verma S, Vesala T, Willson K, Wofsy S (2001) FLUXNET: a new tool to study the temporal and spatial variability of ecosystem-scale carbon dioxide, water vapor, and energy flux densities. Bull Amer Meteorol Soc 82:2415-2434
Berbigier P, Bonefond J-M, Mellmann P (2001) $CO_2$ and water vapour fluxes for 2 years above Euro flux forest site. Agr Forest Meteorol 108:183-197
Black TA, Denhartog G, Neumann HH, Blanken PD, Yang PC, Russell C, Nesic Z, Lee X, Chen SG, Staebler R, Novak MD (1996) Annual cycles of water vapour and carbon dioxide fluxes in and above a boreal aspen forest. Glob Change Biol 2:219-229
Clark KL, Gholz HL, Moncrieff JB, Cropley F, Loescher HW (1999) Environmental controls over net exchanges of carbon dioxide from contrasting Florida ecosystems. Ecol Appl 9:936-948
Curtis PS, Hanson PJ, Bolstad P, Barford C, Randolph JC, Schumid HP, Wilson KB (2002) Biometric and eddy-covariance based estimates of annual carbon storage in five eastern North American deciduous forests. Agr Forest Meteorol 113:3-19
Falge E, Baldocchi D, Tenhunen J, Aubinet M, et al. (2002) Seasonality of ecosystem respiration and gross primary production as derived from FLUXNET measurements. Agr Forest Meteorol 113:53-74
Gamo M (2003) Measurements of net ecosystem production by eddy correlation method. The Tropical Forestry 57:7-16 (in Japanese)
Goulden ML, Munger JW, Fan S-M, Daube BC, Wofsy SC (1996) Exchange of carbon dioxide by a deciduous forest: response to inter-annual climate variability. Science 271:1576-1578
Goulden ML, Munger JW, Fan S-M, Daube Be, Wofsy SC (1996) Measurements of carbon sequestration by long-term eddy covariance: methods and a critical evaluation of accuracy. Glob Change Biol 2:169-182
Goulden ML, Daube BC, Fan S-M, Sutton DJ, Bazzaz A, Munger JW, Wofsy SC (1997) Physiological responses of a black spruce forest to weather. J Geophys Res 102:28987-28996
Goulden ML, Wofsy SC, Harden JW, Trumbore SE, Crill PM, Gower ST, Fries T, Daube BC, Fan S-M, Sutton DJ, Bazzaz A, Munger JW (1998) Sensitivity of boreal forest carbon balance to soil thaw. Science 279:214-217
Grace J, Malhi Y, Lloyd J, Mcintyre J, Miranda AC, Meir P, Miranda HS (1996) The use of eddy covariance to infer the net carbon dioxide uptake of Brazilian rain forest. Glob Change Biol 2:209-217
Granier A, Pilegaard K, Jensen NO (2002) Similar net ecosystem exchange of beech stands located in France and Denmark. Agr Forest Meteorol 114:75-82
Greco S, Baldocchi DD (1996) Seasonal variations of $CO_2$ and water vapor exchange rates over a temperate deciduous forest. Glob Change Biol 2:183-197
Hirano T, Hirata R, Fujinuma Y, Saigusa N, Yamamoto S, Harazono Y, Takada M, Inukai K, Inoue G (2003) $CO_2$ and water vapor exchange of a larch forest in northern Japan. Tellus 55B:244-257
Hollinger DY, Goltz SM, Davidson EA, Lee JT, Tu K, Valentine HT (1999) Seasonal patterns and environmental control of carbon dioxide and water vapour exchange in an ecotonal boreal forest. Glob Change Biol 5:891-902

Jia S, Akiyama T, Mo W, Inatomi M, Koizumi H (2003) Temporal and spatial variability of soil respiration in a cool temperate broad-leaved forest, 1. Measurement of spatial variance and factor analysis. Japanese Journal of Ecology 53:13-22 (in Japanese with an English summary)

Law BE, Thornton PE, Irvine J, Anthoni PM, Van Tuyl S (2001) Carbon storage and fluxes in ponderosa pine forests at different developmental stages. Glob Change Biol 7:755-777

Lee M-S, Nakane K, Nakatsubo T, Koizumi H (2003) Seasonal changes in the contribution of root respiration to total soil respiration in a cool-temperate deciduous forest. Plant Soil 255: 311-318

Lee X, Fuentes JD, Staebler RM, Neumann HH, (1999) Long-term observation of the atmospheric exchange of $CO_2$ with a temperate deciduous forest in southern Ontario, Canada. J Geophys Res 104:15975-15984

Lindroth A, Grelle A, Moren A-S (1998) Long-term measurements of boreal forest carbon balance reveal large temperature sensitivity. Glob Change Biol 4:443-450

Lloyd J, Shibistova O, Zolotoukhine D, Kolle O, Arneth A, Wirth C, Styles JM, Tchebakova NM, Schulze E-D (2002) Seasonal and annual variations in the photosynthetic productivity and carbon balance of a central Siberian pine forest. Tellus 54B:590-610

Malhi Y, Nobre AD, Grace J, Kruijt B, Pereira MGP, Culf A, Scott S (1998) Carbon dioxide transfer over a Central Amazonian rain forest. J Geophys Res 103:31593-31612

Milyukova IM, Kolle O, Varlagin AV, Vygodskaya NN, Schulze E-D, Lloyd J (2002) Carbon balance of a southern taiga spruce stand in European Russia. Tellus 54B:429-442

Mo W, Jia S, Lee M-S, Uchida M, Koizumi H (2003) Temporal and spatial variation of soil respiration in a cool-temperate deciduous broad-leaved forest in Takayama, Japan. - Proceedings of Synthesis Workshop on the Carbon Budget in Asian Monitoring Network (Takayama, Japan), pp 53-57

Nakai Y, Kitamura K, Suzuki S, Abe S (2003) Year-long carbon dioxide exchange above a broadleaf deciduous forest in Sapporo, Northern Japan. Tellus 55B:305-312

Nishimura N, Matsui Y, Ueyama T, Mo W, Saijo Y, Tsuda S, Yamamoto S, Koizumi H (2004) Evaluation of carbon budgets of a forest floor Sasa senanensis community in a cool-temperate forest ecosystem, central Japan. Jpn J Ecol 54 (in press)

Ohtani Y, Mizoguchi Y, Watanabe T, Yasuda Y, Toda M (2001) Carbon Dioxide Flux Above an Evergreen needle leaf forest in a temperate region of Japan. Sixth International Carbon Dioxide Conference, pp 469-472

Ohtsuka T (2003) Biometric based estimates of annual carbon budget in a cool-temperate deciduous forest stand beneath a flux tower. - Proceedings of Synthesis Workshop on the Carbon Budget in Asian Monitoring Network (Takayama, Japan), pp 37-40

Pilegaard K, Hummelshoj P, Jensen NO, Chen Z (2001) Two years of continuous $CO_2$ eddy-flux measurements over a Danish beech forest. Agr Forest Meteorol 107:29-41

Roser C, Montagnani L, Schulze E-D, Mollicone D, Kolle O, Meroni M, Papale D, Marchesine LB, Federici S, Valentini R (2002) Net $CO_2$ exchange rates in three different successional stages of the Dark taiga of central Siberia. Tellus 54B:642-654

Saigusa N, Yamamoto S, Murayama S, Kondo H, Nishimura N (2002) Gross primary production and net ecosystem exchange of a cool-temperate deciduous forest estimated by the eddy covariance method. Agr Forest Meteorol 112:203-215

Schmid HP, Grimmond CSB, Cropley F, Offerlr B, Su H-B (2000) Measurements of $CO_2$ and energy fluxes over a mixed hardwood forest in the mid-western United States. Agr Forest Meteorol 103:357-374

Shimizu T, Shimizu A, Ishizuka S, Daimaru H, Miyabuchi Y, Ogawa Y (2001) Seasonal variation of $CO_2$ exchange and its characteristics over artificial coniferous forest stands in the warm temperate region, Japan. Sixth International Carbon Dioxide Conference, pp 416-419

Tani M, Kosugi Y, et al. (2004) Report of research results, Global environment research fund in FY2003, Ministry of Environment, Japan (in Japanese)

Valentini R, Matteucci G, Dolman AJ, Schulze E-D, et al. (2000) Respiration as the main determinant of carbon balance in European forests. Nature 404:861-865

Valentini R, Angelis PD, Matteucci G, Monaco R, Dore S, Scarasciamugnozza GE (1996) Seasonal net carbon dioxide exchange of a beech forest with the atmosphere. Glob Change Biol 2:199-207

Watanabe T, Yasuda Y, Yamanoi K, Ohtani Y, Okano M, Mizoguchi Y (2000) Seasonal variations in energy and CO2 fluxes over a temperate deciduous forest at Kawagoe, Japan. Cger-Report:ws, pp 11-16

Wofsy SC, Goulden ML, Munger JW, Fan S-M, Bakwin PS, Daube BC, Bassow SL, Bazzaz FA (1993) Net exchange of $CO_2$ in a mid-latitude forest. Science 260:1314-1317

Wang H, Saigusa N, Yamamoto S, Kondo H, Hirano H, Toriyama A, Fujinuma Y (2004) Net ecosystem $CO_2$ exchange over a larch forest in Hokkaido, Japan. Atmos Environ 38:7021-7032

Yamamoto S, Murayama S, Saigusa N, Kondo H (1999) Seasonal and inter-annual variation of $CO_2$ flux between a temperate forest and the atmosphere in Japan. Tellus, 51B:402-413

Yamamoto S, Saigusa N, Murayama S, Kondo H (2001) Long-term results of flux measurement from a temperate deciduous forest site (Takayama). CGER-Report M011:5-10

# 3-D remote sensing of woody canopy height and carbon stocks by helicopter-borne scanning lidar

Kenji Omasa and Fumiki Hosoi

Graduate School of Agricultural and Life Sciences, The University of Tokyo, Yayoi 1-1-1, Bunkyo-ku, Tokyo 113-8657, Japan

**Summary.** A method for 3-D remote sensing of forest canopies using high-resolution, helicopter-borne scanning lidar is presented. The lidar device can scan almost all the ground surface with high resolution because a laser beam with a small footprint and a high scanning rate illuminates the ground surface from a slow-moving helicopter. The method permits the generation of 3-D images such as a Digital Elevation Model, Digital Terrain Model, and Digital Canopy Height Model (DCHM). The validity of the method was demonstrated in two applications. First, we estimated tree height from a DCHM of a forest on a steep slope, and found that errors were within 0.47 m for tree height (0.19 m RMSE). These results show that the estimation of tree height was greatly improved compared with estimates in previous studies. Second, we estimated carbon stocks in each tree and in the stand as a whole. From lidar-derived tree heights and allometric relationships between tree height and carbon stocks, we accurately estimated total carbon stocks of each tree in a coniferous Japanese cedar (*Cryptomeria japonica*) forest as well as carbon stocks at the stand scale.

**Key words.** 3-D remote sensing, Woody carbon stock, DEM, DTM, DCHM, Helicopter-borne scanning lidar, Tree height

## 1. Introduction

Accurate estimation of canopy structures and biomass in forests is crucial for studying the functioning of forests and in studies of the global carbon budget. Remote-sensing techniques have been proven to be reliable tools for assessing environmental changes and their impacts on forests in recent decades. Satellite-based sensors, such as Landsat TM and NOAA AVHRR, can observe wide areas efficiently and have been shown to be effective in many ecological applications (Waring et al. 1995; Goward and Williams 1997). However, these conventional sensors have significant limitations for forestry applications because their sensitivity and accuracy have repeatedly been shown to decrease as aboveground biomass increases (Waring et al. 1995; Turner et al. 1999). In addition, they cannot fully represent the 3-D spatial features of forests.

Lidar (light detection and ranging, a laser-based equivalent to radar) has become a popular, active remote-sensing technology, and its applications to 3-D terrestrial observation have been developing rapidly. Lidar offers many advantages in forestry compared

*Plant Responses to Air Pollution and Global Change*
Edited by K. Omasa, I. Nouchi, and L. J. De Kok ( Springer-Verlag Tokyo 2005 )

with the abovementioned conventional sensors; in particular, it allows measurement of the 3-D distribution of forest canopies. Since the mid-1980s, non-scanning airborne lidar has been used to estimate the average height of forest trees and forest biomass (Nelson at al. 1988). In the mid-1990s, scanning lidar became available, and this technology was soon applied for forest measurement (Nilsson 1996; Næsset 1997; Means et al. 1999). However, ordinary scanning lidar systems with small-footprint lasers were unable to illuminate the entire ground surface because of the high flight speed of the aircraft and the device's low pulse frequency (Flood and Gutelius 1997; Næsset and Økland 2002). As a result, tree tops are not captured exactly by using this system, and tree heights are underestimated. Another airplane-based lidar system, which uses large-footprint lidar with a large scan width, has been used for remote sensing of forests on a regional scale (Nilsson 1996; Means et al. 1999; Harding et al. 2001). The large footprint lidar can cover large areas and fully illuminate the ground surface. However, its spatial resolution is restricted because of its large footprint.

Recently, spatial resolution has been improved by the use of scanning lidar with a small footprint and a high pulse frequency, mounted on a helicopter, which can travel at slower speeds than airplanes can (Omasa et al. 2000, 2003; Brandtberg et al. 2003; Maltamo et al. 2004). These systems are expected to more completely illuminate the ground surface and generate more accurate 3-D data, thereby enhancing the usefulness of scanning lidar. Work using conventional lidar systems has focused on measuring forest attributes such as the mean tree height. However, recent systems based on lidar with high spatial resolution can measure individual trees rather than estimating average values for the stand (Brandtberg et al. 2003). In this paper, remote sensing of forests using a high-resolution, helicopter-borne scanning lidar is presented to reveal the technique's potential.

## 2. Measurement of the forest canopy

Fig. 1 shows a schematic diagram of the helicopter-borne scanning lidar system that we used (Omasa et al. 2000). The elapsed time between the emitted and returned laser pulses was measured, and was used to calculate the distance to the object returning the signal. This system offers two receiving modes: a first-pulse mode (FP-mode) and a last-pulse mode (LP-mode). Laser pulses returned from the outermost canopy surface and from the ground surface were received simultaneously in FP-mode and LP-mode, respectively. In the experiment, more LP-mode beams reached the ground surface through forest canopies because the setting of the footprint for the LP-mode beam was smaller than that of the FP-mode beam. Because of the helicopter's low flight speed (50 km $h^{-1}$) and the lidar's high pulse frequency (25 kHz), each footprint interval nearly equaled the footprint diameter (about 30 cm), and as a result, almost the entire ground surface could be illuminated.

We determined the 3-D geographic position of each data point using a helicopter-borne internal measurement unit (IMU) combined with global positioning system (GPS) devices in the helicopter and on the ground. We generated a Digital Elevation Model (DEM) with a highly precise grid (10×10 cm) by interpolating the lidar data. The canopy surface was described by a DEM created from the FP-mode data, and the ground surface was described as a Digital Terrain Model (DTM) by interpolating the LP-mode data for the ground surface. We then produced a Digital Canopy Height Model (DCHM), which

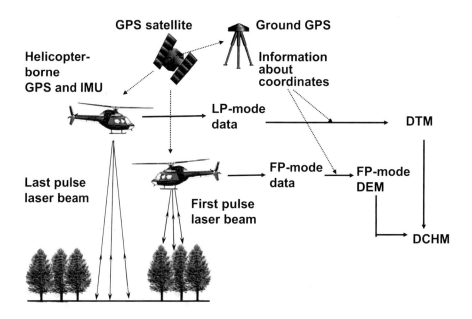

**Fig. 1.** A schematic diagram of the 3-D remote-sensing system based on helicopter-borne scanning lidar.

shows the net tree height after accounting for variations in ground elevation, by subtracting the LP-mode DTM values from the FP-mode DEM values.

## 3. Estimation of ground surface and woody canopy heights

We tested our system at a site located at the foot of a mountain (in Shizuoka, Japan), where the ground surface slopes steeply (Omasa et al. 2000). Fig. 2A shows a 3-D view of the resulting FP-mode DEM. All trees, buildings, and roads can be clearly distinguished. This image shows that our method effectively generates a precise 3-D image of the forest's woody canopy. The elevation of each point in this image is still influenced by the slope of the ground, so tree height cannot be estimated directly from this image; only differences in canopy height can be estimated. In contrast, Fig. 2B shows the DTM produced from the LP-mode data. The ground surface obstructed by the presence of trees was interpolated relative to adjacent ground points measured using the LP-mode. The slope and undulations of the ground can be clearly recognized in this image. Map data and ground-truth measurements confirmed that the DTM was accurate despite of the presence of many trees. Fig. 2C shows the DCHM used to estimate canopy height at the study site after correction for the influence of the ground slope and undulation (i.e., by subtracting the DTM data from the corresponding DEM data). We estimated the errors in the lidar data by comparing the lidar-derived data with the ground-truthed data. The results showed an error within 0.47 m in tree height (0.19 m RMSE; Table 1).

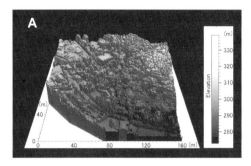

**Fig. 2.** 3-D views of the study site: (A) the DEM created from the FP-mode data; (B) the DTM created from the LP-mode data; (C) the DCHM created by correcting the DEM using the DTM data (Omasa et al. 2000).

**Table 1.** Measurements of the height of the forest canopy.

| Tree species | Ground-truthed tree heights | | Error in the lidar data | |
|---|---|---|---|---|
| | Range (m) | Mean (m) | Range (m) | RMSE (m) |
| Coniferous trees | 11.20 to 19.65 | 15.05 | -0.47 to 0.19 | 0.19 |
| Broadleaved trees | 1.95 to 10.40 | 6.58 | -0.40 to 0.13 | 0.12 |

## 4. Estimation of the forest's carbon stock

Accurate estimation of the carbon stocks of forests is crucial for understanding the global carbon budget and climate change. The height data for a forest canopy obtained from helicopter-borne lidar can be used to accurately estimate forest carbon stocks. For example, Omasa et al. (2003) and Patenaude et al. (2004) successfully used small-footprint lidar to estimate forest carbon stocks. In our study (Omasa et al. 2003), the carbon stocks of each tree in a coniferous Japanese cedar (*Cryptomeria japonica*) forest were estimated by using allometric relationships between lidar-derived tree heights and the corresponding carbon stocks.

We generated the DCHM by subtracting the DTM created from the LP-mode values from the DEM created from the FP-mode values as shown in Fig. 3. Each tree height at the site was automatically determined from the DCHM image by means of filtering to identify the highest point in the canopy surface for each tree. Fig. 4 shows the resulting image for the highest point of trees in the DCHM image. The trees in this image are all Japanese cedars. The position of the highest point of a given tree corresponds with the position of its stem. Therefore, Fig. 4 identifies the position of each tree stem. The error in tree height ranged from –0.25 to –0.42 m, with an RMSE of 0.38 m, compared with the ground-truthed data. Fig. 5 shows the tree height distribution of the study site obtained from the DCHM. Tree heights range from 6.0 to 27.0 m, with an average height of 21.3 m, and most heights (90%) range between 18.5 and 26.0 m.

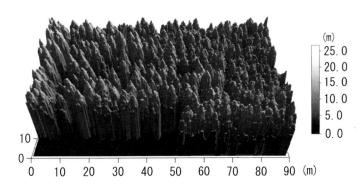

**Fig. 3.** A 3-D view of the DCHM for the Japanese cedar forest (Omasa et al. 2003).

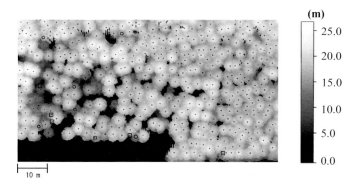

**Fig. 4.** The highest point of each tree in the DCHM image (Omasa et al. 2003).

We developed a regression equation for estimating the stem carbon stocks from tree height, using both field data for our study site and all other available data on Japanese cedar in Japan (212 datasets) (Cannell 1982). We estimated the carbon stocks of a tree stem ($C_{stem}$, kg C tree$^{-1}$) as follows:

$$C_{stem} = 0.0119 H^{2.9696} \ (r^2 = 0.933)$$

where $H$ is tree height (m). We developed another regression equation for estimating the carbon stocks of branches, foliage, and roots from tree height using data collected all over Japan (53 datasets) (Cannell 1982). The carbon stocks of these parts ($C_{BFR}$, kg C tree$^{-1}$) were estimated as follows:

$$C_{BFR} = 0.0075 H^{2.9516} \ (r^2 = 0.864).$$

Using these two equations, we obtained the spatial distribution of carbon stock for each tree at the study site shown in Fig. 6. The gray level inside each outline represents the

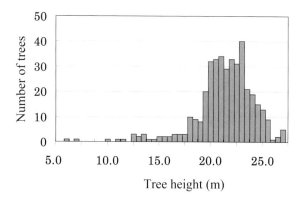

**Fig. 5.** The distribution of tree heights estimated from the DCHM data (Omasa et al. 2003).

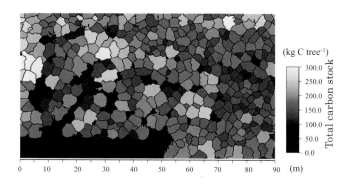

**Fig. 6.** Estimated carbon stock of each tree in the Japanese cedar forest (Omasa et al. 2003).

total carbon stock of that tree and covers the area occupied by that tree's canopy. The maximum carbon storage per tree was less than 300 kg, with most values between 110 and 300 kg. Total carbon storage at the site was estimated at 69 300 kg C, and the average carbon storage was estimated at 175.9 kg C tree$^{-1}$. The results reveal that 3-D remote sensing using helicopter-borne lidar accurately estimated carbon stocks at the stand scale.

## 5. Conclusions

In this paper, we demonstrated two applications of 3-D remote sensing using a high-resolution, helicopter-borne scanning lidar system. In the lidar system, a laser beam with a small footprint and a high pulse frequency illuminates the ground surface from a slow-flying helicopter, and almost all of the ground surface can be scanned at high resolution. The lidar was proven to be effective for obtaining accurate measurements of canopy height and for creating a DTM. Using allometric relationships combined with lidar-derived data, we could also estimate carbon stocks both for individual trees and for the stand as a whole. The 3-D lidar remote-sensing technique thus represents additional progress in the use of remote sensing to accurately estimate various forest properties.

## References

Brandtberg T, Warner TA, Landenberger RE, McGraw JB (2003) Detection and analysis of individual leaf-off tree crowns in small footprint, high sampling density lidar data from the eastern deciduous forest in North America. Remote Sens Environ 85:290–303

Cannell MGR (1982) World forest biomass and primary production data. Academic Press, London, 145-166

Flood M, Gutelius B (1997) Commercial implications of topographic terrain mapping using scanning airborne laser radar. Photogramm Eng Remote Sens 63:327–366

Goward SN, Williams DL (1997) Landsat and earth systems science: development of terrestrial monitoring. Photogramm Eng Remote Sens 63:887–900

Harding DJ, Lefsky MA, Parker GG, Blair JB (2001) Laser altimeter canopy height profiles: methods and validation for closed-canopy, broadleaf forests. Remote Sens Environ 76:283–297

Maltamo M, Eerikäinen K, Pitkänen J, Hyyppä J, Vehmas M (2004) Estimation of timber volume and stem density based on scanning laser altimetry and expected tree size distribution functions. Remote Sens Environ 90:319–330

Means JE, Acker SA, Harding DJ, Blair JB, Lefsky MA, Cohen WB, Harmon ME, McKee WA (1999) Use of large-footprint scanning airborne lidar to estimate forest stand characteristics in the Western Cascades of Oregon. Remote Sens Environ 67:298–308

Næsset E (1997) Determination of mean tree height of forest stands using airborne laser scanner data. ISPRS J Photogramm Remote Sens 52:49–56

Næsset E, Økland T (2002) Estimating tree height and tree crown properties using airborne scanning laser in a boreal nature reserve. Remote Sens Environ 79:105–115

Nelson R, Krabill W, Tonelli J (1988) Estimating forest biomass and volume using airborne laser data. Remote Sens Environ 24:247–267

Nilsson M (1996) Estimation of tree heights and stand volume using an airborne lidar system. Remote Sens Environ 56:1–7

Omasa K, Akiyama Y, Ishigami Y, Yoshimi K (2000) 3-D remote sensing of woody canopy heights using a scanning helicopter-borne lidar system with high spatial resolution. J Remote Sens Soc Jpn 20:394-406

Omasa K, Qiu GY, Watanuki K, Yoshimi K, Akiyama Y (2003) Accurate estimation of forest carbon stocks by 3-D remote sensing of individual trees. Environ Sci Technol 37:1198–1201

Patenaude G, Hill RA, Milne R, Gaveau DLA, Briggs BBJ, Dawson TP (2004) Quantifying forest above ground carbon content using LiDAR remote sensing. Remote Sens Environ 93:368-380

Turner DP, Cohen WB, Kennedy RE, Fassnacht KS, Briggs JM (1999) Relationships between leaf area index and Landsat TM spectral vegetation indices across three temperate zone sites. Remote Sens Environ 70:52–68

Waring RH, Way J, Hunt ER Jr, Morrissey L, Ranson KJ, Weishampel JF, Oren R, Franklin SE (1995) Imaging radar for ecosystem studies. Bioscience 45:715-723

# Assessments of climate change impacts on the terrestrial ecosystem in Japan using the Bio-Geographical and GeoChemical (BGGC) Model

Yo Shimizu, Tomohiro Hajima, and Kenji Omasa

Graduate School of Agricultural and Life Sciences, The University of Tokyo, Yayoi 1-1-1, Bunkyo-ku, Tokyo 113-8657, Japan

**Summary.** This study assessed the future impacts of global climate change on the distribution and functioning of terrestrial ecosystems in Japan using the Bio-Geographical and GeoChemical model (BGGC model). The model enables us to simulate the carbon and nitrogen cycles within ecosystems on the basis of estimated potential natural vegetation distribution. In this study, changes in net primary productivity (NPP) and the distribution of potential natural vegetation were evaluated for assessments of climate change impacts. The GCMs experimental data used for future climate conditions were the CSIRO-Mk2 and ECHAM4/OPYC3 for each of A2 and B2 scenarios in the SRES. Comparison of the averages of simulated NPP under each scenario with the average NPP under current climate conditions showed that the average NPP could increase about 19 to 33 percent by the year 2050s and 25 to 53 percent by the year 2080s with changes in potential natural vegetation type.

**Key words.** BGGC model, Japan, Potential natural vegetation, NPP

## 1. Introduction

Vegetation in terrestrial ecosystems provides not only an important habitat for animals but most importantly it plays a significant role in the cycle of carbon and nutrients. IPCC reports indicated that the changes in global climate caused by the continual increase in greenhouse gases will affect the ecosystem in future. For evaluation of the impact, there are several process-based models to simulate vegetation distribution, carbon and nutrient cycle (Haxeltine and Prentice 1996; Neilson et al. 1998; Ishigami et al. 2002, 2003; Levy et al. 2004; Hajima et al. 2005). The process-based model includes knowledge on the physiological responses of plants to environmental change. This is a suitable model for predicting the effect of climate change on the ecosystem because the plant's physiological and ecological functions must be considered. The objective of this study is to assess the future impacts of global climate change on the distribution and functioning of terrestrial ecosystems in Japan using the Bio-Geographical and GeoChemical model. We evaluated changes in net primary productivity (NPP) and the distribution of potential natural vegetation for assessments of climate change impacts. NPP denotes the net production of organic matter by plants in an ecosystem, and is one of important indicators to provide information on the carbon budget of terrestrial ecosystem.

*Plant Responses to Air Pollution and Global Change*
Edited by K. Omasa, I. Nouchi, and L. J. De Kok ( Springer-Verlag Tokyo 2005 )

## 2. Model structure

The BGGC model consists of two types of process-based model. The one is the bio-geographical model, which puts emphasis on the determination of what kind of vegetation could live in a given location. The other is the bio-geochemical model, which simulates the carbon and nutrient cycles within ecosystems on the basis of given vegetation distribution.

Most of the basic structures of some modules in the model were borrowed from the modified BIOME3 (Ishigami et al. 2002) and sub-model of CENTURY4 (Parton et al. 1993). The BGGC model is divided into two sub-models, the vegetation competition sub-model and the soil organic matter sub-model. The vegetation competition sub-model contains the photosynthesis model and the canopy model. The photosynthesis model estimates the optimized net primary productivity (NPP) and leaf area index (LAI) for each plant functional type (PFT) to satisfy annual moisture and the soil nitrogen. In order to consider the effect of $CO_2$ and nitrogen on photosynthesis, the model contains the Farquhar photosynthesis model as simplified by Collats et al. (1991) and canopy photosynsthesis model developed by Hikosaka (2003). The PFTs were determined by differences in phenological type (evergreen or deciduous), leaf type (broad-leaved or conifer) and the rooting depth. Competition among PFTs is simulated by using the estimated NPP of each PFT as an index of competitiveness. The PFT with the highest NPP is selected as the dominant type. The canopy model takes account of the effect of difference in the canopy structure of the forest vegetation on calculation of daily net photosynthesis. The distribution of potential natural vegetation types was determined by the selected PFT and the estimated NPP and LAI. On the other hand, the soil organic matter sub-model simulates carbon, nitrogen, and water dynamics in the soil ecosystem. The model includes three soil organic matter pools with different potential decomposition rates. The model is linked to the vegetation competition model by the exchange of soil nitrogen and NPP. Fig. 1 shows the schematic diagram of the BGGC model.

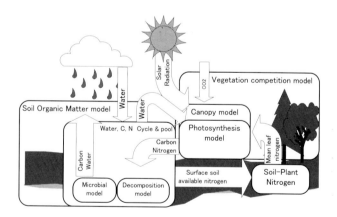

**Fig. 1.** Schematic diagram of BGGC model.

## 3. Data

### 3.1 Climate and soil data

Monthly climate normals data (average for 30 years; 1971 to 2000) and GCMs experimental data prescribed by IPCC-SRES (The Intergovernmental Panel on Climate Change, Special Report on Emissions Scenarios) were used in this study. Current and future climate input data consist of monthly mean temperature, monthly precipitation and solar radiation. The GCMs experimental data were the CSIRO-Mk2 and ECHAM4/OPYC3 for each of A2 and B2 scenarios in the SRES. Since each scenario has different assumptions on global population, gross world product, and technological change, future greenhouse gas (GHG) emissions are different. Although high-spatial resolution GCMs are required to assess the regional climate impacts, grid resolutions of current GCMs are roughly 3° longitude x 3° latitude. Yokozawa et al. (2003) statistically interpolated the GCMs data (based on the IS92a emission scenario) to a 10 x 10 sq. km grid data. This study applied the same method to GCMs data based on SRES scenarios. The number of grids is 4,691. Soil texture data was obtained from Haxeltine and Prentice (1996). All data were arranged in the form of a 10 x 10 sq. km grid system.

Fig. 2 shows boxplots of changes in annual mean temperature between normals and GCMs experimental data (CSIRO-Mk2 and ECHAM4/OPYC3) and those of ratios of

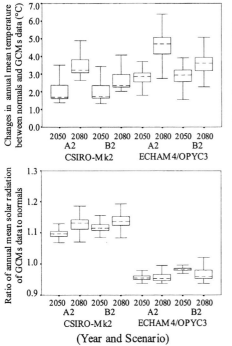

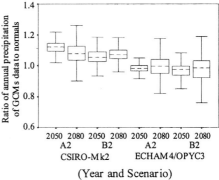

**Fig. 2.** Boxplots of changes in annual mean temperature, annual precipitation and mean solar radiation between normals and GCMs data. The central box indicates the interquartile range, and the dotted line in the box is the 50th percentile.

GCMs annual precipitation and mean solar radiation to normals. In both GCMs data, projected temperatures in the 2080s under the A2 scenario are higher than under the B2. Annual precipitation and solar radiation projected by CSIRO-Mk2 increased over Japan.

### 3.2 $CO_2$ concentration data

$CO_2$ concentration data for A2 and B2 scenarios were obtained from the IPCC WG-1 report (IPCC 2001). $CO_2$ concentrations in 2050 and 2080 for A2 scenario are 537 and 713 ppm respectively, Those for B2 scenarios are 476 and 570 ppm, respectively.

## 4. Results

Fig. 3 shows the estimated NPP and the distribution of potential natural vegetation type under current climatic conditions and $CO_2$ concentration (367 ppm). Comparing with the NPP distribution estimated by our previous model (Ishigami et al. 2002) on the basis of the simulation by Chikugo model (Uchijima and Seino 1985), the BGGC model could simulate it more successfully. The model has correctly estimated the vegetation distribution except for the broad-leaved evergreen forest in the Hokkaido region.

Fig. 4 shows the estimated NPP under CSIRO-Mk2 and ECHAM4/OPYC3 experimental data for each of A2 and B2 scenarios. The simulated NPP under CSIRO-Mk2 was larger than under ECHAM4/OPYC3. It depends on the difference in climatic conditions

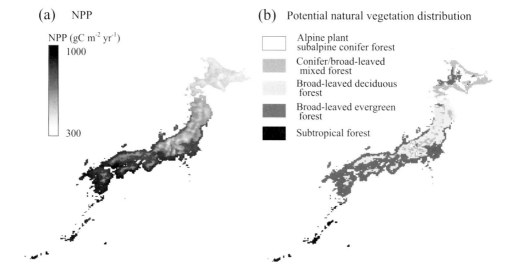

**Fig. 3.** Estimated (a) NPP and (b) distribution of potential natural vegetation type under current climatic conditions and $CO_2$ concentration.

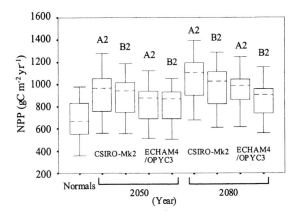

**Fig. 4.** Comparison of simulated NPP under each SRES scenario.

as shown in Fig. 2. Comparing the averages of simulated NPP for each scenario with the average NPP under current climatic conditions, an increase of 19 to 33 percent in the average NPP for year 2050s and about 25 to 53 percent for year 2080s can be observed with changes in potential natural vegetation type.

This study assessed the future impacts of global climate change on the distribution and functioning of terrestrial ecosystems in Japan using the BGGC model. The model has the characteristics of both a bio-geographical model and a bio-geochemical model which enable estimation of the potential distribution of natural vegetation using NPP under future climate conditions.

The model included the processes of $CO_2$ effect on NPP, the responses of NPP to climate that specifically considered plant functional type, and competition among PFTs for light and water. The model is also capable of determining which vegetation type is most suited for given climatic conditions. In order to improve characteristics as a bio-geochemical model, however, there is still a need to consider the dynamic processes involved in vegetation structure due to climate and $CO_2$ effects.

## References

Collatz GJ, Ball JT, Grivet C, Berry JA (1994) Physiological and environmental regulation of stomatal conductance, photosynthesis and transpiration: a model that includes a laminar boundary layer. Agric For Meteorol 54:107-136

Haxeltine A, Prentice IC (1996) BIOME3: An equilibrium terrestrial biosphere model based on ecophysiological constraints, resource availability, and competition among plant functional types. Global Biogeochem Cy 10:693-709

Hajima T, Shimizu Y, Fujita Y, Omasa K (2005) Estimation of net primary production in Japan under nitrogen-limited scenario using BGGC model. J Agric Meteorol 60:1223-1225

Hikosaka K (2003) A model of dynamics of leaves and nitrogen in a plant canopy: an integration of canopy photosynthesis, leaf life span, and nitrogen use efficiency. Am Nat 162:149-164

IPCC (2001) Climate change 2001: impacts, adaptation, and vulnerability. Contribution of working group II to the third assessment report of the intergovernmental panel on climate change. In: McCarthy JJ, Canziani OF, Leary NA, Dokken DJ, White KS (Eds) Cambridge University Press, Cambridge, 1032 p

IPCC (2001) Climate change 2001. The scientific basis: contribution of working group I to the third assessment report of the intergovernmental panel on climate change. In: Houghton JT, Ding Y, Griggs DJ, Noguer M, van der Linden PJ, Dai X, Maskell K, Jonhnson CA (Eds) Cambridge University Press, Cambridge, 881p

Ishigami Y, Shimizu Y, Omasa K (2002) Estimation of potential natural vegetation distribution in Japan using a process model. J Agric Meteorol 58:123-133 (In Japanese with English summary)

Ishigami Y, Shimizu Y, Omasa K (2003) Projection of climatic change effects on potential natural vegetation distribution in Japan. J Agric Meteorol 59:269-276 (In Japanese with English summary)

Levy PE, Cannell MGR, Friend AD (2004) Modelling the impact of future changes in climate, $CO_2$ concentration and land use on natural ecosystems and the terrestrial carbon sink. Global Environ Chang 14:21-30

Neilson RP, Prentice IC, Smith B, Kittel T, Viner D (1998) Simulated changes in vegetation distribution under global warming. In: Watson RT, Zinyowera MC, Moss RH (Eds) The regional impacts of climate change: an assessment of vulnerability. Special report of IPCC working group II. Cambridge University Press, Cambridge, pp 439-456

Parton WJ, Scurlock JMO, Ojima D, Gilmanov TG, Scholes RJ, Schimel DS, Kirchner T, Menaut JC, Seastedt T, Moya EG, Kamnalrut A, Kirchner JI (1993) Observation and modeling of biomass and soil organic matter dynamics for the grassland biome world-wide. Global Biogeochem Cy 7:785-809

Uchijima Z, Seino H (1985) Agroclimatic evaluation of net primary productivity of natural vegetation (1) Chikugo model for evaluating net primary productivity. J Agric Meteorol 40:343-352

Yokozawa M, Goto S, Hayashi Y, Seino H (2003) Mesh climate change data for evaluating climate change impacts in Japan under gradually increasing atmospheric $CO_2$ concentration. J Agric Meteorol 59:117-130

# VII. Air Pollution and Global Change in Asia

# Establishing critical levels of air pollutants for protecting East Asian vegetation – A challenge

Yoshihisa Kohno [1], Hideyuki Matsumura [1], Takashi Ishii [1], and Takeshi Izuta [2]

[1] Environmental Science Research Laboratory, Central Research Institute of Electric Power Industry, Abiko 1646, Abiko City, Chiba 270-1194, Japan
[2] Institute of Symbiotic Science and Technology, Tokyo University of Agriculture and Technology, Saiwai-cho 3-5-8, Fuchu, Tokyo 183-8509, Japan

**Summary.** Critical levels of ozone ($O_3$) and sulfur dioxide ($SO_2$) for protecting European forests are not evaluated to apply to East Asian vegetation. Based on the results obtained from the long-term experimental studies on the effects of chronic exposure to $O_3$ or $SO_2$ on 30 young potted grown tree species using open-top chambers, we analyzed dose-response relationships between the whole-plant dry mass increment during the experiment and concentration of air pollutants to establish Asian critical levels.

While the annual mean concentration of $SO_2$ corresponding to a 10% reduction in the whole-plant dry mass increment for the most sensitive tree species was estimated to be 5-15 nl $l^{-1}$ (ppb), it was over 30 ppb for the less sensitive tree species. The AOT40 of $O_3$ corresponding to a 10% growth reduction for the sensitive tree species was estimated to be 8-15 ppm-h for a growing season (6 months) and over 31 ppm-h for the less sensitive species. Critical level of N load for a sensitive Japanese evergreen broad-leaved tree species was considered to be about 50 kg N $ha^{-1}$ $y^{-1}$. Furthermore, model estimation suggested that N load in a part of southern China have already exceeded this value.

These critical values were estimated by the results obtained from limited experimental studies on the effects of air pollutants or N load on the growth of tree species native to relatively cool temperate zone. Therefore, information about sensitivity of plants grown in tropical to semiarid zones should be accumulated to cover East Asian vegetation.

**Key words.** Air pollution, Critical level, Trees, Sensitivity, Asian vegetation

## 1. Introduction

Natural vegetation, forests and agricultural fields are receptors of $SO_x$, $NO_x$ and photochemical oxidants. In Europe, it has been started to discuss critical levels of air pollutants in the late 1980's and the first provisional critical levels of gaseous air pollutants for agricultural crops and forests were proposed in 1988 (Ashmore and Wilson 1992, CLAG 1994). Current European critical levels of ozone ($O_3$) corresponding to 10% growth reduction was estimated to be 10 ppm-h of AOT40 for forests during April-September with global radiation at and greater than 50 W $m^{-2}$ (Kärenlampi and Skärby 1996, Fuhrer et al. 1997). In Asian countries including Japan, there are few activities to establish critical levels of air pollutants for protecting Asian vegetation except for the RAPIDC

*Plant Responses to Air Pollution and Global Change*
Edited by K. Omasa, I. Nouchi, and L. J. De Kok ( Springer-Verlag Tokyo 2005 )

(http://www.york.ac.uk/inst/sei/rapidc2/rapidc.html) (Emberson et al. 2003). To establish and propose critical levels of air pollutants, it is necessary to accumulate information on the plant sensitivity to air pollutants and their threshold values for the occurrence of adverse effects on the plant growth based on the results obtained from long-term exposure experiments in the region. That will include whether the European values are applicable to East Asia where more complicated vegetation and components developed in the varying climatic regimes rather than that the relatively simple vegetation in the Europe did.

In this paper, we tried to discuss provisional critical levels of air pollutants for protecting East Asian vegetation from the potential threats due to rapid increasing consumption of fossil fuels and its related emissions of pollutants.

## 2. Calculation of critical levels for $SO_2$ or $O_3$

Results obtained from the long-term exposure experiments conducted for multiple growing seasons using open-top chambers to evaluate effects of $SO_2$ or $O_3$ on the tree growth were analyzed to determine critical levels for tree species (Matsumura 2000, 2001; Matsumura and Kohno 1997, 2001, 2003). We evaluated growth responses to $SO_2$ or $O_3$ in 30 young potted tree species (11 deciduous broad-leaved, 9 evergreen broad-leaved and 10 coniferous trees) grown in andisol.

Experiments were conducted in the site at an elevation of 25 m or 540 m above sea levels in the Kanto Plain of the central Japan for 3 growing seasons in 1993-1995 and 2 growing seasons in 1998-1999 and 2000-2001. Sulfur dioxide was daily added to charcoal-filtered air, and it was regulated at a constant concentration of 5-40 ppb for 24 hours. Ozone generated by high electric discharge was added to charcoal-filtered air proportionately adjusted at the same level, +50% or +100% of the ambient concentration.

Even though the most of the tree species did not show significant growth responses to $SO_2$ or $O_3$ after the exposure in the first growing season, they showed significant growth reduction after the consecutive exposure in the 2nd or 3rd growing season. Therefore, we analyzed results obtained from the longest exposure experiments.

In the analysis of dose-response relationships, the independent variable was a dose of $SO_2$ or $O_3$; total 24 hours dose of $SO_2$ or AOT40 of $O_3$ with a global radiation ($\geq 50 Wm^{-2}$) between April and November during the whole period of experiment. The dependent variable was a relative value (%) of whole plant dry mass increment from the initial harvest to the final harvest in exposed plants compared to that of plants grown in the charcoal-filtered air. After linear regression analysis in the individual tree species, doses corresponding to a 10% reduction in the whole-plant dry mass increment of the species were estimated from the slopes. In this paper, a total 24-h dose of $SO_2$ was converted to a mean concentration (ppb), and an AOT40 (ppm-h) of $O_3$ for multiple growing seasons was adjusted to a value per a growing season as a critical level.

## 3. Critical levels of $SO_2$ for individual tree species

Critical levels of $SO_2$ corresponding to a 10 % reduction in increment of whole-plant dry mass were presented in Table 1. Based on the growth responses of the individual species, tree sensitivity was classified into the 3 groups: high (sensitive), moderate and low (tolerant) at which level biomass was significantly reduced by the exposure to $SO_2$ for 2 or 3 growing seasons. Considering the individual plant sensitivity, *Betula platyphylla* var. *japonica* and *Pinus strobes* were most sensitive to $SO_2$ and critical levels were estimated to be 5-6 ppb. Other high-sensitive ones including *Zelkova serrata*, *Quercus myrsinaefolia*, *Pinus densiflora*, *Larix kaempferi* and *Abies homolepis* would be 8-13ppb. In contrast, *Quercus serrata*, *Quercus mongolica* var. *grosseserrata*, *Cinnamomum camphora* and *Chamaecyparis obtusa* in the moderate sensitivity group ranged to be 15-30 ppb and those for low sensitive (tolerant) tree species would be greater than 30 ppb.

**Table 1.** Mean $SO_2$ concentration corresponding to 10% reduction of whole-plant dry mass.

| Sensitivity | Type | Species | Mean (ppb) |
|---|---|---|---|
| High | Deciduous | *Betula platyphylla* var. *japonica*, *Zelkova serrata* | 5-15 |
| | Evergreen | *Quercus myrsinaefolia* | |
| | Coniferous | *Pinus strobus*, *Pinus densiflora*, *Larix kaempferi*, *Abies homolepis* | |
| Moderate | Deciduous | *Quercus serrata*, *Quercus mongolica* var. *grosseserrata* | 15-30 |
| | Evergreen | *Cinnamomum camphora* | |
| | Coniferous | *Chamaecyparis obtusa* | |
| Low | Deciduous | *Populus maximowiczii*, *Fagus crenata*, *Prunus jamasakura* | 30< |
| | Evergreen | *Machilus thunbergii*, *Castanopsis cuspidata* var. *sieboldii*, *Lithocarpus edulis* | |
| | Coniferous | *Pinus thunbergii*, *Cryptomeria japonica* | |

Sensitivity:
High: Whole-plant dry dry mass increment significantly reduced by the exposure of 10 ppb $SO_2$.
Moderate: Whole-plant dry dry mass increment significantly reduced by the exposure of 20 ppb $SO_2$.
Low: Whole-plant dry dry mass increment did not reduce by the exposure of 20 ppb $SO_2$.
$SO_2$: Mean concentration from April to November

**Table 2.** AOT40 of $O_3$ for a growing season corresponding to 10% reduction of whole-plant dry mass.

| Sensitivity | Type | Species | AOT40 (8 months) | AOT40 (6 months) |
|---|---|---|---|---|
| High | Deciduous | *Populus maximowiczii*, *Populus nigra*, *Fagus crenata*, *Zelkova serrata* | 11-20 | 8-15 |
| | Evergreen | *Castanopsis cuspidata* var. *sieboldii* | | |
| | Coniferous | *Pinus densiflora*, *Larix kaempferi* | | |
| Moderate | Deciduous | *Quercus serrata*, *Betula platyphylla* var. *japonica* | 21-40 | 16-30 |
| | Evergreen | *Quercus myrsinaefolia*, *Cinnamomum camphora* | | |
| | Coniferous | *Abies homolepis* | | |
| Low | Deciduous | *Quercus mongolica* var. *grosseserrata* | 41< | 31< |
| | Evergreen | *Lithocarpus edulis*, *Machilus thunbergii* | | |
| | Coniferous | *Pinus thunbergii*, *Cryptomeria japonica*, *Chamaecyparis obtusa* | | |

Sensitivity:
High: Whole-plant dry mass increment significantly reduced by the exposure of ambient level of $O_3$.
Moderate: Whole-plant dry mass increment significantly reduced by the exposure of +50 % to 100% greater than ambient level of $O_3$.
Low: Whole-plant dry mass increment did not reduce by the exposure of +50 % to 100% greater than ambient level of $O_3$.
AOT40 (8 months): Ozone dose from April to November at and greater than 50W/m$^2$ of global radiation.
AOT40 (6 months): Calculated value from AOT40 (8 months).

## 4. AOT40 of $O_3$ for individual tree species

Critical levels of $O_3$ expressed as AOT40 corresponding to a 10% reduction in increment of the whole-plant dry mass were shown in Table 2. Based on the growth responses in the individual species, the sensitivity of trees to $O_3$ was classified into the 3 groups. *Populus maximowiczii* and *Populus nigra* were most sensitive to $O_3$ among the tested species and AOT40 corresponding to a 10% growth reduction for a growing season was estimated to be 11-12 ppm-h for 8 months. The AOT40 for relatively high-sensitive species such as *Zelkova serrata, Castanopsis cuspidata, Quercus serrata, Pinus densiflora* and *Larix kaempferi* was 11-20 ppm-h for 8 months. The AOT40 for moderate and low sensitive (tolerant) species was estimated to be 21-40 ppm-h and greater than 41 ppm-h for 8 months, respectively. These values were equivalent to 16-30 and greater than 31 ppm-h for 6 months, respectively.

## 5. Critical levels of N load for evergreen broad-leaved trees

To evaluate sensitivity of Japanese evergreen broad-leaved trees to N load, two year-old seedlings of *Castanopsis cuspidata, Lithocarpus edulis, Quercus glauca* and *Quercus acuta* those are major component species in warm temperate zone in the southern Japan, were grown in brown forest soil treated with N as $NH_4NO_3$ at 0, 10, 50, 100 and 300 kg N $ha^{-1} y^{-1}$ for 2 growing seasons. Threshold values of N load for the occurrence of growth reduction in *C. cuspidata, L. edulis* and *Q. glauca* were estimated to be approximately 50, 100 and 200 kg $ha^{-1} y^{-1}$, respectively. In contrast, N load even at 300 kg $ha^{-1} y^{-1}$ stimulated plant growth of *Q. acuta*.

Leaf concentrations of N and Mn increased with increasing the N load. In *C. cuspidata, L. eduli* and *Q. glauca*, the relative whole-plant dry mass (the ratio of the whole plant dry mass of plants grown in the N treatments to that of the seedlings grown in the control treatment) was negatively correlated with the concentration ratio of Mn/Mg in the leaves.

Current results suggested that there was a great difference in the sensitivity of Japanese evergreen broad-leaved tree species to N load, and that Mn/Mg concentration ratio in the leaves would be a useful indicator for the negative effects of N load on the growth of Japanese evergreen broad-leaved trees.

## 6. Mapping of air quality in Japan and East Asia

Japanese standard for air pollutants was established for $SO_2$, $NO_2$ and photochemical oxidants ($O_x$), but not for $O_3$. In addition, $O_x$ data was presented as only total excess hours over the standards, but hourly concentration data was not. As several sites recorded both excess hours and hourly concentration data, the relationship between total excess hours and observed AOT40 ($O_x$) was analyzed. Total excess hours were expressed as a sum of excess hours over 0.06 ppm and 0.12 ppm during April to September. Obtained correlation equation was applied to nationwide monitoring data of total excess hours for presenting a distribution of nationwide estimated AOT40 ($O_x$) as presented in the Fig. 1.

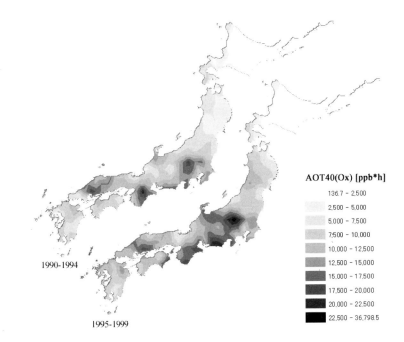

**Fig. 1.** Distribution and its trend of mean AOT40 ($O_x$) in 1990s of Japan.

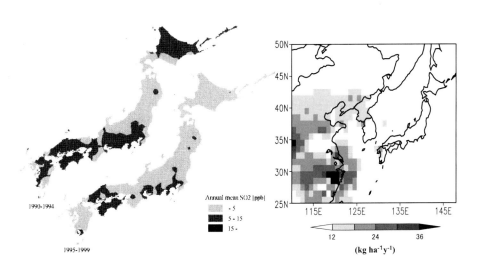

**Fig. 2.** Distribution and its trend of mean $SO_2$ concentration in 1990's of Japan and total (Wet+Dry) deposition of $NH_x$-N in East Asia.

While a whole country map of mean value of AOT40 ($O_x$) in 1990s showed no big change in an overall distribution tendency, high dose areas, especially in the central part of Japan expanded and exceeded 25 ppm-h in the second half of 1990s.

Nationwide mean concentration of $SO_2$ in the ambient air of Japan and total deposition of $NH_x$-N in the East Asia were presented in Fig. 2. Areas ranging 5-15 ppb of $SO_2$ decreased in the second half of 1990s.

The wet and dry deposition of oxidized ($NO_x$) and reduced ($NH_x$) nitrogen compounds in East Asia were calculated on the basis of the various precipitation chemistry data collected between 1980s and 1990s. Estimated total deposition of $NO_x$ was approximately 13 kg ha$^{-1}$ y$^{-1}$ and that of NHx was approximately 9 kg ha$^{-1}$ y$^{-1}$ when the rate of gas and particle in the atmosphere was assumed to be 50:50. The largest deposition flux of $NO_x$ was observed in the northeastern coast of China where the contribution of dry deposition was relatively large. The largest deposition flux of $NH_x$ was observed in the southeastern coast of China with the contribution of wet deposition to total deposition was as large as 60 to 95%.

Currently an average total deposition of N ($NO_x+NH_x$) in East Asia estimated 22 kg ha$^{-1}$ y$^{-1}$ and maximum deposition would be greater than 50 kg ha$^{-1}$ y$^{-1}$. This estimation suggested that certain areas in China might reach a critical level of N load for N sensitive evergreen broad leaved trees.

## 7. Risk assessment of photochemical oxidants for *Pinus densiflora* and *Fagus crenata*

Some mountain and sub-alpine forests in Japan are suffering from unidentified causes and declining as well as those in the western countries. There are many factors affecting tree growth or vitality under the field conditions (Kohno et al. 1998, Manion and Lachance 1992). Some scientists pointed out these phenomena would be in a natural transition or succession process under natural conditions; others suggested effects of anthropogenic pollutants such as acidic deposition, gaseous air pollutants including photochemical oxidants and other related derivatives.

Considering past and current ambient air quality in Japan and actual concentration of primary gaseous air pollutants, especially $SO_2$ and $NO_x$ in the remote mountainous areas with extremely low concentrations, forest and/or tree decline is not likely to be correlated with these pollutants. However, concentration of $O_3$ at high elevation sites is generally higher than that at lower elevation sites. Exposure experiments using open-top chambers with or without filters suggested that current ambient level of $O_3$ has a significant potential threat for $O_3$ sensitive tree species such as Japanese red pine and beech (Matsumura 2000, 2001).

As presented in Table 1 and 2, *Pinus densiflora* (Japanese red pine) is sensitive to both $SO_2$ and $O_3$. *Fagus crenata* (Japanese beech) is relatively tolerant to $SO_2$, but sensitive to $O_3$. The AOT40 of $O_3$ corresponding to 10% growth reduction for these two species was estimated to be 13 ppm-h for 6 months. Assuming that AOT40 of $O_x$ is the same equivalent value to that of $O_3$, plants grown in areas over 15 ppm-h of AOT40 ($O_x$) distributing mostly in the central part of Japan are disposed to have a high risk of $O_3$ impacts on the growth (Fig. 3). Japanese beech in Tanzawa Mountain is just located in the high risk

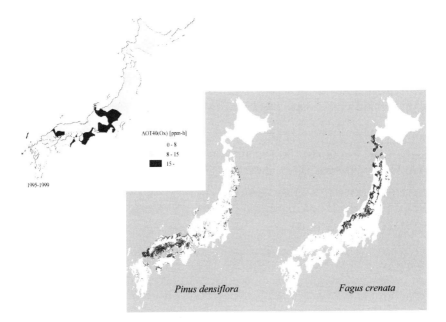

**Fig. 3.** Distribution of *Pinus densiflora* and *Fagus crenata* in Japan and AOT40($O_x$) in the second half of 1990s
Distribution of *P. densiflora* and *F. crenata* was prepared from the National Survey of the Natural Environment, Ministry of the Environment, Japan.

areas facing the Kanto Plain in the central Japan. Takeda and Aihara (2004) demonstrated that ambient level of $O_3$ definitely induced early defoliation of beech seedlings grown in open-top chambers with or without charcoal-filters at the mountain site where the AOT40 of $O_3$ for 6 months was 22.4 ppm-h in 2003.

Direct cause for Japanese red pine decline as pine wilt disease has been identified with pine wood nematode infection carried by Japanese pine sawyer (Mamiya 1983). As well as Japanese beech forest decline may have a potential correlation with photochemical oxidants, decline of Japanese red pine forests is still a matter of controversy among scientists. General trend of distribution pattern of the decline expanded from south west to north in Japan. Considering AOT40 of $O_3$ for Japanese red pine, $O_3$ might have a predisposing role or chronic potential effects on pine plant growth. Plants with reduced vitality under such a predisposing stress for a long term and infected with pathogen or insects would be an epidemic center coupling with current improper forest management circumstances in Japan.

Trials to present exceedance maps of pollutants would help a risk assessment of the impacts on forests health, especially in the remote areas, and would support to discuss reducing emissions for protecting vegetation from direct and subtle effects of anthropogenic pollutants in East Asia. However, it should be accumulated information about sensitivity of plants grown in tropical to semiarid zone in East Asia, as there are few data in this area. In addition, it should be in mind that an extraporation from the experimental results raises scientific arguments whether results from juvenile plants could be applicable to those in mature trees under natural conditions or not.

Acknowledgements. This report was a part of contribution of research program "C-7 Study on impacts of acidic and oxidative substances on vegetation and establishment of tentative critical level for protecting East Asian vegetation" financially supported by Global Environment Research Fund of Ministry of the Environment, Japan.

## References

Ashmore M, Wilson R (1992) Critical levels of air pollutants for Europe. Background papers prepared for the United Nations Economic Commission for Europe Workshop on Critical levels. Egham, UK, 23-26 March, 1992
CLAG (1994) Critiacl loads of acidity in the United Kingdom. Critical load advisory group summary report. Prepared at the request of the Department of the Environment, February 1994
Emberson L, Ashmore M, Murray F (2003) Air pollution impacts on crops and forests. A global assessment. Air pollution reviews vol.4, Imperial College Press, London, UK
Fuhrer J, Skärby L, Ashmore M (1997) Critical levels for ozone effects on vegetation in Europe. Environ Pollut 97:91-106
Kärenlampi L, Skärby L (1996) Critical levels for ozone in Europe: testing and finalizing the concepts. UN/ECE workshop report, University of Kuopio, Finland
Kohno Y, Matsumura H, Kobayashi T (1998) Differential sensitivity of trees to simulated acid rain or ozone in combination with sulfur dioxide. In: Bashkin V, Park S-U (Eds) Acid deposition and ecosystem sensitivity in East Asia. Nova Science Publishers New York, USA, pp143-188
Mamiya Y (1983) Pathology of the pine wilt disease caused by Vursaphelechus xylophilus. Ann Rev Phytopath 21:201-220
Manion P, Lachance D (1992) Forest decline concepts. APS Press, St Paul, Minnesota, USA
Matsumura H (2000) Effects of simulated acid mist and/or ambient ozone on the growth of nine coniferous and five deciduous broad-leaved tree species. CRIEPI Report U99035. Central Research Institute of Electric Power Industry (in Japanese with English summary)
Matsumura H (2001) Impacts of ambient ozone and/or acid mist on the growth of 14 tree species: an open-top chamber study conducted in Japan. Water Air Soil Pollut 130: 959-964
Matsumura H, Kohno Y (1997) Effects of ozone and/or sulfur dioxide on tree species. In: Kohno Y (Ed) Proceedings of CRIEPI international symposium on transport and effect of acidic substances, Nov 28-29, 1996, CRIEPI, Tokyo, Japan, pp 190-205
Matsumura H, Kohno Y (2001) Effects of sulfur dioxide and/or ozone on the growth of Prunus mume, Piunus jamasakura, Quercus serrata, Castanopsis cuspidata and Chamaecyparis obtusa. CRIEPI Report U01028. Central Research Institute of Electric Power Industry (in Japanese with English summary)
Matsumura H, Kohno Y (2003) Effects of sulfur dioxide and/or ozone on Japanese evergreen broad-leaved tree species. CRIEPI Report U02021. Central Research Institute of Electric Power Industry (in Japanese with English summary)
RAPIDC (Regional Air Pollution in Developing Countries) http://www.york.ac.uk/inst/sei/rapidc2
Takeda M, Aihara K (2004) Effects of ambient ozone of Tanzawa on Fagus crenata seedlings – open-top chamber study. Proceedings of the 45th annual meeting of Japan society for atmospheric environment, Akita, 2004, pp 360 (in Japanese)

# Major activities of acid deposition monitoring network in East Asia (EANET) and related studies

Tsumugu Totsuka[1], Hiroyuki Sase[1] and Hideyuki Shimizu[2]

[1] Acid Deposition and Oxidant Research Center, Sowa 1182, Niigata 950-2144, Japan
[2] National Institute for Environmental Studies, Onogawa 16-2, Tsukuba, Ibaraki 305-8506, Japan

**Summary.** In order to act against acid deposition problems in East Asian regions, the Acid Deposition Monitoring Network in East Asia (EANET) was established on the decision at the First Intergovernmental Meeting in 1998 among participating countries including China, Indonesia, Japan, Malaysia, Mongolia, Philippines, Republic of Korea, Russia, Thailand and Viet Nam. Based on the decision, the preparatory phase activities of EANET were started from 1998, and finished successfully in 2000. From January 2001, the regular phase activities were started. Major activities of EANET are, 1) to implement the national monitoring of acid deposition, 2) to develop and implement the QA/QC (quality assurance/quality control) programs, 3) to develop and implement training programs, etc. The acid deposition monitoring covers four environmental media; wet deposition, dry deposition, soil and vegetation, and inland aquatic environment.

In addition to the mentioned activities, Network Center for EANET has joint research projects among participating countries of EANET. As one of them, joint project with Mongolia was performed to accumulate information on plant sensitivity to acid deposition in Ulaanbaatar, which will be reviewed briefly.

**Key words.** EANET, Monitoring, Wet and dry deposition, Soil and vegetation, Plant sensitivity, Joint research, Mongolia

## 1. Results on the acid deposition monitoring

### 1.1 Wet deposition monitoring

During the preparatory phase for 2 and a half year since 1998, wet deposition monitoring for EANET was performed at 38 sites, including 16 remote, 8 rural and 14 urban sites (Interim Scientific Advisory Group of EANET, 2000).

Table 1 shows annual mean pH of precipitation and wet deposition amounts of major components obtained in 2000. The data were collected in urban, rural and remote sites in a whole year round without any trouble. Annual mean pH of precipitation was in the range of 4.22-6.42. Wet deposition amounts in mmol m$^{-2}$ year$^{-1}$ of major components were in the range of 2.16-239 in non-sea-salt $SO_4^{2-}$ (nss-$SO_4^{2-}$), 3.04-92.9 in $NO_3^-$, 0.94-

136 in non-sea-salt $Ca^{2+}$ (nss-$Ca^{2+}$), 3.47-235 in $NH_4^+$, and 0.17-153 in $H^+$, respectively. Underlined data in the table show the extreme value in each column. 'Data Report on the Acid Deposition in the East Asian Region 2000' was published in November 2001 (Network Center for EANET, 2001).

Table 2 indicates the example of difference among urban, rural and remote sites in wet deposition data in China and Russia in 2000. Results of the annual average of rainwater pH and of wet deposition amount of nss-$SO_4^{2-}$ at the monitoring sites in the participating countries of EANET in 2001 are shown in Fig. 1.

**Table 1.** Annual mean pH of precipitation and wet deposition amount of major components in EANET (2000).

| Characteristics of sites | Country | Name of sites | pH | nss-$SO_4^{2-}$ * | $NO_3^-$ | $NH_4^+$ | nss-$Ca^{2+}$ ** | $H^+$ |
|---|---|---|---|---|---|---|---|---|
| | | | | | mmol/m$^2$ | | | |
| Urban | China | Guanyinqiao | 4.33 | 163 | 45.5 | 174 | 67.9 | 50.5 |
| | China | Shizhan | 5.68 | 198 | 48.6 | <u>235</u> | 128 | 1.17 |
| | China | Hongwen | 4.72 | 28.5 | 28.4 | 47.9 | 9.59 | 29.0 |
| | China | Xiang Zhou | 5.15 | 40.3 | 31.8 | 41.7 | 41.3 | 14.2 |
| | China | Zhuxian Cavern | 4.64 | 82.4 | 59.6 | 90.7 | 43.2 | 40.8 |
| | Malaysia | Petaling Jaya | 4.35 | 79.3 | <u>92.9</u> | 149 | 23.0 | <u>153</u> |
| | Philippines | Metro Manila | 5.48 | 87.0 | 48.9 | 143 | 56.0 | 13.2 |
| | Russia | Irkutsk | 5.11 | 15.8 | 11.2 | 17.2 | 14.8 | 4.15 |
| | Thailand | Bangkok | 4.95 | 24.4 | 24.8 | 47.4 | 12.9 | 12.8 |
| | Thailand | Samutprakarn | 4.83 | 24.4 | 15.1 | 31.1 | 10.6 | 14.5 |
| | Vietnam | Hoa Binh | 5.11 | 29.4 | 18.3 | 15.2 | 24.2 | 14.7 |
| Rural | China | Nanshan | <u>4.22</u> | 164 | 52.4 | 132 | 69.3 | 76.2 |
| | China | Weishuiyuan | 6.42 | <u>239</u> | 42.9 | 190 | <u>136</u> | 0.17 |
| | Japan | Ijira | 4.52 | 50.1 | 64.8 | 63.8 | 17.0 | 80.7 |
| | Philippines | Los Banos | 5.44 | 22.4 | 14.3 | 35.9 | 16.3 | 9.58 |
| | Russia | Listvyanka | 5.07 | 6.67 | 8.25 | 7.29 | 5.61 | 3.78 |
| | Thailand | Patumthani | 5.25 | 18.1 | 17.8 | 36.6 | 12.3 | 5.41 |
| | Vietnam | Hanoi | 5.45 | 36.5 | 20.2 | 34.5 | 25.7 | 4.48 |
| Remote | China | Dabagou | 5.42 | 116 | 24.1 | 165 | 102 | 3.14 |
| | China | Xiaoping | 4.91 | 22.0 | 22.8 | 31.0 | 0.94 | 19.1 |
| | Japan | Happo | 4.73 | 28.1 | 24.3 | 31.0 | 8.12 | 40.6 |
| | Japan | Oki | 4.64 | 20.4 | 22.4 | 23.5 | 9.07 | 27.6 |
| | Japan | Yusuhara | 4.71 | 28.4 | 19.7 | 14.2 | 6.80 | 54.7 |
| | Japan | Ogasawara | 5.23 | 6.22 | 4.73 | 4.62 | 4.24 | 11.7 |
| | Japan | Hedo | 5.13 | 17.1 | 16.7 | 21.0 | 2.35 | 21.5 |
| | Malaysia | Tanah Rata | 4.79 | 12.5 | 10.4 | 129 | 6.93 | 50.4 |
| | Russia | Mondy | 5.26 | 2.16 | 3.04 | 3.47 | 1.75 | 1.68 |
| | Thailand | Khao Lam | 5.56 | 2.72 | 5.06 | 6.03 | 4.02 | 2.41 |

*nss-$SO_4^{2-}$: non-sea-salt $SO_4^{2-}$, **nss-$Ca^{2+}$: non-sea-salt $Ca^{2+}$

**Table 2.** Site difference of annual mean pH of rainwater and concentration of major components in $\mu$mol l$^{-1}$ of nss-SO$_4^{2-}$, NO$_3^-$, nss-Ca$^{2+}$, and H$^+$ in Xi'an, China (a) and in Russia (b) in 2000.

(a)

| Site | pH | nss-SO$_4^{2-}$ | NO$_3^-$ | nss-Ca$^{2+}$ | H$^+$ |
|---|---|---|---|---|---|
| Urban (Shizhan) | 5.68 | 353 | 87 | 229 | 2.1 |
| Rural (Weishuiyuan) | 6.42 | 534 | 95.8 | 305 | 0.4 |
| Remote (Dabagou) | 5.42 | 140 | 29.1 | 124 | 3.8 |

(b)

| Site | pH | nss-SO$_4^{2-}$ | NO$_3^-$ | nss-Ca$^{2+}$ | H$^+$ |
|---|---|---|---|---|---|
| Urban (Irkutsk) | 5.11 | 29.6 | 21 | 27.8 | 7.8 |
| Rural (Listvyanka) | 5.07 | 15.1 | 18.7 | 12.7 | 8.6 |
| Remote (Mondy) | 5.26 | 7.1 | 10 | 5.7 | 5.5 |

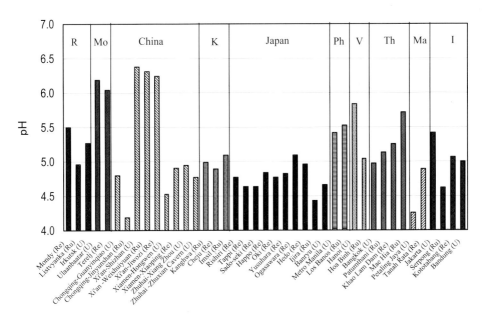

**Fig. 1 a.** Results of the annual average value of rainwater pH (above figure) in 2001 at the monitoring sites in the participating countries of EANET
Country names abbreviated in the figures: R, Russia; Mo, Mongolia; K; Republic of Korea; Ph; Philippines; V, Viet Nam; Th, Thailand; Ma, Malaysia; I, Indonesia.

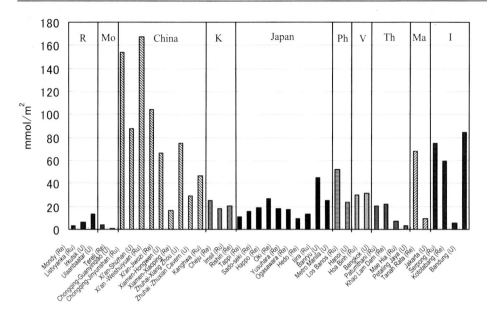

**Fig. 1 b.** The same as in **Fig. 1 a**, but in annual wet deposition amount of nss-$SO_4^{2-}$.

## 1.2 Dry deposition (air concentration) monitoring

For dry deposition monitoring, eight countries including 31 sites (14 remote, 5 rural and 12 urban sites) were participated in 1999. Automatic monitoring method and filter pack method were used to carry out the dry deposition (air concentration) monitoring mainly focusing on the following priority chemical species: First priority: $NO_2$ (urban), $SO_2$, $O_3$, and NO, and particulate mass concentration. Second priority: $NO_2$ (rural and remote), $HNO_3$, $NH_3$, particles ($SO_4^{2-}$, $NO_3^-$, $NH_4^+$, and $Ca^{2+}$). In case of automatic monitoring method, air concentration monitoring for $SO_2$, $NO_2$ (urban), NO, $O_3$ and particulate matter are available, and an hourly average of these components can be reported.

Four-stage filter pack method was used for EANET filter pack monitoring, with which gaseous ($SO_2$, $HNO_3$, $HCl$, $NH_3$) and particulate components ($SO_4^{2-}$, $NO_3^-$, $Cl^-$, $NH_4^+$, $Ca^{2+}$) concentrations were determined. The concentration of parameter obtained by filter packs was calculated from the comparison between extracted solution and standard solution of corresponding chemicals. In 2002, automatic monitoring method was applied in China, Japan, Russia, and Thailand, and filter pack method was used in Indonesia, Japan, Malaysia, Mongolia, Philippines, Republic of Korea, Russia, Thailand and Vietnam. Table 3 shows data of air concentration of $SO_2$, $NO_2$, and $O_3$ in ppb in several monitoring sites in 2000. Fig. 2 shows annual average of $SO_2$ concentration of ambient air in 2002 at 34 monitoring sites in the participating countries.

Data of dry deposition monitoring in 2000, 2001 and 2002 were reported by the Network Center for EANET (2001, 2002, 2003).

**Table 3.** Data of air concentration (ppb) of $SO_2$, $NO_2$ and $O_3$ in 2000 obtained in several monitoring sites, where the monitoring period was partly limited, as shown in the Table. Nos. 1 and 2 attached in the name of site show the difference of automatic monitor (1) and filter pack (2) methods (adapted from the Network Center for EANET, 2001)

| Country | Site | $SO_2$ ppb | $NO_2$ ppb | $O_3$ ppb | Monitoring period |
|---|---|---|---|---|---|
| China | Chongqing[1] | 48.2* | 51.2* | – | *Jan.–Sep. |
| | Xi'an[1] | 6.1 | 6.3 | – | |
| | Xiamen[1] | 5.8* | – | – | *Jan.–Oct. |
| | Zhuhai[1] | 7.1 | – | – | |
| Indonesia | Serpong[1] | 2.1* | 6.3 | – | *Jul.–Dec. |
| Japan | Ijira[1] | 0.1 | 2.1 | 30.0 | |
| | Banryu[1] | 0.6 | 3.6 | 35.6 | |
| Malaysia | Tana Rata[2] | N.D. | – | – | |
| Mongolia | Terelj[2] | 0.4 | – | – | |
| | Ulaanbaatar[2] | 1.5* | – | – | *Apr.–Oct. |
| Philippines | Metro Manila[2] | 4.8* | – | – | *Jun.–Sep., Nov., Dec. |
| | Los Banos[2] | 0.8* | – | – | *Sep., Nov., Dec. |
| Russia | Mondy[2] | 0.1 | – | – | |
| | Listvyanka[2] | 1.0 | – | – | |
| | Irkutsk[2] | 3.4 | – | – | |
| Thailand | Khao Lam[1] | 1.3* | 2.3* | 20.0* | *Mar., Jul., Nov. |
| | Bangkok[1] | 4.3 | – | – | |
| | Samputkarm[1] | 4.3* | 17.8* | – | *May–Dec. |

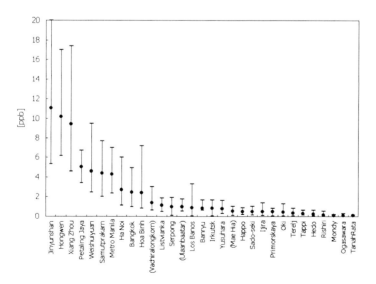

**Fig. 2.** Annual average of $SO_2$ concentration of air (max. and mini. monthly average basis) in 2002 at the monitoring sites in the participating countries. "(site name)" in the figure means its annual data completeness under 75%.

## 1.3 Soil and vegetation monitoring

Forest and (forest) soil monitoring is carried out as a part of the soil and vegetation monitoring in EANET.

Six countries including China, Japan, Malaysia, Philippines, Russia and Thailand performed in 2000 and 2001 their activities on forest and (forest) soil monitoring. In 2002 three countries of Philippines, Russia and Thailand reported the results. One to four areas were selected basically within 50km radius of the acid (wet/dry) deposition monitoring sites in the respective countries. Description of trees (name of species, diameter at breast height and height of trees), understory vegetation survey, and survey of tree decline were carried out in the forest monitoring, and analyses of soil chemical properties such as pH and exchangeable cations were done in the soil monitoring. The surveys are carried out with 3-5 years interval.

As for forest monitoring, most monitoring forests are secondary forest or natural forest, and consist of several tree species. Tree species are varied from coniferous species (e.g. *Pinus sibirica*) in sub-arctic zone to rain forest species (e.g. Dipterocarpaceae family) in tropical zone. Tree decline symptoms have been observed only in the sites of Japan and Russia, however, it seemed that natural environmental factors, such as insect attack, heavy snow, steep slope, and poor soil, might be main causes. No obvious evidence on implication between acid deposition/air pollution and tree decline has reported.

As for soil monitoring, sixteen soil types (by soil unit in the FAO/UNESCO classification) have been identified for the monitoring plots although several soil types (units) have not been identified by the FAO/UNESCO classification. Ten areas have monitoring forests with two (or more) soil types respectively according to the recommendation in the Technical Manual. Monitoring of two soil types (with different chemical properties) in an area would be informative for comparison of their sensitivities to acid deposition. The condition of forest soil in East Asia could be discussed several years later with the data accumulated.

As of 2004, the Data Report 2001 and 2002 including data of twelve monitoring forests during the regular phase which was started from 2001, were disclosed (Network Center for EANET, 2002, 2003), and a part of data in other monitoring forests during the preparatory phase (1998 – 2000) were also published in the previous reports (Network Center for EANET, 2001).

# 2. Joint research project with Mongolia

## 2.1 Field survey

In Bogdkhan Mountain around Ulaanbaatar, Mongolia, decline of larch trees (*Larix sibirica*) has been reported, and it might be due to air pollution from the thermal power plants. A number of useful plant (grass) species also grows in the mountainous area, and effect of the air pollution on those plant species should be studied.

Fig. 3 shows the locality of field survey sites in Ulaanbaatar, Mongolia. Surveys on air concentrations by passive samplers, field observation of tree decline and chemical

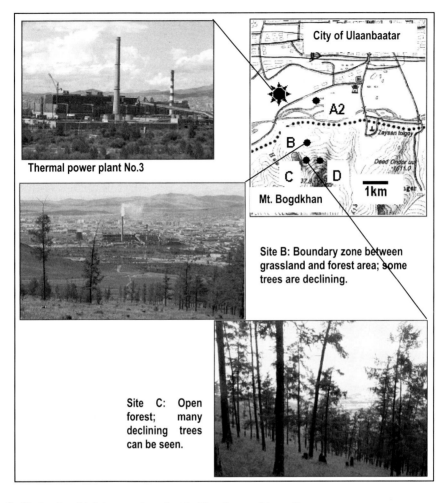

**Fig. 3.** Study site of joint research project in Ulaanbaatar, Mongolia.

properties of needles/soils were carried out in 2001 and 2003 in Bogdkhan Mountain in order to obtain information on implication between the air pollution and chemical/ (eco-) physiological properties on plant and soil. Tree decline was observed not only on the sites B and C in Fig. 3, 2-3 km far from the thermal power plant No. 3 (capacity, 300 MW; no treatment system for $SO_x/NO_x$ emission), but also in reference forests because of insect attack. However, decline symptoms observed on the slope were different from those in the reference forests. In mid summer 2001, concentrations of all the pollutants except $O_3$ were not high; less than 5 ppb (average concentration of two weeks), but $O_3$ concentration was relatively high in mid summer and gradually decreased from autumn to winter (ca. 40 ppb). From autumn to winter, concentrations of $NO_x$ increased probably due to increasing of combustion of coal/wood in winter. Mean concentrations of $SO_2$ and

O₃ during the sampling period (July to November) of the year 2003 were higher at the sites on the slope than at other sites. In addition, sulfur contents of larch needles were also higher on the slope than other sites. It was suggested that the slope facing the thermal power plant have stronger effects than the other areas. Soil acidification due to acid deposition may hardly occur because of the high concentration of base cations; e.g. pH ($H_2O$): 5.8, exchangeable Ca: 22.6 cmol(+) kg$^{-1}$ in a site on the slope. Direct effects of air pollution, especially effects of $O_3$, should be considered as one of the possible causes for the tree decline on the slope as well as natural environmental factors such as insect attack.

It was clarified that air pollutants such as $SO_2$ and $O_3$ were transported to Bogdkhan Mountain, and the slope facing the thermal power plant had larger effects of the pollution than the reference forest area. Concentration of $SO_2$ was very low but $O_3$ was relatively high. Effects of air pollution, especially the effect of $O_3$, should be considered as one of the possible causes for tree decline though effect of insect attack might be the major cause as observed in large extent of Bogdkhan Mountain. As for the surveys, detailed description was reported by Sase et al. (2005).

### 2.2 Gas exposure experiment for plant sensitivity test

Increasing of air pollution level in Ulaanbaatar, Mongolia has caused great concerns in the stability of some semiarid grassland ecosystems. So far, there is no experimental datum available for assessing the effects of air pollutants on the native grass species in this region. In our recent study, five Mongolian semiarid plant species of *Astragalus* spp., *Carex* spp., *Chamaenerion angustifolium*, *Polygonum alopecuroides* and *Sanguisorba officinalis*, were grown for 4 weeks from the seeds collected at Mt. Bogdokhan. They were exposed to either $O_3$ (0.05 ppm in daily average, 0.1 ppm for 3 hours in the daytime) or $SO_2$ (0.1 ppm constant) for 4 weeks in the naturally-lit environment controlled chambers. The chambers were maintained 25/15°C (14hr day/10hr night) and 55/75% (day/night, rel. humidity). So as to investigate the growth responses of these species to $O_3$ and $SO_2$, growth parameters such as leaf number, leaf area, biomass and root/shoot ratio were measured at the final harvest. Different species showed different sensitivities to $O_3$ and $SO_2$. $O_3$ significantly reduced leaf number of *Carex* spp., biomass and root/shoot ratio of *Polygonum alopecuroides*, and root/shoot ratio of *Sanguisorba officinalis*, however $O_3$ had no effect on other parameters of these species or all the parameters of the other species. $SO_2$ significantly reduced leaf number and biomass of *Carex* spp., and had no effects on other parameters for this species and all the parameters of the rest of the four species. Results in detail were reported by Shimizu et al. (2005).

# References

Interim Scientific Advisory Group of EANET 2000. Report on the acid deposition monitoring of EANET during the preparatory phase – Its results, major constraints and ways to overcome them

Network Center for EANET 2001. Data report on the acid deposition in the East Asian region 2000

Network Center for EANET 2002. Data report on the acid deposition in the East Asian region 2001

Network Center for EANET 2003. Data report on the acid deposition in the East Asian region 2002
Sase H, Bulgan T., Batchuhuluun T, Shimizu H, Totsuka T(2005) Tree decline and its possible causes around Mt. Bogdkhan in Mongolia. Phyton (in press)
Shimizu H, An P, Zheng Y. R, Chen L. J, Sase H, Totsuka T, Bulgan T, Zheng,Y (2005) Response to $O_3$ and $SO_2$ for five Mongolian semiarid plant species, Phyton (in press)

# Land degradation and blown-sand disaster in China

Pei-Jun Shi[1], Hideyuki Shimizu[2], Jing-Ai Wang[3], Lian-You Liu[1], Xiao-Yan, Li[1], Yi-Da Fan[4], Yun-Jiang Yu[1,2], Hai-Kun Jia[1], Yanzhi, Zhao[1], Lei Wang[1], and Yang Song[1]

[1]College of Resources Science & Technology, Beijing Normal University, No.19 Xinjiekouwai Street, Beijing 100875, China
[2]National Institute for Environmental Studies, Onogawa 16-2, Tsukuba, Ibaraki 305-8506 Japan
[3]College of Geography and Remote Sensing Science, Beijing Normal University, No.19 Xinjiekouwai Street, Beijing 100875, China
[4]National Disaster Reduction Center of China, Ministry of Civil Affairs, Bai Guang Lu No.7, Beijing 100053, China

**Summary.** China is a country with severe land degradation and blown sand disasters. The arid and semi-arid regions, in which land desiccation, vegetation degeneration, wind erosion, sandification, Gobi-pebblization and salinization occur, take up one third of China's total land area. Vegetation degradation is most serious in lower flood plains of the inland rivers and the semi-arid Agro-pastoral Ecotone due to excessive use of water resources, grassland reclamation, overgrazing and collection of firewood and herbal medicines. Wind erosion features are common around terminal dry lakes, in inland-river fluvial plains, and the semi-arid dry grasslands. Studies by the methods of aeolian sand transport, soil texture analysis, $^{137}$Cs tracing and archaeology confirmed that the rate of wind erosion is normally between 1000 to 2000 ton km$^{-1}$ a$^{-1}$. The gravel Gobi on Mongolian Plateau has been formed to a large degree by wind erosion. The severity of sandification has been manifested by the twelve sandy deserts and lands occupying 710, 000 km$^2$, and the enlargement of sandy land at increasing spreading rates in the past three decades. Salinization has not received enough public attention yet, but soil salinization in Ningxia and Hetao Plains and dry lake basins is unfavorable for crop growth and natural vegetation. Salinization of surface water and ground water in the lower reaches of most inland rivers restricts utilization of insufficient water resources. The exacerbation of sand and dust storms disasters is the ultimate outcome of desertification in China.

**Key words.** Land degradation, Arid and semi-arid regions, China, Blown sand disasters

## 1. Introduction

Land degradation is a global environmental issue on account of its genesic ubiquity, damage severity and subsequent adverse impacts. China is a country with the largest population and the most serious land degradation in the world (Ci 1995). Land degradation in the arid and semi-arid China, the so-called land desertification, is the outcome of long-term interaction between natural and social-economic factors (Shi 1991). It is complex and disastrous due to its multiple genesis, and various damage manifestations (Dong et al. 1987; Yang et al. 1991; Kar and Takeuchi 2004). Relevant research in China has

*Plant Responses to Air Pollution and Global Change*
*Edited by K. Omasa, I. Nouchi, and L. J. De Kok* ( Springer-Verlag Tokyo 2005 )

been chiefly in sandification, or sandy desertification, which is characterized by wind-induced sand actions under dry windy climate, sandy ground conditions and various human disturbances (Wu 1987; Zhu and Chen 2000). Investigations by the Bureau of Forestry indicated that desertification in China has been worsened as a whole, and meliorated at some local parts (Ci 1995). The exacerbation of sand and dust storm disasters in China is generally thought as the aftermath of land degradation in the arid and semi-arid regions (Shi et al. 2000).

Multidisciplinary approaches confirmed that land degradation in China involves with physical, chemical, ecological and human processes and appears as land desiccation, vegetation degeneration, wind erosion, sandification, Gobi pebblization, and salinization. Up to now, the various types of land degradation other than sandy desertification in China have been rarely documented. For a better understanding of the overall status of land degradation in China, we present firstly the distribution of the arid and semiarid areas; then discuss the characteristics of land degradation and blown sand disaster; and finally try to reveal the relationship between land degradation and blown sand disaster.

## 2. Distribution of the arid and semiarid areas

Under influence of the East-Asian Monsoon and continentality, large-scale horizontal climatic zonation takes place from the southeast humid to the northwest arid regions in China (Fig.1). Vegetation, soils, landforms and social-economic features differentiate correspondingly in a similar spatial pattern. By thermal conditions, the arid and semiarid regions in China could be classified into: (1) The temperate arid and semiarid, and (2) the cold high arid and semiarid zones. The arid and semiarid region with humidity index <0.5 covers 3.15 million km$^2$, and makes up 32.78% of the total territorial area of China (Ci 1995). Due to insufficient rainfall, sparse vegetation, and frequent strong wind, land degradation occurs mainly in the arid and semiarid regions.

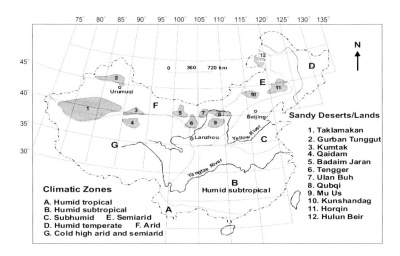

Fig. 1. Distribution of the climatic zones and major sandy deserts and sandy lands in China.

## 3. Distributions and tendency of land degradation

### 3.1 Land desiccation

The aridity of northwestern China is chiefly due to its inland locality and far from the world oceans (Cooke et al. 1993). With the uplift of the Qinghai-Tibetan Plateau, northwest China has experienced a long history of intensified desiccation (Zhang 1992). In the past several decades, further land desiccation was induced due to both climatic fluctuation and human activities. In the semiarid area, annual rainfall change significantly with monsoon fluctuation, periodical land desiccation takes place due to frequent droughts. In the arid regions, irrational reclamation and excessive use of limited water resources in the middle and upper reaches of inland rivers, such as in Gansu, Xinjiang, led to large-area land desiccation, vegetation degeneration, and severe wind erosion in the lower reaches. The migrating lake Lop Nor in the terminal of Tarim River became completely desiccated in 1972 (Yan et al. 1998). In the Heihe River in Western Inner Mongolia, the terminal lake Gashun Nor, exposed its bottom in 1962, while the Sorgo Nor, shared the same fate in 1992 (Zhang et al. 1998).

### 3.2 Vegetation degeneration

Vegetation degeneration is the most important indicator of land degradation (Fig. 2). In the inland river basin, natural forest in the lower reaches degrades as a result of water shortage. Over the past twenty years the underground water table has descended at least 5 m. The root systems of natural desert vegetation can hardly absorb any moisture and died

**Fig. 2.** Degeneration of vegetation due to (a) land desiccation, (b) salinization, (c) over-grazing, and (d) random traffic.

due to land desiccation. The area of *Populus Euphratica* in Ejina reduced from 50,000 ha in 1949 to 23,000 ha in 2002, and about 100 species of vegetation became extinct. In the rangeland, vegetation deteriorates due to reclamation, grazing and traffic. Since the 1950s about 87 M ha grassland in China has been degraded (Li 1993).

### 3.3 Wind erosion

Wind erosion takes place in succession to land desiccation and vegetation degeneration. Wind erosion features are extensive in China (Fig. 3). Numerous streamlined Yadangs occupy vast area of the Qaidam Basin. In the vicinity of Lop Nor, the White Dragon Mounds more than 20m-high, which were described in the Han Dynasty 2000 years ago, might have a history of 20,000 years (Huntinton 1907). Even the 1-2m-high clay mesas took its shape in about one thousand years. The rates of wind erosion on bare land in the arid region vary significantly due to the difference in lithology, moisture, land cover, and wind regimes. It has been estimated through different methods that 0.5 to 5mm thick topsoil could be removed by wind every year (Hedin 1905; Liu 1999; Dong et al. 2000). Wind erosion in the semiarid regions could be more intense due the accelerating effects of various human activities, such as reclamation, over-grazing and traffic disturbance (Liu et al. 2003; Li et al. 2004).

### 3.4 Sandification

Sandification is the major form of the land degradation in China (Fig. 1). Sand deserts and sandy lands total 710,000 km$^2$ in area (Zhu et al. 1980). There are 11 provinces and

**Fig. 3.** Wind eroded features in China: (a) mega Yadangs in Qaidam Basin, (b) the White Dragon Mounds, east Lop Nor, (c) clay mesas in western Inner Mongolia, and (d) sand abraded grooves along traffic tracks, Xilin Gol Grassland, Inner Mongolia.

212 counties impacted by sandification. In the arid area to the west of Helan Mount, sandification is characterized by shifting sand seas. The Taklamakan Desert has an area of 337,600 km$^2$. Sand deserts distribute in huge tectonic basins where alluvial, fluvial and lacustrine sediments are transported from the surrounding mountains. In the semiarid area, sandification is distributed as fixed or semi-fixed sandy lands (Fig. 4). The area of sandification increased due to climatic change and human disturbances. Sandified land expanded at a rate of 1570 km$^2$ a$^{-1}$ from the end-1950s to mid-1970s, 2100 km$^2$ a$^{-1}$ from mid-1970s to mid-1980s, 2460 km$^2$ a$^{-1}$ from mid-1980s to mid-1990s, and 3450 km$^2$ a$^{-1}$ from mid-1990s to the end-1990s (Wang 2000).

## 3.5 Gobi-pebblization

Gobi pebblization represents a desert pavement, a surface concentrated with pebbles lagged by wind deflation or wetting and drying cycles. The coarse pebbles, which prevent the land from further erosion, can be easily disturbed. In northern China, the area of Gobi desert totals 569,500 km$^2$ (Zhu et al. 1980). It is mainly distributed in Xinjiang, Inner Mongolia and Gansu Province (Fig.5).

**Fig. 4.** Sandification (a) the Sand Mount near Dunhuang, (b) sandified grassland in Tibet.

**Fig. 5.** Gobi pebblization: (a) farm land texture coarsening, (b) the black Gobi in Inner Mongolia.

## 3.6 Salinization

Salinization of soils, surface water and ground water is common in the arid and semiarid area. There are 99.13 M ha salinized soils in China (Wang et al. 1993). Salinized soil could be classified into recent (36.93 M ha), residual (44.87 M ha) and potential salinized soils (17.33 M ha). Due to intense evaporation and climatic fluctuation, soil salification is characterized by alternate intense surface salt accumulation and desalinization. Salinized soils in the arid and semiarid areas in China are distributed mainly in Inner Mongolia, Gansu, Xinjiang, Qianghai and Tibet. The water salinity in Qinhai Lake, Bosten Lake, and Dalai Nor is 13.13, 1.58, and 5.55 g l$^{-1}$, respectively.

# 4. Distribution of sand and dust storms in relation to land degradation

Accompanied by intense Siberian/Mongolian cold fronts, sand and dust storms in eastern Asia move out of the Gobi and sand desert regions of China along varied southeastward trajectories (Qiu et al. 2001). The annual average frequency of sand and dust storms in the arid and semiarid regions vary from 1 to 37 days with a general increase from southeast to northwest. Strong sand and dust storms occurs mainly in southern Xinjiang, in the middle and western Inner Mongolia. About 70 percent of the sand and dust storms occur in Spring.

Sand and dust storms occur under strong wind conditions, and sand entrainment, dust emission vary significantly from one land surface type to another. The ratio of sand and dust storm days to strong wind days at different land surfaces indicates the susceptibility of blown sand disasters on different degraded lands. The data in Fig. 6 and Table 1 imply that sand and dust storms can take place most easily in the sand desert and farmland areas, moderately in the grassland and gobi area, but most rarely in the salinized land.

Suspended particles carried in sand and dust storms from the arid and semiarid regions of China have caused extensive consequential effects on the atmospheric environment of Northern China, Eastern Asia and even the northern hemisphere (Duce et al. 1980; Liu et al. 1985). Air quality monitoring indicated that $PM_{10}$ was the chief pollutant of the major cities in northern China (Shi 2000). The seasonality of sand and dust storms coincide with the severest air pollution in the major cities in China (Fig. 7).

# 5. Conclusions

Land degradation occurs widely in the arid and semiarid regions due to insufficient rainfall, sparse vegetation, and frequent strong wind. Land degradation in China involves with physical, ecological, chemical and anthropological processes. Land desiccation is the ultimate reason for vegetation degeneration, salinization, wind erosion, sandification, and Gobi pebblization. Suspended particles carried in sand and dust storms from the arid and semiarid regions have caused extensive effects on the atmospheric environment. Sand and dust storms occur most readily from the sand desert and farmland regions.

Acknowledgements. This work was supported by NKBRSF Project (G2000018604), NSFC project (40471014) and a China-Italy joint project (2003DFB00040).

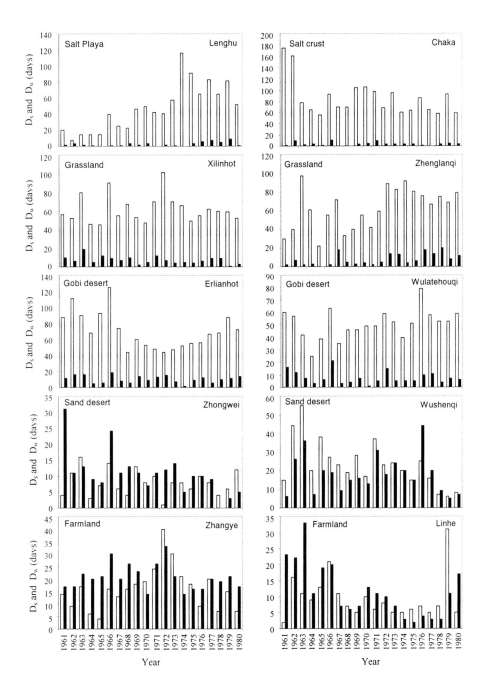

**Fig. 6.** Relationship between: ■ the days of sand and dust storm ($D_s$), and ☐ the days of strong wind ($D_w$) at stations under different land cover and use conditions.

**Table 1.** The days of sand and dust storm (Ds) and strong wind (Dw) under different land surfaces.

| Land surfaces | Stations | $D_s / D_w$ | Period |
|---|---|---|---|
| Salinized land | Lenghu | 1/15.0 | 1961-1980 |
| | Chaka Salt Lake | 1/23.0 | 1961-1980 |
| Grassland | Xilinhot | 1/8.5 | 1961-1980 |
| | Zhenglanqi | 1/8.1 | 1961-1980 |
| Gobi Desert | Erlianhot | 1/6.7 | 1961-1980 |
| | Wulatehouqi | 1/6.6 | 1961-1980 |
| Sand Desert | Zhongwei | 1/0.7 | 1961-1980 |
| | Wushenqi | 1/1.3 | 1961-1980 |
| Farmland | Zhangye | 1/0.8 | 1961-1980 |
| | Linhe | 1/0.8 | 1961-1980 |

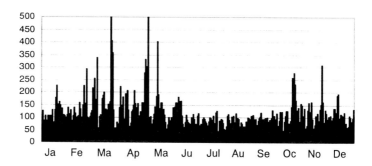

**Fig. 7.** Air pollution index affected by dust storms in Beijing, 2002.

# References

Ci LJ (1995) Influence of global change on China's desertification. Sci Technol Rev 1:61-64

Cooke RU, Warren A, Goudie AS (1993) Desert Geomorphology. UCL Press, London

Dong GR (1983) The Quaternary fossil aeolian sand on Ordos Plateau. Acta Geographic Sinica 38:341-346

Dong GR, Li CZ, Jin J, Gao SY, Wu D (1987) Some results of the simulating experiment on soil wind erosion. Chinese Sci Bull 32:1703-1709

Dong ZB, Wang XM, Liu LY (2000) Wind Erosion in Arid and Semiarid China: an overview. J Soil Water Conservation 55:439-444

Hedin S (1905) LOP-NOR: Scientific Results of A Journey in Central Asia (1899-1902). Stockholm: Scientific Report, Vol. 2

Huntinton E (1907) The Pulse of Asia. Boston and New York: Houghton Mifflin Company

Kar A, Takeuchi K (2004) Yellow dust: an overview of research and felt needs. J Arid Environments 59:167-187

Li XY, Liu LY, Wang JH (2004) Wind tunnel simulation of aeolian sandy soil erodibility under human disturbance. Geomorphology 59:3-11

Li YT (1993) Prevention and Cure of China's Land Deterioration. Beijing: Science and Technology Press

Liu DS (1985) Loess and environment. Beijing, Science Press

Liu LY (1999) The quantity and intensity of regional aeolian sand erosion and deposition: the case of Shanxi-Shaanxi-Nei Monggol region. Acta Geographica Sinica 54:59-68

Liu LY, Shi PJ, Zou XY, Gao SY, Li XY, Wang JH, Ta WQ (2003) Short-term Dynamics of Wind Erosion of Three Newly Cultivated Grassland Soils in Northern China. Geoderma 115:55-64

Qiu XF, Zeng Y, Miao QL (2001) Temporal and spatial distribution as well as tracks and source areas of sand-dust storms in China. Acta Geographica Sinica, 56:316-322

Shi PJ (1991) Theory and practice in the study of geographical environment evolution-Study on geographical environment evolution since late Quaternary in the Ordos region. Beijing: Science Press, pp 182

Shi PJ, Yan P, Gao SY, Wang YM, Ha S, Yu YJ (2000) Disaster of sand and dust storms in China: Research progress and prospects. J. Natural Disaster 9:71-77

Wang T (2000). Research on desertification and control to its calamity in the large-scale development of the western China. J Desert Res 20:345-348

Wang ZQ (1993) Saline soils in China. Beijing: Science Press, pp 573

Wu Z (1987) Aeolian Geomorphology. Beijing: Science Press, 316 p

Yan S, Mu GJ, Xu YQ (1998) Quaternary environment evolution of the LOP-NOR region, China. Acta Geographica Sinica 4:332-340

Yang GS, Di XM, Huang ZH (1991) Land desertification in Northern Loess Plateau region. Beijing: Science Press, pp125-186

Zhang LY, Jiang ZL (1992) Discussion on the causes of climatic aridity in northwestern China. Arid Land Geography 15:1-12

Zhang ZK, Wu RJ, Wang SM (1998) Environmental changes recorded by lake sediments from east Juyanhai Lake in Inner Mongolia during the past 2600 years. Journal of Lake Sciences 2:44-51

Zhu ZD, Chen GT (2000) Sandy desertification in China. Beijing: Science Press, pp 250

Zhu ZD, Wu Z, Liu S (1980) Introduction to Sand Deserts in China. Beijing, Science Press

# Impact of meteorological fields and surface conditions on Asian dust

Seiji Sugata, Masataka Nishikawa, Nobuo Sugimoto, Ikuko Mori, and Atsushi Shimizu

National Institute for Environmental Studies, Onogawa 16-2, Tsukuba, Ibaraki, 305-8506, Japan

**Summary.** Dust aerosol in eastern Asia (yellow sand, kosa) is one of the most notable topics in atmospheric environment problems. The number of days of observational dust event was high in 2001 and 2002, and low in 2003 in downstream areas in eastern Asia, such as Japan. The aerosol is simulated by the combination of the meteorological model, RAMS, and the transport model, CMAQ, in this study. Simulated surface concentrations show good agreement with those observed by laser particle counters at some surface observation sites. We divided dust aerosol in the transport model into six different tracers according to their emission areas; Kazakhstan, Mongolia, three parts of Xinjiang, and Inner Mongolia. As for dust concentration at surface in Beijing, dust from Mongolia accounts for more than half, that from Inner Mongolia does one third, and dust from Xinjiang does the other approximately 15%. It is found that not only reduction of the amount of emission in Mongolia but also the change of transport from Mongolia toward downstream areas is main reason for less dust events in 2003. Large snow cover in 2003 cannot account for all the reduction of the emission in Mongolia compared to the other two years.

**Key words.** Dust, Numerical simulation, Temporal variation of dust emission, Contribution of emission area

## 1. Introduction

Interest has grown to understand the fugitive dust originating from eastern Asia. One of the most focused topics on the dust is whether the amount of the dust transported into downstream areas, such as eastern China, Korean peninsula, Japan, has recently been increasing or decreasing. It should depend on how the amount of the emission itself in emission areas changes and also on how the amount and the route of the transport of the dust from the emission areas toward the downstream areas changes. The former is attributed to natural and anthropogenic changes of surface conditions in the emission areas, meteorological changes of the strengths of surface wind, and so on. The latter is attributed to the change of stream lines of the transport of dust due to wind field change, the change of characteristics of planetary boundary layer, which controls vertical transport of dust, and so on. As for the numbers of accumulated observation day in Japan, it was very high in 2001 and 2002 and was very low in 2003. This study simulates the dust in the eastern Asia for these three years to investigate what causes the difference of the amount of the dust among the years.

*Plant Responses to Air Pollution and Global Change*
Edited by K. Omasa, I. Nouchi, and L. J. De Kok ( Springer-Verlag Tokyo 2005 )

First, we briefly describe the model used in this study and configuration of calculations. Second, we compare the results of the calculations with observed time series of dust concentration at some sites to demonstrate the model performance. Third, we investigate which part of emission area gives large contribution for surface concentration at downstream areas. Finally, we discuss causes making difference of yearly averaged dust concentrations.

## 2. Model description

We use the Community Multiscale Air Quality (CMAQ) modeling system (Byun and Ching, 1999), which has been developed by U.S. EPA, version 4.4 with the meteorological data provided by the Regional Atmospheric Modeling System (RAMS) version 4.3. The former is a comprehensive air quality modeling system that consists of the CMAQ Chemical Transport Model (CCTM) and several interface processors such as Meteorology-Chemistry Interface Processor (MCIP). One of key design objectives of CMAQ was to achieve flexibility that enables linkage of different science processors and modules to build appropriate air quality models to meet user's needs. It uses efficient modular structure with minimal data dependency and a set of a generalized chemistry solver module and chemical mechanism reader to handle multi-pollutant problems. The CMAQ modeling system utilizes the Mesoscale Model Generation 5 (MM5) as the default meteorological driver. RAMS meteorological data is used in this study with the modified CMAQ system (Sugata et al. 2001) where some of the interface programs both in the RAMS and CMAQ systems were modified to allow the linkage.

The CCTM represents the particle size distribution of aerosols as the superposition of three lognormal modes. As for the soil-derived aerosol, it is represented by only one mode for a PM10 component. Explicit treatment of emission for the soil-derived aerosol is not implemented in the model as its default setting. Therefore a reading module for aerosol emissions was modified to read a prepared dust emission in this study.

Dust emission is estimated according to soil and landuse of each grid cell. The World Soils for Global Climate Modeling by Zobler is used as the data for soil, which has 27 categories in one degree square latitude/longitude grid cells, and USGS Global Land Cover Characterization data is used as the data for landuse, which has 100 categories with 30s resolution. Each grid is supposed to be erodible when its soil is yermosol (desert soil) or xerosol (semi-desert soil) and its landuse is bare desert, semi desert, semi desert shrub, or low sparse grassland. Following Gillette and Passi (1988), the amount of emission is expressed as,

$$E = CU_*^4 (1 - \frac{U_{*t}}{U_*}) \qquad (U_* > U_{*t})$$
$$= 0 \qquad (U_* < U_{*t})$$

where $E$ is the mass of dust emitted, $U_*$ is friction velocity, $U_{*t}$ is a threshold of $U_*$, and $C$ is a constant to be determined by evaluation. The threshold is assumed according to surface characteristics in Table 1 and the other combination of soil and landuse produces no emission. Effect of snow cover is also considered, where the mass emitted is reduced in proportion to SSM/I snow cover fraction.

**Table 1.** Dependence of the threshold value of friction velocity for the dust emission on surface characteristics. Units are in m s$^{-1}$.

| Landuse Soil | Bare desert | Semi desert | Semi desert shrub | Low sparse grassland |
|---|---|---|---|---|
| Yermosol | 0.25 | 0.3 | 0.4 | 0.7 |
| Xerosol | 0.3 | 0.35 | 0.45 | 0.75 |

## 3. Configuration for simulations

To provide meteorological data for CMAQ simulations, RAMS was run for five mouths from January to May for 2001, 2002, and 2003. The modeling domain is 6,300 x 2,700 km$^2$ on the rotated polar-stereographic map projection centered at the (43 N, 110 E) with 50 km mesh. The model extends vertically up to approximately 18 km, which is represented with the 20 layers in the sigma-z coordinates. For example, lowest four vertical layers are placed at elevations 57.3, 188.7, 346.5, and 535.8 m. RAMS options used for the run were; non-hydrostatic dynamics, simplified Kuo for cloud parameterization, Mellor-Yamada 2.5 for vertical diffusion, and Louis (1979) surface flux parameterizations. The RAMS was run with large-scale meteorological data provided by the ECMWF analysis with 2.5 degree horizontal resolution and 4 times a day. Then, the RAMS output is fed into MCIP to generate all the necessary meteorological parameters for the CMAQ chemical transport model (hereafter CCTM) simulations. The amount of dust emission is calculated based on calculated friction velocity and surface characteristics data described in the previous section including snow cover data. The CCTM was run for three months from March to May for each year.

Numerical simulation was carried out besides the standard simulations, where all the snow cover data were ignored, in order to estimate the impact of snow cover on the amount of dust emission.

## 4. Results and discussions

Distribution of dust emission for three month from March to May averaged for three years from 2001 to 2003 is shown in Fig. 1. Two large source areas located in Taklimakan Desert and Gobi desert, whose peak values are comparable to each other. We divided dust aerosol to six different tracers according to area that they were emitted. Area 1 is for outside of China and Mongolia, mainly for Kazakhstan, area 2 is for Mongolia, area 3 is for northern Xinjiang area, area 4 is for western Xinjiang area corresponding to main part of Taklimakan Desert, area 5 is for eastern Xingjian, and area 6 is for Inner Mongolia in China.

Table 2 shows dependence of dust emission for three month from March to May on each year and each source area. As shown in the distribution in Fig. 1, area 2 and area 4 show largest emissions, which are comparable to each other. The second largest emission is recorded in area 6 and the other areas give smaller emission. You see that area 2 shows very small amount in 2003 compared to the other two years, approximately 60 %. The table also shows result of no-snow-cover simulation. The difference of values between the no-snow-cover simulation and the standard simulation presents how much snow cover in each source area reduces the emission. It is show that the effect of the cover is large in area 2 and 5, is not small in area 6, and is very small in area 1, 3, and 4, where there is very small snow fall every spring. It is noticed that the emission amount would have been smaller in 2003 than the other years in area 2 even without the snow cover, which means the small emission is not due to larger snow cover in Mongolia in the year.

Figure 2 illustrates temporal variation of concentrations of dust, total concentrations accumulated for all source area, for March 2001 and March 2002 at two sites in China; Beijing and Erenhot. The figure compares surface concentrations observed by a laser

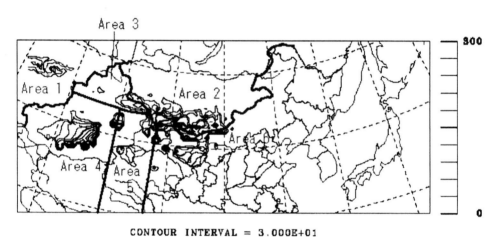

**Fig. 1.** Distribution of dust emission averaged for three years. Unit is ton 3 months$^{-1}$ km$^{-2}$. Six areas in order to investigate contribution of each emission are also shown. Area 1, Kazakhstan; area 2, Mongolia; area 3, northern Xinjiang area; area 4, Taklimakan Desert; area 5, eastern Xingjian; and area 6, Inner Mongolia.

**Table 2.** Dust emission calculated for three months in each year from each source area. Units are $10^6$ ton per three months. Numbers in parenthesis are the amount with no-snow cover assumption.

|      | Area 1 | Area 2 | Area 3 | Area 4 | Area 5 | Area 6 |
|------|--------|--------|--------|--------|--------|--------|
| 2001 | 6 (6)  | 38(42) | 7(8)   | 35(36) | 10(15) | 23(25) |
| 2002 | 12(12) | 41(52) | 7(8)   | 41(42) | 13(21) | 30(35) |
| 2003 | 4 (5)  | 23(29) | 5(7)   | 44(46) | 12(21) | 23(26) |

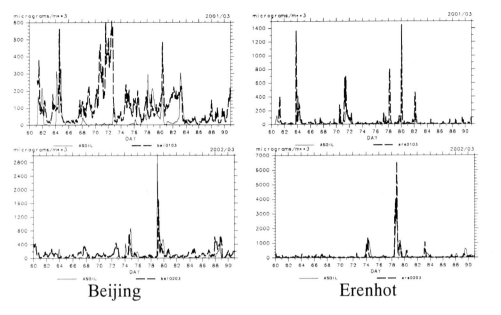

**Fig. 2.** Temporal change for one month at Beijing (left) and Erenhot (right) in 2001 (upper) and 2002 (lower). Thin solid line is calculation in the lowest layer and thick broken line is observation at the surface.

particle counter with simulated concentrations in the lowest layer. Note that the unit of the former is count per minutes and that of the latter is microgram per cubic meter, therefore it is not possible to compare them quantitatively. However their variations can be compared to each other. In Erenhot located very close to source areas in Mongolia and Inner Mongolia, almost all peaks of observational concentration are reproduced by the simulation. In Beijing, some of observational peaks are not shown in the simulation while the others are well simulated. It was clarified that the peaks not reproduced by the simulation correspond to observational high concentrations due to air pollution at Beijing by observational study with lidar (figures not shown).

Contribution of six source area on surface concentration in Beijing was investigated. Table 3 shows surface dust concentration simulated in Beijing averaged for three months for each year for each source area. You see that dust from area 2, Mongolia, occupies more than half concentration, area 6, Inner Mongolia, does one third, and area 3, 4, and 5, Xinjiang areas, does the other 15 % for total three years average. Total concentration is very low in 2003, whose reason is attributed to the reduction of dust from area 2. The concentration from area 2 in 2003 is approximately one third of that in 2001 and much less than one half of 2002. Taking into account that the emission from area 2 itself is more than half compared to the other years in Table 2, the reduction of the concentration originated from area 2 is not only from the reduction of the emitted amount but also the change of transport from the area to Beijing. Future study is necessary to investigate the variation of the transport from area 2 to Beijing and other downstream areas.

**Table 3.** Calculated surface concentrations of dust aerosol in Beijing with each generated area in microgram per cubic meter. Numbers in parenthesis shows percentages of contribution of each area for each year.

|         | Area 1 | Area 2 | Area 3 | Area 4 | Area 5 | Area 6 | Total |
|---------|--------|--------|--------|--------|--------|--------|-------|
| 2001    | 0.2    | 90.1   | 7.3    | 1.8    | 5.7    | 50.3   | 155.5 |
|         | (0.1%) | (58.0%)| (4.7%) | (1.2%) | (3.7%) | (32.4%)| (100%)|
| 2002    | 0.4    | 80.4   | 4.9    | 2.6    | 6.2    | 39.0   | 133.4 |
|         | (0.3%) | (60.2%)| (3.6%) | (1.9%) | (4.7%) | (29.2%)| (100%)|
| 2003    | 0.1    | 30.0   | 4.3    | 5.6    | 7.2    | 31.4   | 78.6  |
|         | (0.1%) | (38.2%)| (5.5%) | (7.1%) | (9.2%) | (39.9%)| (100%)|
| average | 0.2    | 66.8   | 5.5    | 3.3    | 6.4    | 40.2   | 122.5 |
|         | (0.2%) | (54.6%)| (4.5%) | (2.7%) | (5.2%) | (32.8%)| (100%)|

## 5. Summary

Dust aerosol is simulated in eastern Asia from March to May in 2001, 2002, and 2003. Calculated concentrations are well corresponding to observations with lazar particle counter in some sites in China. Contribution of source areas on surface concentration in downstream area is investigated. Mongolia shows the largest contribution on the surface concentration in Beijing, more than half, and Inner Mongolia does the second largest, approximately one third, and Xinjiang area does the other. By comparing concentration in the three years, it is clarified that small concentration in 2003 is attributed to small concentration whose source is in Mongolia, whose reason is partly small emission in Mongolia itself and partly the change of the transport to downstream areas.

## References

Gillette D, Passi R (1988) Modeling dust emission caused by wind erosion. J Geophysical Res 93:14233-14242

Sugata S, Byun D, Uno I (2001) Simulation of sulfate aerosol in East Asia using Models-3/CMAQ with RAMS meteorological data. In: Gryning SE, Schiermeier FA (Eds) Air Pollution Modeling and Its Application XIV, pp 267-275

# A case study on combating desertification at a small watershed in the hills-gully area of loess plateau, China

Junliang Tian[1], Puling Liu[1], Hideyuki Shimizu[2], and Shinobu Inanaga[3]

[1]Institute of Soil and Water Conservation, Chinese Academy of Sciences and Ministry of Water Resources, 26# Xinong Rd. Yangling, Shaanxi 712100, China
[2]National Institute for Environment Studies, Onogawa 16-2, Tsukuba, Ibaraki 305-8506 Japan
[3]Arid Land Research Center, Tottori University, Hamasaka 1390, Tottori 680-0001, Japan

**Summary.** Soil erosion and drought are twin problems in hills-gully area of the Loess Plateau, China, and they are main causes of desertification and poverty in the area. To combat desertification and improve rural economy in the area, a case study for reconstruction of the degenerated eco-environment in a small watershed has been conducted since 1997. The agriculture structure was changed by construction of terraces and conversion of cropping land to forest/grass land. The reconstruction results show that decreasing cropping land and rationally rearrangement of land use are key points to reduce soil and water loss in a watershed scale. The primary experiment results have shown that the reconstruction models based on the eco-environment condition are successful not only in restoration of vegetation but also in increasing farmer income. As a result, the severe soil erosion and the degradation of vegetation began to be controlled.

**Key words.** Loess plateau, Soil erosion control, Vegetation restoration, Combating desertification

## 1. Introduction

The Loess Plateau with an area of 0.62 million km$^2$, located in the middle reaches of the Yellow River in China, is the cradle of Chinese civilization. Cultivation in this region started 6000 years ago. The Loess Plateau as a part of the west area of China has been caused a great attention in recent years with the implement of the stratagem of the west development launched by Chinese central government. However, the fragile environment in the Loess Plateau is a serious problem to prevent the area from development.

In the plateau, 67.7% of the total area suffers from soil erosion, and severe soil erosion area has 0.28 million km$^2$, mainly in hills-gully area of the plateau. About more than half of the annual sediment discharge of 1.6 billion tons in Yellow River is from the hills-gully area.

To improve the eco-environment of the plateau, several eco-environment reconstruction projects supported by Chinese central government have been on going in recent years. One of the projects called conversion cropland to forest/grass land launched in 1999, and cropland with a total area of more than 2.49 million hectares has been converted by the end of 2002 (The Ministry of National Land Resources of China, 2003). Actually, since 1950's, the central government of China has paid a great attention to restoration of the vegetation in the plateau. In the past 50 years, the mass campaign of tree planting on the Loess Plateau, for an example, has never stopped and the aggregated area of the afforestation drive over the years is considerable. Except in a few limited key project areas, however, less than 10% of the total artificial afforestation area now is estimated to have significant tree cover (Tian 2003). Many factors, such as technical and natural as well as economic factors, are responsible for this kind of failed reconstruction of ecosystem. How to avoid the situation to recur in the eco-environment reconstruction projects on going is an important topic to be studied in China.

This case study aimed to the main environment problems in hills-gully area of the Loess Plateau and the issues caused the failed reconstruction to provide scientific and technical bases for reconstruction of the eco-economy system in a watershed scale, and to build a demonstration model in which the development of agriculture is to be in harmony with the improvement of the eco-environment.

## 2. Study site and the main environment problems

The study site locates in Yan'gou Watershed, a typical watershed in the hills-gully area of the Loess Plateau (Fig. 1,2). The watershed has a total area of 46.88 km$^2$ in latitude N 36°28' ~ 36°32', longitude N 109°20' ~ 109°35', and includes 14 villages with 693 families and a population of 2932. The mean elevation of the watershed is 1197m varying from 986 to1245m. Mean gradient of hill slopes in the watershed is 22°. The watershed locates in a transition zone of vegetation from forest to typical steppe. It basically belong to semi-arid climate region with mean annual air temperature 9.8°C and a multi-year average precipitation 558.4 mm. The soil type mainly is Huangmiantu developed from the loess, a poor soil in fertility with organic mater less than 1%.

Because of excess cropping and overgrazing before the reconstruction in 1997, severe loss of soil and water, wide area degeneration of vegetation and degradation of land were main environment problems in the watershed.

The area of soil and water loss in Yan'gou watershed was 42.55 km$^2$, which accounted for 90.7% of total area. Erosion modules reached 6000-9000 t km$^{-2}$ a$^{-1}$. And only 23.9% of the total soil erosion area was in control (Tian, Liang and Liu 2003).

The watershed had very sparse vegetation, 67% of the total area had NDVI between 0 and 0.2. The primary zonal forest was completely destroyed, the cover rate of survived secondary forest lower than 10%. The pasture on the hill slopes had a very low capacity to carry cattle and sheep, less than one sheep equivalent per hectare.

There was 1831.3 hm$^2$ of arable land, of which 90% was slope land on hills. Mean yield was 1095 kg hm$^{-2}$. Farmer's life in the watershed mainly relied on planting cropping with a lower input to the farmland. The annual average of farmer net income was 763 yuan (RMB) per capita, much lower than the average of farmer income in China.

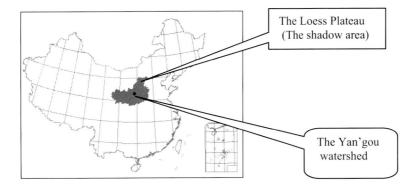

**Fig. 1.** Location of the Yangou watershed.

**Fig. 2.** The distribution map of village and water system of Yan'gou watershed.

With the interaction between the fragile eco-environment and the poverty of farmer life, an evil circle between the rural ecology and economy had been formed. In the circle, to meet the demands for farmer to live on, the forest land was disafforested and reclaimed for cropping, caused vegetation degeneration and soil erosion and land degradation, and the decreases of land productivities as well the farmer's income subsequently followed. Thus repeated time to time around the circle, more and more forest land disafforested and reclaimed. And as a result the poorer the farmer's life was and the worse the environment changed. How to break off the evil circle is the main task of the reconstruction in the case study.

## 3. The reconstruction of the watershed

The case study on reconstruction of eco-environment in Yangou watershed has been conducted since 1997.

### 3.1 Strategy of the reconstruction

According to the historical lessons of the failing reconstruction in the hills-gully region of the Loess Plateau in the past 50 years, strategy of the reconstruction in the watershed is that the improvement of farmer's income was to be considered at the same time with vegetation restoration. Based on our studies on restriction factors for vegetation restoration, natural restoration should put at the most important position. Building high quality terraces with importing and integrating planting techniques for increasing unit yield of the farmland was taken into account as a key measure to break off the evil circle in the area.

### 3.2 Main measures employed

#### 3.2.1 Regeneration of vegetation

In the period of this case study, the conversion program was also implemented in the study site. To restore the vegetation, the government provided subsidy for farmers to stop cropping on slopes and to plant trees/grasses on the cropland. This is greatly effective for the restoration of vegetation in the watershed.

According to our experiments and long period monitoring, degenerated vegetation in this area is able to be restored naturally as long as the disturbing from human activities is stopped. Experiment results in this area show that the vegetation on hill slope could be naturally restored from a coverage of 35-40% of herbaceous community to 80-90% with the inbreak of shrub species after forbidding cropping and grazing for 3 years. In addition, the vegetation restored naturally follows the rules of native vegetation succession so that it is good for development of stable vegetation. The man made vegetation in this area is hardly to be stable if the distribution of forestation or species is irrational. Furthermore, development of the drying soil layer in artificial afforestation land resulted from unbalance compensating of soil water resources is much graver compared with the soil in the land of natural vegetation restoration (Yang and Tian 2004). The development of the drying soil layer is a common restriction factor for forestation in the semiarid region of the Loess Plateau (Liang 2003). So that the restoration of vegetation in the reconstruction mainly relied on natural restoration, especially in areas where the farmland or wasteland is on steep hill slopes.

Afforestation also was carried into execution in the construction according to the requirement of the cropland conversion program. A great attention was paid on the land condition. The lands to plant trees were mainly selected in the area where the land has moderate gradient of hill slops, and it is necessary that measures for harvesting runoff water on slope were used on the forestation land to increase the surviving rate of the trees planted. A planning of restoration of vegetation in the watershed was mapped out first based on the land type distribution, for an example, grasses or shrubs are main types planned on the top of hills or upper part of hill slopes where the soil moistures as a rule are lower. It is in the highest priority that local native plant species were selected for the afforestation, and the specific species used was under a consideration of the potential of the vegetation succession (Xue et al. 2000).

### 3.2.2 Forbidding reclaiming and grazing on hill slopes

To protect the vegetation restoration from abortion, the local government enacted a regulation to forbid reclaiming and grazing on the hills. This made the natural restoration of vegetation on the barren slope land possible. The barren slope land occupies an area more than 20% of the total watershed area. The result from investigation of the natural restoration has shown that number of herbage community increased 7-73%, and vegetation coverage increased 81-167% in a hill slope forbidden for 7 years.

### 3.2.3 Increase farmer income

To provide enough food for farmer self-consumption and feeding stuff for stock raising after the decrease of cropland, it is necessary to build high quality terraces as cropland instead of the hill slope land. The unit yield of corn production in the terrace land, for an example, could reach to as 4-5 times high as the yield in hill slope land because of the good soil water condition in the terraces and new crop species and integrated plating techniques imported.

About 10-20% of the farmer income was from stock raising in the watershed before the reconstruction, mainly grazing goats on the barren land of the hill slope. To maintain the share of stock raising in the total GDP of the area after the forbidding grazing on the hills, shed-breeding (feeding goats or sheep in sheepfolds) was developed in the watershed. The grassland developed in the conversion program provided part of forage for the breeding.

Furthermore, to increase farmer income, cash trees and vegetable production with sunlight greenhouses are also developed in the area.

## 4. Results and discussion

The effects from the reconstruction have been gradually achieved in recent years.

### 4.1 Land use and cover changes in Yangou watershed

After the reconstruction, the land use structure has been changed greatly (Table 1). The slope cropland sharply decreased from 34.50 % of the total watershed area to 0.26%, and the area of high quality level terraces increased from 4.56% to 12.89%. The total area for plating crop reduced from 39.06% of the total watershed area to 13.15%. As a result, most slope cropland was converted to forest or grassland. The total area of artificial

**Table 1.** Land use and cover changes in Yangou watershed.

| Year | Slope land | Quality farmland (terrace etc.) | Natural forest | Artificial arbors and shrubs | Artificial grasses | Cash trees | Barren slopes |
|---|---|---|---|---|---|---|---|
| 1997 | 34.50% | 4.56% | 9.27% | 10.83% | 0 | 3.71% | 26.11% |
| 2003 | 0.26% | 12.89% | 9.27% | 26.65% | 7.46% | 11.31% | 20.99% |

vegetation including planted trees, shrubs and grasses has reached 45.42%. Land utilization structure tends to be rational (Table 1).

### 4.2 Benefits from the reconstruction

Including the area of artificial afforestation and the natural restoration area, the vegetation coverage in the watershed has reached above 70% in 2003.

The engineering and biological measures, including building terraces and increase of vegetation coverage, have greatly decreased soil and water loss. The area where the soil and water loss was in control has reached to 73.3% of the total soil and water loss area in the watershed. According to the monitoring results of a hydrological station at the mouth of Yan'gou watershed, the silt transport module has greatly decreased year by year (Fig. 3).

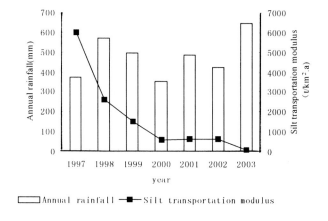

**Fig. 3.** Changes of silt transportation modulus of the watershed.

Table 2. The indices of reconstruction benefits in Yan'gou watershed.

| Year | Controlling degree(%) | Vegetation Coverage (%) | Silt transportation Modulus (t km$^{-2}$ a$^{-1}$) | Income per capita (RMB) | Yield (kg hm$^{-2}$) |
| --- | --- | --- | --- | --- | --- |
| 1997 | 23.9 | 20.1 | 6000 | 763 | 1095 |
| 2003 | 73.3 | 54.7 | 79 | 1968 | 3695 |

Through the measures like strengthen orchard management and technical training and importing new cropping techniques, the unit crop yield and income per capita have been rapidly increased. From 1997 to 2003, mean income per farmer is increased from 763 to 1968 yuan (RMB) and the unit yield of cropland from 1095 to 3695 kg hm$^{-2}$ (Table 2). Mean annual increase rate is as high as 17.1%. The income mainly comes from crops, fruits as well as animal husbandry. The unit yield in dam land and terrace increased 27.9% and 12.8% respectively. Total crop yield of the watershed are stable at the level of 1600t. Per capita owned food reaches 511kg.

According to the evaluation results by using single item and different levels of benefit analysis, the benefit from the eco-environmental reconstruction shows that, between the eco-environment and economy in the watershed, the evil circle has been broken and the eco-economy system stepped into a good cycling (Xu et al. 2004, Wang 2003).

### 4.3 Challenge to the further study on the reconstruction

How to synthetically assess the reconstruction impacts on environment is to be a topic for further studies. For an example, the redistribution of the precipitation as well as the run-off water resulted from the land use and cover changes will have certain impacts on watershed hydrology circle. A further study also is needed to assess if the new agro-forestry system constructed in this case study is to be a sustainable production system with the potential of replacing present agricultural patterns.

## 5. Conclusion

To improve agriculture condition and farmer income is an important guaranteed term for reconstruction of a degenerated eco-environment in poverty rural area. Soil drying layer development is a common problem of vegetation restoration in semi-arid area, reasonable distribution of forestation is a key point to avoid its aggravation. Natural restoration of vegetation is the best way for management of the degenerated eco-system in the semiarid area.

## References

Liang Yimin (2003) Vegetation construction on the loess plateau. Huanghe Publishing House, Zhengzhou, pp 113-116

The Ministry of National Land Resources of China (2003) The report of land resources of China in 2002. The Geology Publishing House, pp 107-108

Tian Junliang (2003) Restoring the eco- environment in conformity to natural law - some considerations on the vegetation restoration on the loess plateau. Bull Chin Acad Sci 17:101-106

Tian Junliang, Liang Yi-min, Liu Puling (2003) Exploration about the eco-agriculture in middle scale field in the loess hills and gully region. Huanghe Publishing House, Zhengzhou, pp 243-244

Wang Ji-jun (2003) A study on evaluating index system of eco-construction benefits on a middle scale region. Res Soil Water Conserv 7:10-13

Xu Yong, Tian Jun-liang, Liu Pu-ling (2004) The evaluating method of eco-environment restoration patterns: a case study of the loess hilly-gully region. Acta Geographica Sinica 59:621-628

Xue Zhi-de, Yang Guang, Liang Yi-min (2000) Research of artificial vegetation building patterns and rapidly constructing in Yan'gou watershed. Res Soil Water Conserv 7:128-132

Yang Wenzhi, Tian Junliang (2004) An investigation of soil drying on the loess plateau. Pedology Acta 41:1-6

# A recipe for sustainable agriculture in drylands

Shinobu Inanaga[1], A. Egrinya Eneji[1], Ping An[1], and Hideyuki Shimizu[2]

[1] Arid Land Research Center, Tottori University, Hamasaka 1390, Tottori, 681-0001, Japan
[2] National Institute for Environmental Studies, Onogawa 16-2, Tsukuba, Ibaraki 305-8506, Japan

**Summary.** About 47% of the earth's land area is classified as dryland, wherein the farming system is characterized by an annual rainfall of approximately 300 – 500mm, with much of the rainfall in the spring and early summer. The low rainfall, which is not only insufficient but irregular, constitutes a major challenge to profitable farming in dry areas. Meeting this challenge in the years ahead will require a more sustainable product per unit area of land, conservation and rational use of natural resources, protection of the environment, improved management practices, cost-effective technologies, and favorable government policies and incentives for farmers to increase their productivity. In this review, we discuss the major ingredients essential for sustainable dryland agriculture which include: efficient use of water, seeding at rates corresponding to the soil water supply, management practices that minimize water loss and soil erosion, managing soil fertility and organic matter, selection of suitable crop species and management of pests. We conclude that an approach integrating scientific technologies with indigenous farming methods offers the best hope for sustainable agriculture in the drylands of the world.

**Key words.** Dryland, Environment, Sustainable farming, Technologies

## 1. Introduction

Drylands are found in climate regimes that are not very favorable to crop production. Low total rainfall (300 – 500 mm per annum or less) and high variability in rainfall patterns present particularly difficult challenges for growing crops. Nevertheless, local populations depend on these lands for producing food. Drylands are inhabited by more than two billion people worldwide. In terms of numbers and percent, Asia has the largest population living in drylands, with over 1.4 billion people, or 42 percent of the region's population (White and Nackoney 2003). Africa has nearly the same percent of people living in drylands but the total number (270 million) is less than that of Asia's while South America has 30 percent of its population or approximately 87 million people in drylands. The drylands cover about 54 million square kilometers of the globe (UNSO/UNDP 1997) of which semi-arid areas are the most extensive (18%) followed by arid areas (12%), dry sub-humid lands (10%) and hyperarid lands (7.5%). These aridity zones spread across all continents, with Asia and Africa predominating. Interestingly, various land cover types

*Plant Responses to Air Pollution and Global Change*
Edited by K. Omasa, I. Nouchi, and L. J. De Kok ( Springer-Verlag Tokyo 2005 )

are found in drylands, ranging from shrubland, forests, croplands and urbanized settlements.

Since water is the limiting factor of agricultural production, the primary problem is the most effective means of storage of the natural precipitation in the soil. Of nearly equal importance is how to retain the water in the soil until it is needed by plants. In attempting to sustainably improve dryland agriculture to satisfy the needs of the growing population, the relation of crops to the prevailing conditions of drylands must be taken into account. Some plants require much less water than others. Others mature early, and in that way become desirable for dryland farming. Some crops grown under humid conditions, may easily be adapted to dryland conditions, if the correct methods are employed. The individual characteristics of each crop should be known in relation to low rainfall and dry conditions. Rain-fed farming as currently practiced in drylands, is a system of low inputs combined with soil and water conservation practices and risk reducing strategies. The system can be sustainable if practiced properly. Although water shortage is the main limiting factor, a successful dryland farming in rain-fed conditions should also maintain reasonable practices to minimize other limiting factors such as poor nutrient status, weeds and biotic stresses, which can reduce the efficiency by which the crop uses the limited moisture.

## 2. The concept of sustainable agriculture

During the $20^{th}$ century, agricultural production increased tremendously, especially in the developed countries, due to new technologies, mechanization, increased use of agrochemicals, specialization and favorable government policies. While these changes had many positive effects and reduced many risks associated with farming, there have also been significant costs such as loss of topsoil, groundwater contamination, increasing costs of production, and the breakdown of economic and social conditions in rural areas. When the production of food and fiber degrades the natural resource base, the ability of future generations to produce and flourish decreases. The decline of ancient civilizations in the Middle East and other regions is thought to have been strongly influenced by natural resource degradation from non-sustainable farming and forestry practices. Water is the principal resource that has helped agriculture and society to prosper, and it has been a major limiting factor when mismanaged.

In view of the foregoing problems, the concept of sustainable agriculture has been proposed to address many environmental and social concerns, and offer innovative and economically viable opportunities for farmers, consumers, policymakers and many others in the entire food system. Sustainable agriculture guarantees environmental health, economic profitability, and social and economic equity. Sustainability is based on the principle that we must meet the needs of the present without compromising the ability of future generations to meet their own needs. Therefore, careful management of both natural and human resources is of great importance. A sustainable land use is simply one that is able to continue without degrading the land it is using (UNEP 1997). In the drylands, this will involve maintaining or enhancing land and environmental quality for the long term. At present, more research and development are needed to apply sustainable agriculture concepts to the dry areas of the world. However, this will require innovative, practically relevant and demonstrable research.

## 3. Requirements for sustainable agriculture in drylands

The development of sustainable farming in drylands would be based on several principles: selection of a suitable site, improved soil and water management practices, good crop management, weed control, fallow management, efficient and rational use of inputs and consideration for indigenous knowledge and practices.

### 3.1 Land selection and preparation

When site selection is an option, factors such as soil type and depth, previous crop history, and topography should be taken into account before planting. A clay loam soil is preferred under dry conditions. Other soils may be equally productive, but are more difficult to cultivate.

### 3.2 Soil management

The key soil characteristics affecting crop growth in drylands are shown in Table 1. Soils in dryland regions have low organic matter contents due the characteristically low plant biomass, and are thus predominantly mineral soils. Many of the soils have a lower clay content than those in more humid regions. In practice, it is very rare to find soils with ideal texture, reaction, fertility and organic content in drylands. Therefore there is need to manage and improve dry soils so that they can perform to their full potential. In sustainable systems, the soil is viewed as a delicate and living medium that must be protected and nurtured to ensure its long-term productivity and stability. A healthy soil is a key component of sustainability as it produces healthy crops with optimum vigor which are less susceptible to environmental stress. Proper soil management can help prevent some pest problems caused by crop stress or nutrient imbalance. Improved soil and farm management can also significantly increase the amount of water a soil can store for the next growing season.

Soil erosion continues to be a serious threat to crop production in drylands. Numerous practices have been developed to keep soil in place, which include reducing or eliminating tillage, use of cover crops and managing irrigation to reduce runoff. It is commonly assumed amongst many environmentalists that erosion can be stopped by planting trees. However, this depends on the way the trees are planted and managed, as benefits in soil and water protection do not accrue automatically by having trees on the land (Douglas, 1998). It is the litter below the trees rather than the tree canopy itself that provides the bulk of the protection against erosion. If the litter is removed for mulch, fodder, fuel etc, then the conservation benefits from planting trees are seriously reduced. As shown in Table 2, trees are not always more efficient at protection than annual crops which can provide adequate cover within 30-45 days and pastures within 2-6 months (Sanchez 1987). When mulched, or managed with low tillage, annual crops give the same results for soil loss as do secondary forests (Lal 1977).

In the context of good watershed management, well managed rotational cropping or well managed pasture may be preferable alternatives to poorly managed forest land use (Shaxson 1992). The risk of soil loss by water and wind erosion is also reduced if the soil

**Table 1.** Major soil characteristics influencing the growth of dryland crops and pastures based on wheat critical limits (from NSW Department of Primary industries - agriculture, Australia).

| Process | Soil property | Specification |
|---|---|---|
| Seed germination | clod diameter adjacent to seed 0.5-2mm | 100% |
| | temperature | 5-35°C (optimum = 23 °C) |
| | electrical conductivity (EC) | <5dS/m |
| | pH | 6-7 |
| Root growth | macroporosity (percentage of pores >30$\mu$m) | >15% |
| | penetrometer resistance | <1MPa |
| | clod diameter 0.5-10mm | 100% |
| | temperature | 10-35°C (optimum= 26 °C) |
| | EC | <6dS/m |
| | pH | 6-7 |
| Activity of earthworms and microbes | organic residues | 10t/ha/year |
| Erosion control | clod diameter >2mm on the soil surface | 100% |
| | organic residues | >30% |

**Table 2.** Soil conservation potential of contour hedgerows on sloping lands (from Eneji et al. 2004).

| Location | Slope (%) | Treatments | Soil loss (t ha$^{-1}$ yr$^{-1}$) |
|---|---|---|---|
| Yurimaguas (Peru) | 15-20 | Hedgerow intercropping | 53 |
| | | Annual crop | <1 |
| Ibadan (Nigeria) | 8 | Hedgerow intercropping | <1 |
| | | Annual crop | 4 |
| Los Banos (Phillipines) | 14-19 | Hedgerow intercropping | 3 |
| | | Annual crop | 127 |
| Chiang Rai (Thailand) | 20-50 | Hedgerow intercropping | 65 |
| | | Annual crop* | |
| Ntcheu (Malawi) | 44 | Hedgerow intercropping | 44 |
| | | Annual crop | 2 |

* Farmers' practice

surface is protected by straw or gravel mulch, which should cover at least 30-35% of the soil surface.

### 3.3 Water management

Dryland agriculture in many areas of the world is highly dependent on precipitation, both snow and rainfall. The water received as rain or snow can be easily lost before it can be used by a crop. Water use by weeds and evaporation are the two most negative pathways of water loss that must be avoided if improvements in precipitation use efficiency are to be accomplished. Where economically feasible, irrigation is the most direct means for combating drought conditions and intensifying agricultural production. But for sustainability, irrigation must be practiced to avoid such hazards as soil erosion, soil salinization, soil leaching and soil disease infection. A sustainable irrigation must be based on knowledge of the crop, the soil properties and the potential evapo-transpiration of the specific crop at the site. This information can also be used to estimate dryland crop water use and deficit at any given time during the crop cycle, which is actually an index of crop drought stress.

Supplemental irrigation (SI) is defined as the application of a limited amount of water to rainfed crops when precipitation fails to provide the essential moisture for normal plant growth (Oweis et al. 2000). This practice has shown potential in alleviating the adverse effects of unfavorable rain patterns and thus improving and stabilizing crop yields (Perrier and Salkini 1991; Oweis et al. 1998). Studies at ICARDA showed that applying two or three irrigations (80–200 mm) to wheat increased crop grain yield by 36 to 450%, and produced similar or even higher grain yields than in fully irrigated conditions (Perrier and Salkini 1991). Supplemental irrigation is widely practiced in Syria, and in southern and eastern Mediterranean countries.

Water harvesting is a broad term describing various methods of collecting runoff from large contributing areas and concentrating it for use in a smaller crop area. Gravel mulching is a method involving dense covering of the soil surface with gravels (Inanaga 2002). It is effective in reducing evapo-transpiration and surface runoff of rain water, thereby increasing its percolation to the soil for use by crops. This is an indigenous practice already adopted in many arid or semi-arid regions. Water saving polymers have also been formulated and manufactured to provide better moisture management capabilities and longer lasting effects on crop performance. Polymer crystals are incorporated into the soil preplant or at planting. These crystals absorb moisture and transform into gel-like nuggets of water and nutrients to meet the needs of plants when root-zone conditions turn dry. The polymers expand hundreds of times their original weight, retaining moisture and water-soluble nutrients until plants need them. However, a major potential limitation to the use of polymers is the high cost.

## 3.4 Crop management

Good crop management systems often result in less inputs of water, nutrients, pesticides, and/or energy for tillage to maintain yields. After a crop has been chosen, skill and knowledge are needed in the proper seeding, cultivation, and harvesting of the crop. New, improved varieties, especially adapted to soils with moisture deficit are ideal. Many profitable and sustainable crop rotations are currently practiced in dry land farming as a replacement to the wheat/fallow method. The most frequently used crops in these rotations are winter wheat, corn, millet and fallow. In these crop rotations the main objective is to minimize the amount of tillage with residue management. Through this process of minimum tillage, water is saved and farmers can produce crops more regularly.

Sustainable crop production practices involve a variety of approaches. Specific strategies must take into account topography, soil characteristics, climate, pests, local availability of inputs and the individual farmer's goals. Diversification of cropping is an economically and ecologically resilient approach for sustainability. By growing a variety of crops, farmers spread economic risk and are less susceptible to the sudden price fluctuations associated with changes in supply and demand. Diversification of cropping to reduce risk is especially important under dryland conditions. It can be achieved either by spatial diversification of fields or crop diversification. For spatial diversification, the farmer's land is divided into several fields which may differ in their topography, soil and hydraulic properties. Some fields may be prone to flooding while others do not hold water. Certain fields may be on a warmer slope while others may be on a cooler one. The different field conditions make it possible to achieve a better fit between the crop and the environment and to reduce the likelihood of stress affecting the farmer. Optimum diversity may be obtained by integrating both crops and livestock in the same farming operation. This is a common indigenous practice in many dry zones.

Crop diversification takes an advantage of the generally low correlation between the performance of crops when grown in a single stress environment. Crops differ in their response to a given environment and this difference is used to reduce the risk associated with monoculture. Mixed cropping or intercropping is an indigenous and successful approach to crop diversification on a single piece of land, where two or more crops are grown together in various possible configurations. If for some reason only one crop is grown, some level of risk reduction can be achieved by varietal diversification. Planting of several crop varieties may offer a better probability for reducing loss due to environmental stress, as compared with growing one variety only.

## 3.5 Weed management

Weeds compete with crops for sunlight, nutrients, and water. They sometimes harbor pathogens, nematodes, insects, or vertebrates that can invade or spread to the crop soon after planting. There is a suggestion that arid and semi-arid communities altered by invasion of non-native weeds may never recover, even with the cessation of all anthropogenic disturbance (Brandt and Rickard 1994). A weed-infested farm is doomed to failure, but weed control can enhance water storage in soils. Weeds should be removed and destroyed as soon as they emerge or before they become serious competitors with crops for water and nutrients. If problem weeds are present in significant numbers, the best strategy is to rotate to a crop in which they can be successfully controlled. Control of weeds after planting is most critical during the seedling stage. An integrated weed control program which relies on several management methods – cultural, biological, mechanical and chemical, to keep weed populations at tolerable levels may be the best option.

## 3.6 Fallow practices

The fallow system in dry areas is aimed at conserving soil moisture from one year to the other, depending on climate and crop. Increasing storage of soil moisture by the fallow system with or without conservation tillage is standard agricultural practice in dryland farming. There is also evidence that a fallow that involves some leguminous shrubs is beneficial to the re-establishment of native plant species threatened with extinction (Isola et al. 1995). Dao et al. (2003) showed that the greatest number of useful plants were

**Table 3.** The number of useful plants in fallows of different ages in Daka and Xishuangbanna in Yunnan, China (from Dao et al. 2003).

| Fallow age (yrs) | 1 | 2 | 4 | 5 | 6 | 17 | 33 |
|---|---|---|---|---|---|---|---|
| Building wood | 1 | - | 3 | 2 | 3 | 5 | 5 |
| Fuelwood | 2 | 5 | 2 | 1 | 5 | 2 | 4 |
| Medicinal plants | 4 | 7 | 2 | 5 | 5 | 14 | 6 |
| Animal feed | 1 | - | - | - | 1 | - | 2 |
| Wild vegetables | 6 | 2 | 2 | 1 | 3 | 4 | 3 |
| Wild fruits | 1 | - | 2 | 3 | 3 | 2 | 3 |
| Other uses | 2 | 2 | 1 | 4 | 1 | 4 | 2 |
| Useful species | 16 | 15 | 10 | 13 | 19 | 24 | 25 |
| Other species | 5 | 15 | 6 | 4 | 3 | 9 | 10 |

found in older fallows (Table 3), with species richness indices in one area (Daka) of Yunnan, China varying from 3.4 to 9.5. The benefit of fallow to increasing crop available soil moisture depends on soil water-holding capacity, climate, topography and management practices. At the beginning of fallow, the surface cover should be about 65-70% to allow for breakdown of stubble during fallow.

## 3.7 Efficient use of inputs

Although many inputs and practices used by conventional farmers are also used in sustainable agriculture, sustainable farmers, maximize reliance on natural, renewable, and on-farm inputs. Equally important are the environmental, social, and economic impacts of a particular system. Adopting sustainable practices in drylands does not mean simple input substitution. Rather, enhanced management and scientific knowledge are employed in place of conventional inputs, especially chemical inputs that harm the farming and living environment in rural communities. The goal should be to develop efficient, biological systems which do not need high levels of material inputs.

## 3.8 Use of indigenous knowledge

Examples of some indigenous soil conservation practices are shown in Table 4. The bulk of the food consumed in many arid and semi-arid regions is produced by peasant farmers, owning 1 to 2 hectares of land. Since they are resource-poor, they are seldom able or willing to adopt standard conservation and land use practices as most of the recommendations are costly and involve sacrificing short-term benefits for long-term sustainability. However, from experience, the farmers are aware of the need to conserve their soil and environment, and have their indigen- ous practices for this purpose. The guiding principle in indigenous farming in dry areas is risk minimization, which has resulted over the years, not only to survival strategies from season to season, but also in ways of conserving the long-term productivity of the soil. Indigenous farming practices contain a wealth of environmental knowledge, conservation practices and flexibility which have not been duly recognized to date (Eneji 1999). Indigenous farming systems and the conservation

**Table 4.** Some indigenous soil and crop management practices (from Eneji 1999).

|  | Manual tillage | Oxen tillage | Conservation effectiveness |
|---|---|---|---|
| (a) Direct production practices |  |  |  |
| Field preparation: |  |  |  |
| Contour plowing | - | x | moderate slopes |
| Ridging (tuber crops) | x | x | moderate slopes |
| Crop management: |  |  |  |
| Manuring | x | x | indirect but effective |
| Crop rotation | x | x | depends on rotation component |
| Fallow (and ley) | x | x | but compaction by grazing |
| Ratooning | x | x | less tillage |
| Relay cropping | x | x | continuous ground cover |
| Intercropping | x | x | better cover |
| Alley cropping | x | x | more organic matter |
| Tree planting | x | - | depends on layout |
| Crop residue and range management: |  |  |  |
| Mulch incorporation | x | x |  |
| Tethered grazing | x | x |  |
| (b) Indirect production oriented practices |  |  |  |
| Excessive water disposal: |  |  |  |
| Drainage, furrowing | x | x | mostly on moderate slopes |
| Ridging | x | x | mostly on moderate slopes |
| Diagonal cutoff (hill side drain) | x | x | very effective protection |
| Cambered beds | limited | limited | slopes only |
| Water conservation: |  |  |  |
| Contour plowing | x | x | moderate slopes |
| Tie ridging, traditional basin | basins | basins | effective on moderate slopes |
| Level bunds (traditional) | limited | limited | effective on moderate slopes |
| (c) Long-term conservation practices |  |  |  |
| Bench terraces (earthen, stone wall) | x | x |  |

practices integral to them are infinitely varied and ultimately compatible in their undisrupted forms with local environmental conditions. Many indigenous conservation practices will need adaptation to stay relevant to rapidly changing circumstances. Therefore, in the search for a sustainable agricultural practice in dry areas, the existing indigenous systems need to be studied and understood. Based on such studies, the various underlying processes that make them sustainable can be identified and improved through research.

## 4. Conclusion

Sustainable agriculture in the drylands of the world should be practiced through integration of modern, research-proven technologies with indigenous farming systems. Such technologies and systems should be simple, cheap, easy-to-use and compatible with native customs and beliefs in respective regions.

## References

Brandt CA, Rickard WH (1994) Alien taxa in the North American shrub-steppe four decades after cessation of livestock grazing and cultivation agriculture. Biol Cons 68:95-105

Dao Z, Guo H, Chen A, Fu Y (2003) Agrodiversity in China. In: Brookfield H, Parsons H, Brookfield M (Eds) Agrodiversity. United Nations press, Tokyo, 343p

Douglas M (1998) The concept of conservation effectiveness. ENABLE – Newsletter of the Association for Better Land Husbandry No. 9, December 1998, 21-29

Eneji AE (1999) The use of indigenous knowledge for controlling soil degradation in Africa. Jpn J Trop Agr 43(3):199-205

Eneji AE, Irshad M, Inanaga S (2004) Agroforestry as a tool for combating soil and environmental degradation: examples from the tropics. Sandune Res 51(1):47-56

Inanaga S (2002) Aiming to realize sustainable dryland farming. Farming Japan 36(3):16-20

Isola TO, Eneji AE, Agboola AA (1995) Influence of hedgerow species on the establishment of volunteer woody species in Ayepe, South-western Nigeria. J Trop For Res Vol.11:37-45

Lal R (1977) Soil management systems and erosion control. In: Lal R, Greenland DJ (Eds) Soil conservation and management in the humid tropics. John Wiley Chichester UK

Oweis T, Pala M, Ryan J (1998) Stabilizing rainfed wheat yields with supplemental irrigation in a mediterranean-type climate. Agron J 90:672-681

Oweis T, Zhang H, Pala M (2000) Water use efficiency of rainfed and irrigated bread wheat in a mediterranean environment. Agron J 92:231-238

Perrier ER, Salkini AB (1991) Supplemental irrigation in the Near East and North Africa. Kluwer Acad Publ, Netherlands

UNEP (1997) World atlas of desertification. In: Middleton N, Thomas D (Eds) John Wiley and Sons, Inc, New York, 182 p

UNSO/UNDP: Office to combat desertification and Drought (1997) Aridity zones and dryland populations: an assessment of population levels in the world's drylands. UNSO/UNDP, New York, 23p

Sanchez PA (1987) Soil productivity and sustainabillty in agro forestry systems. In: Steppler HA, Nair PK (Eds) Agroforestry: a decade of development. ICRAF, Nairobi Kenya

Shaxson TF (1992) Crossing some watersheds in conservation thinking. In: Tato K, Hurni H (Eds) Soil conservation for survival. Soil and Water Conservation Society, Ankeny, Iowa, USA

White RP, Nackoney J (2003) Drylands, people, and ecosystem goods and services: a web-based geospatial analysis:http://pubs.wri.org/pubs_description.cfm?PubID=3813

# Index

**A**

abiotic stresses, 33
acclimation, 76
acid deposition monitoring network in East Asia (EANET), 251
acidification, 39
adaptation, 54
adult forest trees, 22
agriculture, 81
agriforests, 173
air concentration of gaseous and particulate components, 254, 255
air pollutant, 101
air pollution index, 268
air pollution, 45, 243
air velocity, 185
*Allium cepa*, 5
altitudinal gradient, 39
antioxidant levels, 24
antioxidants, 38
AOT40 index, 118
AOT40, 54, 243
AOT40 modification, 53
arabidopsis, 129
arctic, 159
arid and semi-arid regions, 262
ascorbate, 40
Asian vegetation, 243
assessment indicators, 46
assimilation activity, 50
atmosphere-ocean mixing, 176
atmospheric $CO_2$, 73
atmospheric deposition, 113, 115

AtMPK6, 129

**B**

bark-beetle outbreaks, 112, 120
Beijing, 268
benefits, 282
BGGC model, 235
biochemical responses, 32
bio-geographical and geo chemical model, 235
biomass, 74
Biome-BGC model, 211
biometric NEP, 221
biosphere 2 laboratory, 173
birch, 30
blown sand disasters, 266
boundary layer resistance, 185
branch autonomy, 25
branch-bag experiments, 22
*Brassica oleracea*, 14

**C**

C - cycles, 173
$CaCO_3$, 174
*Calluna*, 66
calvin cycle, 141
carbon budget, 221
carbon dioxide, 81, 84
carbon flow, 146
carbon stocks, 227
carboxysome, 153
*cbb* operons, 166
cell death, 126

cell division, 92
cell expansion, 93
cell wall thickening, 93
challenge, 283
China, 261
chlorophyll, 39
chlorotic mottling, 42
*Chromatium vinosum*, 154
chronic $O_3$ stress, 27
citrus, 74
climate change, 81, 101
climatic zones, 262
closed system attributes, 175
$CO_2$ and needle abscission, 106, 107
$CO_2$ concentration, 192
$CO_2$ enrichment, 195
$CO_2$ exhalations, 195
$CO_2$ fixation, 165
$CO_2$ springs, 195
$CO_2$ vents, 196
$CO_2$ and $O_3$ interactions, 107
$CO_2$, 73, 101, 105, 150, 185
combating desertification, 277
configuration for simulations, 273
contribution in Beijing, 275
contribution of emission area, 271
coral reefs, 176
coralline organisms, 175
crassulacean acid metabolism ecosystem responses, 180
critical level, 54, 243
critical loads, 113
crops, 82
cyanobacteria, 149

**D**

damaging factors, 45
DCHM, 228
defence, 25

DEM, 228
desertification, 262
diatom, 159
diffuse radiation, 208
division of source area, 275
dose-response, 244
down-regulation, 94
drought, 30, 177
dry deposition monitoring, 254, 255
dryland, 285
DTM, 228
dust, 271

**E**

ecological fitness, 65
ecosystem competition, 63
ecosystem fluxes, 211
ecosystem stability, 111
eco-system, 280
eddy covariance, 215
elements disbalance, 47
elevated $CO_2$, 89, 176, 198
emission distribution, 273
emission impacts, 51
emission scheme, 272
emission, 271, 276
environment, 160, 278, 290
Erica, 66
ethylene, 128
evaluation with observations, 274
exceedance maps of pollutants, 249

**F**

filter pack method, 254, 255
forest canopy, 228
forest decline, 249
forest health status, 113, 114
forest monitoring, 256
forest sites, 215

forest's carbon stock, 230
forestry, 34
free-air $O_3$ fumigation system, 22
frost, 30
fructose-,16-/sedoheptulose-1,7-bisphosphatase, 142
fructose-1,6-bisphosphatase, 144

## G
gas exchange, 32
gas exposure experiment for plant sensitivity, 258
GCM, 237
gene expression under elevated $CO_2$, 180
genetic engineering, 87
genetic potential, 65, 83
global change, 73
glutathione, 40
Gobi pebblization, 265
grasses, 66
gross primary production, 207
ground surface, 229
growth enhancement, 138
growth responses, 15
growth ring width, 90
growth, 31, 75, 133

## H
$H_2S$ toxicity, 6
$H_2S$, 4
helicopter-borne scanning lidar, 227
hemeroby, 55
high elevations, 37
His-Asp phosphorelay, 167

## I
IctB conservation, 134
IctB, 134
inorganic carbon, 134

integrated NEP, 218
interactions, 33
internal $CO_2$, 136
in-vitro plantlets, 189
isoprene emission, 179

## J
Japan, 235
joint research, 256

## K
kanamycine, 154

## L
land cover and use, 267
land degradation, 261
land desiccation, 263
leaf area index (LAI), 236
leaf area index, 188
leaf longevity, 23
leaf temperature, 190
location, 279
loess plateau, 277, 279

## M
MAPK, 126
MAPKK, 130
marine carbon fluxes, 176
marine, 161
measures, 281
microgravity, 190
micropropagation, 189
mitogen-activated protein kinases, 126
model description, 272
mofettes, 195
Mongolia, 257
monitoring, 251
monsoon, 215
moors, 66

mosses, 66
multivariate patterns, 42
mutant, 154

**N**
N analysis, 103
N availability, 102
N- metabolism, 65
N resorption, 105
N retranslocation, 105
N, 101
$N_2O$, 177
native grass species, 258
needle abscission, 105
needle age class, 103
needle N, 104
needles elements content, 45
N-enhanced fluxes, 18
net ecosystem production, 216
net primary productivity (NPP), 236
Newtonian-Darwinian divide, 174
$NH_3$ deposition, 13
$NH_3$ metabolism, 14
nighttime NEP, 220
nitrate uptake, 17
nitrogen pollutants, 13
nitrogen pollution, 63
nitrogen supply, 63, 84
nitrogen, 30
North American Carbon Program, 206, 209
Northern conditions, 29
Norway spruce forests, 53
Norway spruce, 53
numerical simulation, 273
nutrients, 13, 77

**O**
ontogenetic stages, 26

open-field experiments, 30
open-top chamber, 74
orange, 74
outdoor chambers, 102
oxidative stress, 37
oxygenase, 157
ozone exposure maps, 58
ozone injury index, 43
ozone interpolation model, 55
ozone levels, 59
ozone modelling(approaches), 56
ozone passive samplers, 113, 117
ozone risk modelling, 55
ozone risk, 53
ozone trends, 57
ozone vertical transect, 118
ozone($O_3$), 21, 29, 37, 101, 112, 125, 258

**P**
parameters connection, 49
park grass experiment, 66
PAS motifs, 167
pH, 68
photochemical oxidants, 246
photo-oxidative stress, 37
photorespiration, 151
photosynthesis, 75, 83, 133, 141, 150, 159, 186,
photosynthetic activity, 135
photosynthetic bacteria, 165
phytotrons, 173
pine treestands, 46
plant adaptation, 68
plant breeding, 82
plant dry matter, 67
plant ecology, 63
plant functional type (PFT), 236
plant growth, 64, 82
plant nutrition, 64, 84

plant-insect interactions, 180
plants, 64, 82
plant sensitivity, 258
pollution, 63
ponderosa pine, 101
poplar forests at elevated $CO_2$, 179
poplar, 130
potential natural vegetation, 235
problems, 278
process modeling, 208
productivity, 74
prokaryote, 150
proteins, 77

## R

rainwater pH, 252, 253
reconstruction study, 283
reconstruction, 280, 281, 282
regional carbon balance, 210
remote sensing of chlorophyll fluorescence, 178
remote sensing, 209
resorption, 101
respiration, 83, 211
response patterns, 24
retranslocation, 102
*Rhodopseudomonas palustris*, 167
rice, 84
risk assessment, 22, 248
ROS, 125
RubisCO activity, 137
Rubisco, 143, 150, 157
RuBP regeneration, 145
RuBP, 152

## S

salinization, 266
sand and dust storms, 266
sandification, 264

scaling levels, 22
seasonal variation, 216
secondary sulfur compounds, 5
sedoheptulose-1,7-bisphosphatase, 144
sensitivity, 245
SIPK, 127
sites, 216
snow-cover, 274
$SO_2$, 4
soil and vegetation, 256
soil erosion control, 277
soil monitoring, 256
soil respiration stimulated by elevated $CO_2$, 180
soil respiration, 208
species richness, 67
specificity factor, 158
SRES, 237
stable isotopes, 178
stomatal conductance, 76
structure of growth ring, 94
sugar beet, 84
sulfate accumulation, 9
sulfolipids, 7
sulfur metabolism, 5
sulfur, 3
sulphur dioxide, 38
sustainable farming, 287
synthetic analysis, 218

## T

Tatra Mountains, 112
technology, 86, 286
temperature effects on respiration, 179
temperature, 83
temporal variation of dust emission, 271
terrestrial ecosystem, 235
The Taklamakan Desert, 265
threshold, 244

three dimensional (3-D) remote sensing, 227
tobacco, 127
tocopherol, 41
total deposition, 248
toxin, 17
transformation, 150
transgenic plants, 142
transpiration, 186
transport, 271, 276
tree decline, 257
tree height, 227
tree, 73, 244
treestands state, 48
tropical forest ecosystems, 177
tropical forests, 176
two-component signal transduction system, 167

### U
unifying theory of $O_3$ sensitivity, 23, 26
Urtica, 66

### V
vegetation degeneration, 263
vegetation restoration, 280

vertical gradient, 111
vessel dimensions, 94
visible injuries, 31

### W
water vapor pressure, 192
water, 85
water-use efficiency, 186, 207
weather extremes, 114, 119
wet and dry deposition, 251
wet deposition monitoring, 252, 253
wheat, 84
wildfire carbon emissions, 210
wind erosion, 264
windfalls, 112
WIPK, 128
wood density, 90
wood formation, 90
woody canopy height, 227
woody plants, 89

### X
xanthophyll cycle, 41

### Y
yield, 74